Lecture Notes in Computer Science   11992

More information about this series at http://www.springer.com/series/7412

Alessandro Crimi · Spyridon Bakas (Eds.)

Brainlesion:
Glioma, Multiple Sclerosis, Stroke and Traumatic Brain Injuries

5th International Workshop, BrainLes 2019
Held in Conjunction with MICCAI 2019
Shenzhen, China, October 17, 2019
Revised Selected Papers, Part I

 Springer

Editors
Alessandro Crimi 🆔
University Hospital of Zurich
Zurich, Switzerland

Spyridon Bakas 🆔
University of Pennsylvania
Philadelphia, PA, USA

ISSN 0302-9743   ISSN 1611-3349 (electronic)
Lecture Notes in Computer Science
ISBN 978-3-030-46639-8   ISBN 978-3-030-46640-4 (eBook)
https://doi.org/10.1007/978-3-030-46640-4

LNCS Sublibrary: SL6 – Image Processing, Computer Vision, Pattern Recognition, and Graphics

This Springer imprint is published by the registered company Springer Nature Switzerland AG
The registered company address is: Gewerbestrasse 11, 6330 Cham, Switzerland

Preface

This volume contains articles from the Brain-Lesion workshop (BrainLes 2019), as well as the (a) International Multimodal Brain Tumor Segmentation (BraTS 2019) challenge, (b) Computational Precision Medicine: Radiology-Pathology Challenge on Brain Tumor Classification (CPM-RadPath 2019) challenge, and (c) the tutorial session on Tools Allowing Clinical Translation of Image Computing Algorithms (TACTICAL 2019). All these events were held in conjunction with the Medical Image Computing for Computer Assisted Intervention (MICCAI 2019) conference during October 13–17, 2019, in Shenzhen, China.

The papers presented describe research of computational scientists and clinical researchers working on glioma, multiple sclerosis, cerebral stroke, trauma brain injuries, and white matter hyper-intensities of presumed vascular origin. This compilation does not claim to provide a comprehensive understanding from all points of view; however the authors present their latest advances in segmentation, disease prognosis, and other applications to the clinical context.

The volume is divided into four parts: The first part comprises the paper submissions to BrainLes 2019, the second contains a selection of papers regarding methods presented at BraTS 2019, the third includes a selection of papers regarding methods presented at CPM-RadPath 2019, and lastly papers from TACTICAL 2019.

The aim of the first chapter, focusing on BrainLes 2019 submissions, is to provide an overview of new advances of medical image analysis in all of the aforementioned brain pathologies. Bringing together researchers from the medical image analysis domain, neurologists, and radiologists working on at least one of these diseases. The aim is to consider neuroimaging biomarkers used for one disease applied to the other diseases. This session did not make use of a specific dataset.

The second chapter focuses on a selection of papers from BraTS 2019 participants. BraTS 2019 made publicly available a large (n = 626) manually annotated dataset of pre-operative brain tumor scans from 19 international institutions, in order to gauge the current state of the art in automated brain tumor segmentation using multi-parametric MRI modalities, and comparing different methods. To pinpoint and evaluate the clinical relevance of tumor segmentation, BraTS 2019 also included the prediction of patient overall survival, via integrative analyses of radiomic features and machine learning algorithms, as well as experimentally attempted to evaluate the quantification of the uncertainty in the predicted segmentations, as noted in: www.med.upenn.edu/cbica/brats2019.html.

The third chapter contains descriptions of a selection of the leading algorithms showcased during CPM-RadPath 2019 (www.med.upenn.edu/cbica/cpm2019-data.html). CRM-RadPath 2019 used corresponding imaging and pathology data in order to classify a cohort of diffuse glioma tumors into two sub-types of oligodendroglioma and astrocytoma. This challenge presented a new paradigm in algorithmic challenges, where data and analytical tasks related to the management of brain tumors were combined to arrive at a more accurate tumor classification. Data from both challenges were obtained from The Cancer Genome Atlas/The Cancer Imaging Archive (TCGA/TCGA) repository and the Hospital of the University of Pennslvania.

The final chapter comprises two TACTICAL 2019 papers. The motivation for the tutorial on TACTICAL is driven by the continuously increasing number of newly developed algorithms and software tools for quantitative medical image computing and analysis towards covering emerging topics in medical imaging and aiming towards the clinical translation of complex computational algorithms (www.med.upenn.edu/cbica/miccai-tactical-2019.html).

We heartily hope that this volume will promote further exiting research about brain lesions.

March 2020

Alessandro Crimi
Spyridon Bakas

Organization

BrainLes 2019 Organizing Committee

Spyridon Bakas	University of Pennsylvania, USA
Alessandro Crimi	African Institutes for Mathematical Sciences, Ghana
Keyvan Farahani	National Institutes of Health, USA

BraTS 2019 Organizing Committee

Spyridon Bakas	University of Pennsylvania, USA
Christos Davatzikos	University of Pennsylvania, USA
Keyvan Farahani	National Institutes of Health, USA
Jayashree Kalpathy-Cramer	Harvard University, USA
Bjoern Menze	Technical University of Munich, Germany

CPM-RadPath 2019 Organizing Committee

Spyridon Bakas	University of Pennsylvania, USA
Benjamin Bearce	Harvard University, USA
Keyvan Farahani	National Institutes of Health, USA
Jayashree Kalpathy-Cramer	Harvard University, USA
Tahsin Kurc	Stony Brook University, USA
MacLean Nasrallah	University of Pennsylvania, USA

TACTICAL 2019 Organizing Committee

Spyridon Bakas	University of Pennsylvania, USA
Christos Davatzikos	University of Pennsylvania, USA

Program Committee

Ujjwal Baid	Shri Guru Gobind Singhji Institute of Engineering and Technology, India
Jacopo Cavazza	Instituto Italiano di Tecnologia (IIT), Italy
Guray Erus	University of Pennsylvania, USA
Anahita Fathi Kazerooni	University of Pennsylvania, USA
Hugo Kuijf	Utrecht University, The Netherlands
Jana Lipkova	Technical University of Munich, Germany
Yusuf Osmanlioglu	University of Pennsylvania, USA
Sadhana Ravikumar	University of Pennsylvania, USA
Zahra Riahi Samani	University of Pennsylvania, USA

Aristeidis Sotiras   Washington University in St. Louis, USA
Anupa Vijayakumari   University of Pennsylvania, USA
Stefan Winzeck   University of Cambridge, UK

Sponsoring Institutions

Center for Biomedical Image Computing and Analytics, University of Pennsylvania, USA

Contents – Part I

Brain Tumor Image Segmentation

Contents – Part II

Combined MRI and Pathology Brain Tumor Classification

Tools AllowingClinical Translation of Image Computing Algorithms

Brain Lesion Image Analysis

Convolutional 3D to 2D Patch Conversion for Pixel-Wise Glioma Segmentation in MRI Scans

Mohammad Hamghalam[1,2], Baiying Lei[1(✉)], and Tianfu Wang[1]

[1] National-Regional Key Technology Engineering Laboratory for Medical Ultrasound, Guangdong Key Laboratory for Biomedical Measurements and Ultrasound Imaging, School of Biomedical Engineering, Health Science Center, Shenzhen University, Shenzhen 518060, China
m.hamghalam@gmail.com, {leiby,tfwang}@szu.edu.cn
[2] Faculty of Electrical, Biomedical and Mechatronics Engineering, Qazvin Branch, Islamic Azad University, Qazvin, Iran

Abstract. Structural magnetic resonance imaging (MRI) has been widely utilized for analysis and diagnosis of brain diseases. Automatic segmentation of brain tumors is a challenging task for computer-aided diagnosis due to low-tissue contrast in the tumor subregions. To overcome this, we devise a novel pixel-wise segmentation framework through a convolutional 3D to 2D MR patch conversion model to predict class labels of the central pixel in the input sliding patches. Precisely, we first extract 3D patches from each modality to calibrate slices through the squeeze and excitation (SE) block. Then, the output of the SE block is fed directly into subsequent bottleneck layers to reduce the number of channels. Finally, the calibrated 2D slices are concatenated to obtain multimodal features through a 2D convolutional neural network (CNN) for prediction of the central pixel. In our architecture, both local inter-slice and global intra-slice features are jointly exploited to predict class label of the central voxel in a given patch through the 2D CNN classifier. We implicitly apply all modalities through trainable parameters to assign weights to the contributions of each sequence for segmentation. Experimental results on the segmentation of brain tumors in multimodal MRI scans (BraTS'19) demonstrate that our proposed method can efficiently segment the tumor regions.

Keywords: Pixel-wise segmentation · CNN · 3D to 2D conversion · Brain tumor · MRI

1 Introduction

Among brain tumors, glioma is the most aggressive and prevalent tumor that begins from the tissue of the brain and hopefully cannot spread to other parts of the body. Glioma can be classified into low-grade glioma (LGG) and high-grade glioma (HGG). LGGs are primary brain tumors and usually affect young

© Springer Nature Switzerland AG 2020
A. Crimi and S. Bakas (Eds.): BrainLes 2019, LNCS 11992, pp. 3–12, 2020.
https://doi.org/10.1007/978-3-030-46640-4_1

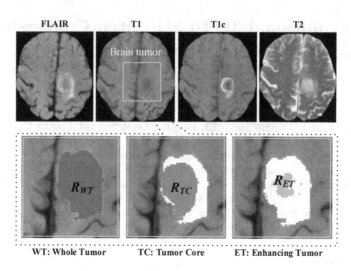

Fig. 1. Structural MRI provides a non-invasive method to determine abnormal changes in the brain for clinical purpose. Four MRI modalities (FLAIR, T1, T1c, and T2) along with brain lesion: WT (all internal parts), TC (all except edema), and ET (enhancing tumor).

people compared to HGGs. Multimodal MR sequences comprised of FLAIR, T1, T1c, and T2 are usually used to segment internal parts of the tumor, i.e., whole tumor (WT), tumor core (TC), and enhancing tumor (ET) as depicted in Fig. 1. Since the shape and location of tumors are unpredictable, it is difficult to identify exactly type of brain tumor by studying the brain scans. On the other hand, the low tissue contrast in the lesion regions makes the tumor segmentation a challenging task. Moreover, manual annotation of these tumors is a time-consuming and often biased task. Thus, automatic segmentation approaches are a crucial task in diagnosis, analysis, and treating plane.

Many segmentation methods have been proposed to segment tissue of interest based on traditional [5,6,19,20] and modern machine learning methods [7] in medical application. The brain tumor segmentation methods [8,14] can be roughly categorized into the pixel-wise [9,15] and region-wise [11,12,16,18,21] techniques. The former predicts only the central pixel of each input patch while the latter predicts labels of the most pixels inside the input patches. The region-wise methods are usually based on 3D [11,16,21] and 2D [12,18] fully convolutional networks (FCNs). Wang *et al.* [21] applied the cascaded framework with three stages to segment WT, TC, and ET on each stage, respectively. Isensee *et al.* [11] employed a U-Net-like architecture [17] that was trained on the BraTS training dataset [3,13] along with a private clinical dataset with some augmentations. In another work, Pereira *et al.* [16] introduced two new blocks to extract discriminative feature maps: recombination-recalibration (RR) and segmentation squeeze-and-excitation (SegSE) blocks.

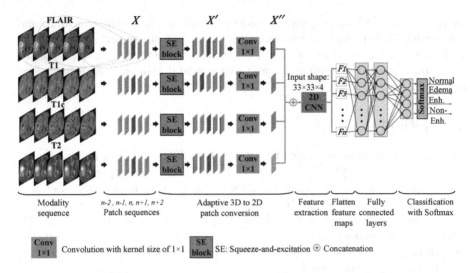

Fig. 2. Illustration of the proposed 3D to 2D patch conversion model for the pixel-wise segmentation of glioma. The 3D to 2D model includes the SE block and 1×1 convolution as the bottleneck layer. The input data are 3D multimodal image patches, and the outputs are four types of clinical subregions, i.e., edema, non-enhancing, enhancing, and healthy tissue.

In 2D structures, Shen *et al.* [18] utilized a multi-task FCN framework to segment tumor regions. Additionally, Le *et al.* [12] introduced deep recurrent level set (DRLS) based on VGG-16 with three layers: convolutional, deconvolutional, and LevelSet layer. In the pixel-wise networks [9,15], the authors established the 2D CNN-based model to predict a single class label for the central pixel in the 2D multimodal patches. However, the intra-slice features are not used in their segmentation frameworks.

Although the 3D FCN models can capture 3D information from MRI scans, 3D architectures are too computationally expensive because of the complicated network structure, including the 3D kernels, 3D input patches, and input dimensions. Notably, the size of the image patches is the most notable memory factor in convolutional nets, especially in the multimodal BraTS scans with four sequences. In the case of multimodal 3D scans, we have 5-dimensional tensors, including batch size, width, length, depth, and the number of modality concatenation. These tensors require much more memory for training and testing compared to 2D FCN.

The focus of the current study is to develop a 3D to 2D conversion network for the pixel-wise segmentation. The conversion block employs squeeze-and-excitation (SE) block to adaptively calibrate slices in the input patch sequence by explicitly modeling the interdependencies between these slices. The bottleneck layer is applied to encode the 3D patches to 2D ones to decrease the number of input channel to the following feature extraction block. We use multimodal 2D output patches for segmentation through the 2D-CNN network. Particularly, we utilize the 3D feature between consecutive slices while using convolutional layers with 2D kernels in our framework. The rest of our paper is organized as follows. In Sect. 2, we describe 3D to 2D conversion method. Section 3 explains the databases used for evaluation and experimental results. Some conclusions are drawn in Sect. 4.

2   Method

Our goal is to segment an input MR volume, $\mathbf{I} \in \mathbb{R}^{H \times W \times D}$, according to manual labels $\mathbf{S} \in \{1, 2, ..., c\}^{H \times W \times D}$, where c is the number of output classes. Also H, W, and D are the spatial height, width, and depth, respectively. Let $\mathbf{x} \in \mathbb{R}^{\omega \times \omega \times L}$ denotes the cropped 3D input patch on the central voxel, $x^{\frac{\omega}{2}, \frac{\omega}{2}, \frac{L}{2}}$. We need to predict the label of central voxels in each extracted 3D patch via 2D-CNN network. Figure 2 demonstrates an overview of the proposed method. We first introduce the adaptive 3D to 2D conversion module, and then 2D-CNN architecture will be discussed.

2.1   Convolutional 3D to 2D Patch Conversion

We extend the SE block [10] to deal with the calibration of input 3D patches. Our model squeezes the global spatial information in each slice by computing average in each slice as:

$$z_l = F_{sq}(\mathbf{x}_l) = \frac{1}{\omega \times \omega} \sum_{i=1}^{\omega} \sum_{j=1}^{\omega} \mathbf{x}_l(i, j). \tag{1}$$

where z_l is the global embedded information in the slice of l. The second operation called 'excitation' is applied to capture slice-wise dependencies with a sigmoid (σ) and ReLU (δ) activation, respectively. Thus we have:

$$\mathbf{u} = F_{ex}(\mathbf{z}, W) = \sigma(W_2 \delta(W_1 \mathbf{z})) \tag{2}$$

where $W_1 \in \mathbb{R}^{r \times w^2}$ and $W_2 \in \mathbb{R}^{w^2 \times r}$ are the weight matrices of two fully-connected with reduction ratio r. At last, the scalar u_l and input slice $\mathbf{x}_l$ are multiplied to obtain the calibrated 3D patch, $\mathbf{x}' \in \mathbb{R}^{\omega \times \omega \times L}$.

Our bottleneck layer is a block that contains one convolutional layer with the kernel size of 1×1 to represent calibrated 3D slices as 2D with nonlinear dimensionality reduction, $\mathbf{x}''$. Each 2D patch thus forms a 3D-like representation of a part (n consecutive slices) of the MR volume. This model allows incorporating some 3D information while bypassing the high computational and memory requirements of the 3D CNN.

2.2   Classifier Block for Pixel-Wise Prediction

The output slices from four 3D to 2D blocks are concatenated and fed into classifier block to predict the label of voxel where is located at the center of its cropped patch. The proposed network allows jointly capturing contextual features from FLAIR, T1, T1c, and T2 modality. For feature extraction, we rely on CNN block to learn from ground truth scores. Our feature extractor consists of two levels of 3×3 convolutions along with max-pooling layers. The number of kernels in each level is 32, 32, 32, 64, 64, and 64, respectively. The fully-connected layers are composed of 64 and 32 hidden neurons, respectively, followed by the final Softmax layer. Finally, we optimize cross-entropy loss between the predicted score, $F_{seg}(\mathbf{x}^{FLAIR}, \mathbf{x}^{T1}, \mathbf{x}^{T1c}, \mathbf{x}^{T2}; \mathbf{W})$, and the ground truth label, $s^{\frac{w}{2}, \frac{w}{2}, \frac{L}{2}}$, with ADADELTA optimizer [22] as:

$$\underset{\mathbf{W}}{\arg\min} \ - \sum_i^c s^{\frac{w}{2}, \frac{w}{2}, \frac{L}{2}} . \log(F_{seg}(\mathbf{x}^{FLAIR}, \mathbf{x}^{T1}, \mathbf{x}^{T1c}, \mathbf{x}^{T2}; \mathbf{W})) \qquad (3)$$

where c is the class number and $\mathbf{W}$ is the trainable parameter of the model.

3 Experimental Results

3.1 Implementation Details

We implement the proposed method using the KERAS and TensorFlow with 12 GB NVIDIA TITAN X GPU. We have experimentally found that volumes of seven have the best compromise between accuracy and complexity. Thus, the input MR volumes are partitioned into $33 \times 33 \times 7$ patches at the center of each label, then the concatenated patches from four modalities are considered as training data. For efficient training and class imbalance in brain tumor, we perform augmentation in the number of patches for the small sample size classes. The model is trained using the ADADELTA [22] optimizer (learning rate = 1.0, $\rho = 0.95$, epsilon=1e−6) and cross-entropy as the loss function. Dropout is employed to avoid over-fitting during the training process ($p_{drop} = 0.5$).

3.2 Datasets

The performance of the proposed pixel-wise method is evaluated on BraTS [1–4,13] dataset to compare with other segmentation methods based on the pixel. BraTS'13 contains small subjects, i.e., 30 cases for training and 10 cases for the Challenge. We additionally evaluate the proposed technique on BraTS'19, which has two publicly available datasets of multi-institutional pre-operative MRI sequences: Training (335 cases) and Validation (125 cases). Each patient is contributing $155 \times 240 \times 240$ with four sequences: T1, T2, T1c, and FLAIR. In BraTS'19, it identifies three tumor regions: non-enhancing tumor, enhancing tumor, and edema. Evaluation is performed for the WT, TC, and ET. The evaluation is assessed by the SMIR[1] and CBICA IPP[2] online platforms. Metrics computed by the online evaluation platforms in BraTS'19 are Dice Similarity

---

[1] https://www.smir.ch/BRATS/Start2013.
[2] https://ipp.cbica.upenn.edu.

Table 1. Impact of the 3D to 2D conversion block in segmentation: we perform experiments using the same setting to evaluate performance with and without proposed block.

Model	DSC			HD95 (mm)		
	ET	WT	TC	ET	WT	TC
With 3D to 2D block	83.15	91.75	92.35	1.4	3.6	1.4
Without 3D to 2D block	80.21	89.73	88.44	3.1	4.2	1.5

Coefficient (DSC) and the 95th percentile of the Hausdorff Distance (HD95), whereas, in BraTS'13, the online platform calculates DSC, Sensitivity, and Positive Predictive Value (PPV). DSC is considered to measure the union of automatic and manual segmentation. It is calculated as $DSC = \frac{2TP}{FP + 2TP + FN}$ where TP, FP, and FN are the numbers of true positive, false positive, and false negative detections, respectively.

3.3 Segmentation Results on BRATS'13

Ablation Study. To investigate the effect of the proposed adaptive 3D to 2D block, we perform experiments with and without considering the 3D to 2D block. For the latter, we directly apply multimodal 3D volume into the 3D plain CNN model. We train both models with the 320 K patch for an equal number of the patch in each group and validate on ten unseen subjects. Also, Dropout is employed to avoid over-fitting during the training process ($p_{drop} = 0.5$). As presented in Table 1, the results with 3D to 2D block increase the accuracy of segmentation in terms of standard evaluation metrics compared to the 3D baseline.

Comparison with State-of-the-Arts. We also compare the performance of the proposed method with the well-known pixel-wise approach [9,15] and 2D region-wise ones [18] on BraTS'13 Challenge. Table 2 shows DSC (%), Sensitivity, and PPV for EN, WT, and TC, respectively. Moreover, it can be seen that the proposed method outperforms others in DSC for WT.

3.4 Segmentation Results on BRATS'19

One limitation of pixel-wise methods is the time complexity at inference time due to pixel by pixel prediction. Specifically, we have to process about 9M voxels per channel for each patient. Although we eliminate voxels with the value of zero in testing time, the pixel-wise prediction still needs longer time compared to region-wise ones. This issue limits our method for evaluation on BraTS'19 with 125 validation samples. To decrease the inference time, we use a plain 3D U-Net model to solely predict WT as an initial segmentation, which further allows us to compute a bounding box concerning tumor region for our pixel-wise method.

Table 2. Comparison of proposed 3D to 2D method with others on BraTS'13 Challenge dataset.

Method	DSC			Sensitivity			PPV		
	EN	WT	TC	EN	WT	TC	EN	WT	TC
Shen [18]	0.76	0.88	**0.83**	**0.81**	**0.90**	0.81	0.73	0.87	**0.87**
Pereira [15]	**0.77**	0.88	**0.83**	**0.81**	0.89	0.83	**0.74**	0.88	**0.87**
Havaei [9]	0.73	0.88	0.79	0.80	0.87	0.79	0.68	0.89	0.79
Proposed method	0.74	**0.89**	0.80	0.78	0.86	**0.86**	0.73	**0.92**	0.76

Table 3. DSCs and HD95 of the proposed method on BraTS'19 Validation set (training on 335 cases of BraTS'19 training set).

	Dice			Sensitivity			Specificity			HD95 (mm)		
	ET	WT	TC	ET	WT	TC	ET	WT	TC	ET	WT	TC
Mean	72.48	89.65	79.56	73.25	90.60	79.57	99.87	99.45	99.69	5.4	7.8	8.7
Std.	29.47	8.968	21.62	26.61	08.91	24.77	0.23.5	0.58	0.36	9.2	15.5	13.5
Median	84.46	92.19	89.17	83.20	93.66	91.14	99.94	99.64	99.82	2.2	3.1	3.8
25 quantile	70.99	88.31	74.63	67.73	87.83	72.88	99.84	99.26	99.56	1.4	2.0	2.0
75 quantile	89.22	94.72	93.39	88.71	96.58	96.04	99.98	99.81	99.93	4.2	5.3	10.2

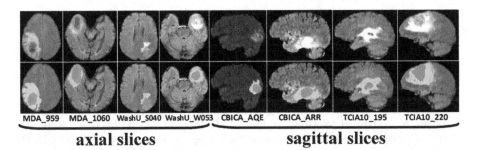

MDA_959 MDA_1060 WashU_S040 WashU_W053 CBICA_AQE CBICA_ARR TCIA10_195 TCIA10_220

axial slices **sagittal slices**

Fig. 3. Segmentation results are overlaid on FLAIR axial and sagittal slices on BraTS'19 Validation Data. The yellow label is edema, blue color means enhancing tumor, and the green one presents the necrotic and non-enhancing tumor core. Each column displays one slice of different Subject IDs of BraTS'19. (Color figure online)

In this way, the segmentation of the internal part of the tumor area is performed inside the bounding box. The results in Table 3 show that our method achieved competitive performance on automatic brain tumor segmentation. Results are reported in the online processing platform by BraTS'19 organizer.

Moreover, Fig. 3 shows examples for glioma segmentation from validation slices of BraTS'19. For simplicity of visualization, only the FLAIR image is shown in the axial and sagittal view along with our segmentation results. The subject IDs in each column are related to the validation set.

4   Conclusion

This paper provides a framework that adaptively converts 3D patch into 2D to highlight discriminative pixels for the label prediction of central voxels. The converted 2D images are fed into the classifier block with 2D kernels for the predication. This conversion enables incorporating 3D features while bypassing the high computational and memory requirements of fully 3D CNN. We provided ablation study to examine the effect of our proposed conversion block on the segmentation performance. Results from the BraTs'13 and BraTS'19 dataset confirm that inter and intra-slice features effectively improve the performance while using 2D convolutional kernels. Though pixel-wise methods have limitation in inference time, we can take advantage of pre-trained network for classification purpose through fine-tuning with MRI training set. Future works will concentrate on 3D to 2D patch conversion with an attention mechanism.

Acknowledgment. This work was supported partly by National Natural Science Foundation of China (Nos. 61871274, 61801305, and 81571758), National Natural Science Foundation of Guangdong Province (No. 2017A030313377), Guangdong Pearl River Talents Plan (2016ZT06S220), Shenzhen Peacock Plan (Nos. KQTD2016053112 051497 and KQTD2015033016 104926), and Shenzhen Key Basic Research Project (Nos. JCYJ20170413152804728, JCYJ20180507184647636, JCYJ20170818142347 251, and JCYJ20170818094109846).

References

1. Bakas, S., Akbari, H., Sotiras, A., Bilello, M., Rozycki, M., Kirby, et al.: Segmentation labels and radiomic features for the pre-operative scans of the TCGA-GBM collection. Cancer Imaging Arch. (2017). https://doi.org/10.7937/K9/TCIA.2017. KLXWJJ1Q
2. Bakas, S., Akbari, H., Sotiras, A., Bilello, M., Rozycki, M., Kirby, et al.: Segmentation labels and radiomic features for the pre-operative scans of the TCGA-LGG collection. Cancer Imaging Arch. (2017). https://doi.org/10.7937/K9/TCIA.2017. GJQ7R0EF
3. Bakas, S., et al.: Advancing the cancer genome atlas glioma MRI collections with expert segmentation labels and radiomic features. Nat. Sci. Data **4** (2017). https://doi.org/10.1038/sdata.2017.117. Article no. 170117
4. Bakas, S., Reyes, M., Jakab, A., Bauer, S., Rempfler, M., Crimi, A., et al.: Identifying the best machine learning algorithms for brain tumor segmentation, progression assessment, and overall survival prediction in the BRATS challenge. arXiv preprint arXiv:1811.02629 (2018)
5. Hamghalam, M., Ayatollahi, A.: Automatic counting of leukocytes in giemsa-stained images of peripheral blood smear. In: 2009 International Conference on Digital Image Processing, pp. 13–16 (2009). https://doi.org/10.1109/ICDIP. 2009.9
6. Hamghalam, M., Motameni, M., Kelishomi, A.E.: Leukocyte segmentation in giemsa-stained image of peripheral blood smears based on active contour. In: 2009 International Conference on Signal Processing Systems, pp. 103–106 (2009). https://doi.org/10.1109/ICSPS.2009.36

7. Hamghalam, M., et al.: Brain tumor synthetic segmentation in 3D multimodal MRI scans. arXiv preprint arXiv:1909.13640 (2019)
8. Hatami, T., et al.: A machine learning approach to brain tumors segmentation using adaptive random forest algorithm. In: 2019 5th Conference on Knowledge Based Engineering and Innovation (KBEI), pp. 076–082 (2019). https://doi.org/10.1109/KBEI.2019.8735072
9. Havaei, M., et al.: Brain tumor segmentation with deep neural networks. Med. Image Anal. **35**, 18–31 (2017)
10. Hu, J., Shen, L., Sun, G.: Squeeze-and-excitation networks. In: 2018 IEEE/CVF Conference on Computer Vision and Pattern Recognition, pp. 7132–7141 (2018)
11. Isensee, F., Kickingereder, P., Wick, W., Bendszus, M., Maier-Hein, K.H.: Brain tumor segmentation and radiomics survival prediction: contribution to the BRATS 2017 challenge. In: Crimi, A., Bakas, S., Kuijf, H., Menze, B., Reyes, M. (eds.) BrainLes 2017. LNCS, vol. 10670, pp. 287–297. Springer, Cham (2018). https://doi.org/10.1007/978-3-319-75238-9_25
12. Le, T.H.N., Gummadi, R., Savvides, M.: Deep recurrent level set for segmenting brain tumors. In: Frangi, A.F., Schnabel, J.A., Davatzikos, C., Alberola-López, C., Fichtinger, G. (eds.) MICCAI 2018. LNCS, vol. 11072, pp. 646–653. Springer, Cham (2018). https://doi.org/10.1007/978-3-030-00931-1_74
13. Menze, B.H., Jakab, A., Bauer, S., Kalpathy-Cramer, J., Farahani, K., et al.: The multimodal brain tumor image segmentation benchmark (BRATS). IEEE Trans. Med. Imaging **34**(10), 1993–2024 (2015). https://doi.org/10.1109/TMI.2014.2377694
14. Najrabi, D., et al.: Diagnosis of astrocytoma and globalastom using machine vision. In: 2018 6th Iranian Joint Congress on Fuzzy and Intelligent Systems (CFIS), pp. 152–155 (2018). https://doi.org/10.1109/CFIS.2018.8336661
15. Pereira, S., Pinto, A., Alves, V., Silva, C.A.: Brain tumor segmentation using convolutional neural networks in mri images. IEEE Trans. Med. Imaging **35**(5), 1240–1251 (2016)
16. Pereira, S., Alves, V., Silva, C.A.: Adaptive feature recombination and recalibration for semantic segmentation: application to brain tumor segmentation in MRI. In: Frangi, A.F., Schnabel, J.A., Davatzikos, C., Alberola-López, C., Fichtinger, G. (eds.) MICCAI 2018. LNCS, vol. 11072, pp. 706–714. Springer, Cham (2018). https://doi.org/10.1007/978-3-030-00931-1_81
17. Ronneberger, O., Fischer, P., Brox, T.: U-Net: convolutional networks for biomedical image segmentation. In: Navab, N., Hornegger, J., Wells, W.M., Frangi, A.F. (eds.) MICCAI 2015. LNCS, vol. 9351, pp. 234–241. Springer, Cham (2015). https://doi.org/10.1007/978-3-319-24574-4_28
18. Shen, H., Wang, R., Zhang, J., McKenna, S.J.: Boundary-aware fully convolutional network for brain tumor segmentation. In: Descoteaux, M., Maier-Hein, L., Franz, A., Jannin, P., Collins, D.L., Duchesne, S. (eds.) MICCAI 2017. LNCS, vol. 10434, pp. 433–441. Springer, Cham (2017). https://doi.org/10.1007/978-3-319-66185-8_49
19. Soleimany, S., et al.: A novel random-valued impulse noise detector based on MLP neural network classifier. In: 2017 Artificial Intelligence and Robotics (IRANOPEN), pp. 165–169 (2017). https://doi.org/10.1109/RIOS.2017.7956461
20. Soleymanifard, M., et al.: Segmentation of whole tumor using localized active contour and trained neural network in boundaries. In: 2019 5th Conference on Knowledge Based Engineering and Innovation (KBEI), pp. 739–744 (2019). https://doi.org/10.1109/KBEI.2019.8735050

21. Wang, G., Li, W., Ourselin, S., Vercauteren, T.: Automatic brain tumor segmentation using cascaded anisotropic convolutional neural networks. In: Crimi, A., Bakas, S., Kuijf, H., Menze, B., Reyes, M. (eds.) BrainLes 2017. LNCS, vol. 10670, pp. 178–190. Springer, Cham (2018). https://doi.org/10.1007/978-3-319-75238-9_16
22. Zeiler, M.D.: ADADELTA: an adaptive learning rate method. CoRR abs/1212.5701 (2012)

TBI Lesion Segmentation in Head CT: Impact of Preprocessing and Data Augmentation

Miguel Monteiro[1]([✉]), Konstantinos Kamnitsas[1], Enzo Ferrante[2],
Francois Mathieu[3], Steven McDonagh[1], Sam Cook[3], Susan Stevenson[3],
Tilak Das[3], Aneesh Khetani[3], Tom Newman[3], Fred Zeiler[3], Richard Digby[3],
Jonathan P. Coles[3], Daniel Rueckert[1], David K. Menon[3],
Virginia F. J. Newcombe[3], and Ben Glocker[1]

[1] Biomedical Image Analysis Group, Imperial College London, London, UK
miguel.monteiro@imperial.ac.uk
[2] sinc(i), FICH-Universidad Nacional del Litoral, CONICET,
Santa Fe, Argentina
[3] Division of Anaesthesia, Department of Medicine, University of Cambridge,
Cambridge, UK

Abstract. Automatic segmentation of lesions in head CT provides key information for patient management, prognosis and disease monitoring. Despite its clinical importance, method development has mostly focused on multi-parametric MRI. Analysis of the brain in CT is challenging due to limited soft tissue contrast and its mono-modal nature. We study the under-explored problem of fine-grained CT segmentation of multiple lesion types (core, blood, oedema) in traumatic brain injury (TBI). We observe that preprocessing and data augmentation choices greatly impact the segmentation accuracy of a neural network, yet these factors are rarely thoroughly assessed in prior work. We design an empirical study that extensively evaluates the impact of different data preprocessing and augmentation methods. We show that these choices can have an impact of up to 18% DSC. We conclude that resampling to isotropic resolution yields improved performance, skull-stripping can be replaced by using the right intensity window, and affine-to-atlas registration is not necessary if we use sufficient spatial augmentation. Since both skull-stripping and affine-to-atlas registration are susceptible to failure, we recommend their alternatives to be used in practice. We believe this is the first work to report results for fine-grained multi-class segmentation of TBI in CT. Our findings may inform further research in this under-explored yet clinically important task of automatic head CT lesion segmentation.

1 Introduction

Traumatic brain injury (TBI) is a pathology that alters brain function caused by trauma to the head [1]. TBI is a leading cause of death and disability worldwide with heavy socio-economic consequences [2]. Computed tomography (CT)

© Springer Nature Switzerland AG 2020
A. Crimi and S. Bakas (Eds.): BrainLes 2019, LNCS 11992, pp. 13–22, 2020.
https://doi.org/10.1007/978-3-030-46640-4_2

allows rapid assessment of brain pathology, ensuring patients who require urgent intervention receive appropriate care [3]. Its low acquisition time allows for rapid diagnosis, quick intervention, and safe application to trauma and unconscious patients. Research on automatic segmentation of TBI lesions in magnetic resonance imaging (MRI) [4,5] has shown promising results. However, MRI is usually reserved for imaging in the post-acute phase of brain injury or as a research tool. Since CT is routinely used in clinical care and CT voxel intensities are approximately calibrated in Hounsfield units (HUs) across different scanners, effective computational analysis of CT has the potential for greater generalisation and clinical impact than MRI. Prior work on automatic analysis of pathology in head CT is limited, mostly focusing on image-level detection of abnormalities [6,7], feature extraction for outcome prediction [8], or image-level classification [9] instead of voxel-wise semantic segmentation. Previous works on segmentation employ level-sets for the segmentation of specific haemorrhages and haematomas [10,11]. Recently, [12] applied deep learning for binary segmentation of contusion core grouped with haematomas. We present the first multi-class segmentation of contusion core, blood (haemorrhages and haematomas), and oedema. This task is important for patient management and a better understanding of TBI.

State-of-the-art automatic segmentation relies on convolutional neural networks (CNNs) [13]. These models are effective in many biomedical imaging tasks [14]. Neural networks are theoretically capable of approximating any function [15]. This result commonly translates in the expectation that networks are able to extract any necessary pattern from the input data during training. As a result, attention is mainly focused on further development of network architectures, while disregarding other parts of the system, such as data preprocessing. Contrary to popular belief that networks generalise well by learning high-level abstractions of the data, they tend to learn low-level regularities in the input [16]. In practice, the learned representations are largely dependent on a stochastic, greedy and non-convex optimisation process. We argue that appropriate data preprocessing and augmentation can be as important as architectural choices. Preprocessing can remove useless information from the input and help training start in a "better" region of the feature space. Data augmentation can help optimisation by reducing the risk overfitting to training samples. It can also learn a model invariant to information not useful for the task (e.g., rotation).

Motivated by these observations, we present an extensive ablation study of different preprocessing and data augmentation methods commonly found in the literature but whose usage is rarely empirically justified. Our goal is to establish the most appropriate methodological steps for segmentation of TBI lesions in CT. We explore the effects of spatial normalisation, intensity windowing, and skull-stripping on the final segmentation result. We also explore the effects of spatial normalisation vs. spatial data augmentation. We demonstrate that using intensity windowing can replace skull-stripping, and spatial data augmentation can replace spatial normalisation. We show the difference these methodological choices can make is up to 18% in the Dice similarity coefficient (DSC). To the best of our knowledge, this work is the first to report results for multi-class segmentation of contusion core, blood and oedema in TBI head CT.

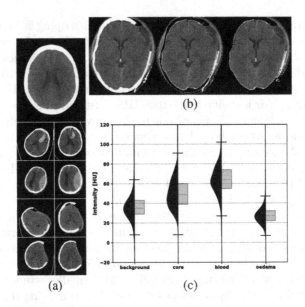

Fig. 1. Left (a): Effect of affine spatial normalisation to atlas space, CT atlas (top), before and after normalisation (left and right). Top right (b): Skull-stripping methods, from left to right: no skull-stripping; thresholding out the skull; level-set method. Bottom right (c): Intensity distribution of classes inside the brain mask.

2   Methods

Dataset: We use 98 scans from 27 patients with moderate to severe TBI. We split the data into training and test (64/34), ensuring images from the same patient are in the same set. All scans have been manually annotated and reviewed by a team of experts to provide reference segmentations. We consider four classes: background; core; blood; and oedema. The core class includes contusion cores and petechial haemorrhages. The blood class includes subdural and extradural haematomas as well as subarachnoid and intraventricular haemorrhages.

Spatial Normalisation: Since CNNs are not scale invariant, it is standard practice to resample all images to have the same physical (isotropic) resolution (e.g., $1 \times 1 \times 1$ mm). Given that brain CT is often highly anisotropic with high in-plane and low out-of-plane resolution, we may opt to resample to anisotropic resolution (e.g., $1 \times 1 \times 4$ mm) without loss of information while saving memory in the CNN's activations. Another preprocessing option is to perform spatial normalisation via registration to a reference frame, e.g., an atlas. Using affine transformations, we can remove inter-subject variability in terms of rotation and scaling which may be beneficial for the CNN. We investigate the effect of resampling and registration using three different settings: **1)** isotropic resolution of 1 mm; **2)** anisotropic resolution of $1 \times 1 \times 4$ mm; **3)** affine-to-atlas registration with 1 mm isotropic resolution. The atlas has been constructed from 20 normal CT scans that show no disease using an iterative unbiased atlas construction scheme [17], and subsequent alignment to an MNI MRI atlas (Fig. 1a).

Skull-Stripping and Intensity Windowing: Skull-striping is commonly used in brain image analysis to eliminate unnecessary information by removing the skull and homogenising the background. Unlike MRI, CT intensities are roughly calibrated in HUs and have a direct physical interpretation related to the absorption of X-rays. A specific intensity value reflects the same tissue density regardless of the scanner. Air is defined as -1000 HUs, distilled water as 0, soft tissue ranges between -100 and 300, while bone has larger values than soft tissue. Consider a lower and an upper bound that define the range of soft tissue. We test three different skull-stripping methods: **1)** no skull-stripping: we set intensities below the lower bound and above the upper bound to the lower and upper bound respectively; **2)** thresholding out the skull: we set values below the lower bound and above the upper bound to the lower bound; **3)** a level-set method (geodesic active contours [18]) to remove the skull followed by thresholding. Figure 1b shows the effect of the three skull-stripping methods. We performed a visual check on all images to make sure the level-set method is not removing parts of the brain. We test two different intensity windows for the bounds and normalisation: **1)** a larger window range $[-100, 300]$; **2)** a smaller window $[-15, 100]$. Figure 1c shows that intensity values of soft tissue fall well inside these windows. After skull-stripping and windowing, we normalise the intensity range to $[-1, 1]$. As seen in Fig. 1b, the brain-mask it not perfect in cases with a craniectomy, yet, this is a realistic scenario when we calculate automatic brain-masks for large datasets.

Data Augmentation: We test the following settings: no augmentation; flipping the x axis; flipping the x and y axes; flipping the x and y axes combined with fixed rotations (multiples of $90°$) of the same axes; flipping and fixed rotations of all axes; random affine transformations (scaling $\pm 10\%$; rotating xy randomly between $\pm 45°$; rotating xz and yz randomly between $\pm 30°$) combined with flipping the x axis; random affine transformations combined with flipping the x and y axes. We test these settings for both isotropic and affine spatial normalisation to study the effects of spatial augmentation vs. spatial normalisation.

Model architecture: Our main goal is to study the effects of preprocessing and hence we use the same architecture for all experiments[1]. We employ DeepMedic [5], a 3D CNN, with 3 parallel pathways that process an image at full resolution, three and five times downsampled. We use the residual version [19], but using 20 less feature maps per convolution layer. To see which results are model specific we re-run a subset of experiments with a 3D U-Net [20].

3   Results and Discussion

To assess which type of preprocessing is most effective we test all combinations of the aforementioned alternatives for spatial normalisation, intensity windowing

---

[1] For the experiments with $1 \times 1 \times 4$ resolution, we turn some of the isotropic kernels into anisotropic $3 \times 3 \times 1$ kernels in order to obtain approximately the same receptive field as in the experiments with isotropic resolution. This did not affect the performance.

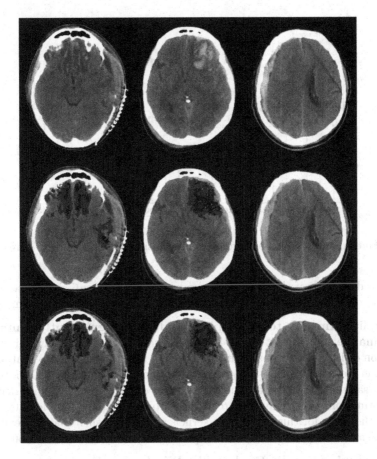

Fig. 2. Visual comparison for three cases. Top to bottom: image, manual and predicted segmentation. Red is contusion core, green is blood, and blue is oedema. (Color figure online)

and skull-stripping. For evaluation, we use the DSC calculated after transforming the prediction back into the original native image space (where the expert segmentation is defined). Thus, we guarantee an accurate comparison with the expert's segmentation. Figure 3 shows the DSC of the foreground class for all preprocessing pipelines (flipping the x axis for augmentation). The foreground class consists of all lesion classes merged into one (after training) for an overall evaluation and comparison. Figure 2 presents a visual comparison between manual segmentation and the prediction made by the best performing model. We can see that the TBI lesions have large inter-subject variability and intra-subject complexity, making for a difficult segmentation problem. Figure 4 shows the result of paired Wilcoxon signed-rank tests to determine which performance differences are statistically significant ($p < 0.05$). Figure 5a presents the per class DSC. Figure 5b presents a subset of experiments replicated with a different

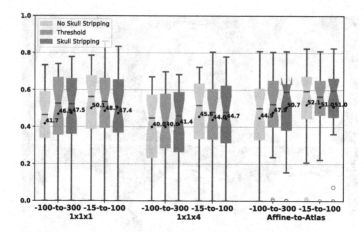

Fig. 3. Foreground DSC box plots for all preprocessing pipelines. (Color figure online)

model to determine if the results are model specific. Figure 6 shows the results of spatial data augmentation for comparison with spatial normalisation.

Resampling Images to Isotropic Resolution Significantly Improves Performance. From Figs. 3 and 4 (purple) we observe that using anisotropic resolution of $1 \times 1 \times 4$ is consistently worse than using the other two spatial normalisation methods which use isotropic resolution. This contradicts the intuition that oversampling the out-of-plane axis is not necessary, even though it does not add information. CNNs have an architectural bias towards isotropic data which is likely the cause for this result. The kernels are stacked such that it is expected the same amount of information to be present in each physical direction.

Skull-Stripping Can Be Replaced by Windowing. From Fig. 4 (yellow) we can see that skull-stripping significantly helps performance when combined with a large intensity window. However, when combined with a small intensity window, skull-stripping does not offer a benefit over no skull-stripping. We observe the same for the second model (Fig. 5b) where the difference is also not statistically significant ($p = 0.2$). This indicates that what is important for the CNN is to remove intensity ranges that are not of interest (e.g., hyper-intense skull, either via windowing or skull stripping), allowing it to focus on the subtle intensity differences between lesions and healthy tissue. Therefore, skull-stripping may be replaced by an intensity window that limits extreme intensity values. Level-set based skull-stripping is susceptible to failure. Although we performed visual checks to ensure quality, these are unfeasible on large scale settings such as the deployment of an automated segmentation pipeline. Conversely, windowing is more robust, hence it should be used instead.

Affine-to-Atlas Registration Can Be Replaced by Data-Augmentation. Although Figs. 3, 4 (green) could lead us to believe that affine registration provides a small benefit over simply using isotropic resolution, when we look at Fig. 6 we see this is not the case. When we add more spatial augmentation to

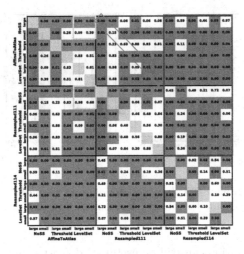

Fig. 4. Results of paired difference tests comparing prepossessing methods. For a p-value of 5% the y-label is statistically significantly better (red) or worse (blue) than the x-label. Yellow boxes: level-set skull stripping only helps performance when the large intensity window is used. Green box: controlling for other preprocessing steps, affine-to-atlas can provide small benefits. Purple box: anisotropic resolution consistently underperforms when compared to isotropic resolution. (Color figure online)

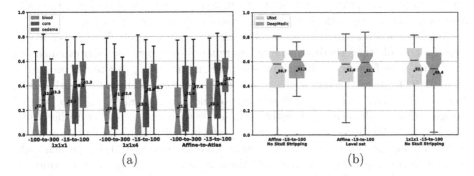

Fig. 5. Left (a): Per class DSC box plots preprocessing pipelines with no skull-stripping. Right (b): Foreground DSC box plots for two different models.

the two spatial normalisation methods (besides only flipping the x axis) their performance becomes comparable. Moreover, we see that this small benefit may not translate to different models (Fig. 5b). We conclude that we can make the network more robust to spatial heterogeneity with data augmentation instead of homogenising the input data. Like skull-stripping, affine-to-atlas registration can fail unpredictably during deployment of automatic pipelines. In contrast, spatial augmentation applied only during model training, and thus it does not constitute a possible point of failure after deployment. As a result, we recommend spatial data augmentation to be used instead of affine-to-atlas registration. Regardless,

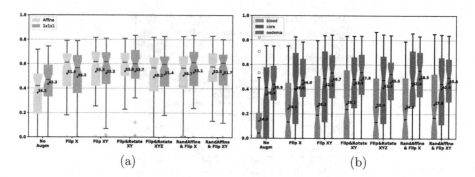

Fig. 6. Left (a): Foreground DSC box plots for all data augmentation methods. Right (b): Per class DSC box plots for all data augmentation methods.

affine-to-atlas registration can still be useful for downstream analysis tasks such as studying the location of lesions across a patient cohort.

Additional Findings. From Fig. 6 we observe that even though random affine augmentation serves its purpose, it did not perform better than fixed rotations. These transformations incur on a computational cost due to interpolation, and hence fixed rotations may be a better choice. We also observe that too much augmentation can start to hinder performance since flipping and rotating all axes performs worse than doing the same on just x and y.

We achieve a maximum DSC of $53.9 \pm 23.0\%$ (foreground). We can see from Figs. 5a and 6b that there is a large discrepancy between the performance of each class. The blood class has the worst performance likely due to the presence of hard to segment lesions such as subarachnoid haemorrhages. Surprisingly, the model performs best for oedema, one of the hardest lesion types to detect visually. Our results are not directly comparable with ones reported in the literature [10–12]. We use a different dataset, perform multi-class segmentation, and our labels include hard to segment lesions such as petechial and subarachnoid haemorrhages. Although the DSC obtained is not as high as in other similar applications (e.g. brain tumour segmentation on MRI), this goes to show the challenging nature of the problem and need to focus more effort on difficult tasks. Importantly, this is the first work reporting fine-grained multi-class segmentation of contusion core, blood and oedema in CT for patients with TBI.

4   Conclusion

We present an in-depth ablation study of common data preprocessing and augmentation methods for CT and show these methodological choices are key for achieving better segmentation performance with CNNs, with a difference of 18% DSC between the worst and best settings. Based on our results we make the following recommendations: **1)** using isotropic resolution is key **2)** choosing the correct intensity window for context and normalisation is superior to skull-stripping since it is simpler and more robust; **3)** affine-to-atlas registration can give small improvements, however, spatial data augmentation can achieve the same benefits while being more robust.

We hope our study will serve as a useful guide and help the community to make further progress on the clinically important task of head CT lesion segmentation. While our results on fine-grained TBI lesion segmentation are promising, we believe this study also shows that this task remains an open challenge and new approaches may be required to tackle this difficult problem. In the future, we aim to apply our findings to a larger dataset and to further fine-grain the segmented classes by separating SDH, EDH and SAH into separate classes.

Acknowledgments. This work is partially funded by a European Union FP7 grant (CENTER-TBI; Agreement No: 60215) and by the European Research Council (ERC) under the European Union's Horizon 2020 research and innovation programme (Agreement No: 757173, project MIRA, ERC-2017-STG). EF is supported by the AXA Research Fund and UNL (CAID-50220140100084LI). KK is supported by the President's PhD Scholarship of Imperial College London. VFJN is supported by a Health Foundation/Academy of Medical Sciences Clinician Scientist Fellowship. DKM is supported by funding from the National Institute for Health Research (NIHR) through a Senior Investigator award and the Cambridge Biomedical Research Centre at the Cambridge University Hospitals National Health Service (NHS) Foundation Trust. The views expressed are those of the authors and not necessarily those of the NHS, the NIHR, or the Department of Health and Social Care.

References

1. Menon, D.K., Schwab, K., Wright, D.W., Maas, A.I.: Position statement: definition of traumatic brain injury (2010)
2. Maas, A.I.R., Menon, D.K., Adelson, P.D., et al.: Traumatic brain injury: integrated approaches to improve prevention, clinical care, and research. Lancet Neurol. **16**, 987–1048 (2017)
3. Coles, J.P.: Imaging after brain injury. BJA **99**, 49–60 (2007)
4. Rao, A., Ledig, C.: Contusion segmentation from subjects with Traumatic Brain Injury: a random forest framework. In: ISBI (2014)
5. Kamnitsas, K., et al.: Efficient multi-scale 3D CNN with fully connected CRF for accurate brain lesion segmentation. MIA **36**, 61–78 (2017)
6. Chan, T.: Computer aided detection of small acute intracranial hemorrhage on computer tomography of brain. CMIG **31**, 285–298 (2007)
7. Arbabshirani, M.R., et al.: Advanced machine learning in action: identification of intracranial hemorrhage on computed tomography scans of the head with clinical workflow integration. NPJ Digit. Med. **1**, 1–7 (2018)
8. Koikkalainen, J.R., Lötjönen, J.M.P., Ledig, C., Rueckert, D., Tenovuo, O.S., Menon, D.K.: Automatic quantification of CT images for traumatic brain injury. In: ISBI (2014)
9. Chilamkurthy, S., et al.: Deep learning algorithms for detection of critical findings in head Ct scans: a retrospective study. Lancet **392**, 2388–2396 (2018)

10. Liao, C.-C., Xiao, F., Wong, J.-M., Chiang, I.-J.: Computer-aided diagnosis of intracranial hematoma with brain deformation on computed tomography. CMIG **34**, 563–571 (2010)

11. Prakash, K.N.B., Zhou, S., Morgan, T.C., Hanley, D.F., Nowinski, W.L.: Segmentation and quantification of intra-ventricular/cerebral hemorrhage in CT scans by modified distance regularized level set evolution technique. IJCARS **7**, 785–798 (2012). https://doi.org/10.1007/s11548-012-0670-0

12. Jain, S., et al.: Automatic quantification of computed tomography features in acute traumatic brain injury. J. Neurotrauma **36**, 1794–1803 (2019)

13. LeCun, Y., et al.: Backpropagation applied to handwritten zip code recognition. Neural Comput. **1**, 541–551 (1989)

14. Litjens, G., et al.: A survey on deep learning in medical image analysis. MIA **42**, 60–88 (2017)

15. Hornik, K.: Approximation capabilities of multilayer feedforward networks. Neural Netw. **4**, 251–257 (1991)

16. Jo, J., Bengio, Y.: Measuring the tendency of CNNs to learn surface statistical regularities, arXiv preprint (2017)

17. Joshi, S., Davis, B., Jomier, M., Gerig, G.: Unbiased diffeomorphic atlas construction for computational anatomy. NeuroImage **23**, S151–S160 (2004)

18. Caselles, V., Kimmel, R., Sapiro, G.: Geodesic active contours. IJCV **22**, 61–79 (1997). https://doi.org/10.1023/A:1007979827043

19. Kamnitsas, K., et al.: DeepMedic for brain tumor segmentation. In: Crimi, A., Menze, B., Maier, O., Reyes, M., Winzeck, S., Handels, H. (eds.) BrainLes 2016. LNCS, vol. 10154, pp. 138–149. Springer, Cham (2016). https://doi.org/10.1007/978-3-319-55524-9_14

20. Ronneberger, O., Fischer, P., Brox, T.: U-Net: convolutional networks for biomedical image segmentation. In: Navab, N., Hornegger, J., Wells, W.M., Frangi, A.F. (eds.) MICCAI 2015. LNCS, vol. 9351, pp. 234–241. Springer, Cham (2015). https://doi.org/10.1007/978-3-319-24574-4_28

Aneurysm Identification in Cerebral Models with Multiview Convolutional Neural Network

Mingsong Zhou[1], Xingce Wang[1(✉)], Zhongke Wu[1], Jose M. Pozo[2], and Alejandro F. Frangi[2]

[1] Information Science and Technology College, Beijing Normal University, Beijing, China
wangxingce@bnu.edu.cn
[2] Centre for Computational Imaging and Simulation Technologies in Biomedicine (CISTIB), School of Computing & School of Medicine, University of Leeds, Leeds, UK

Abstract. Stroke is the third most common cause of death and a major contributor to long-term disability worldwide. Severe stroke is most often caused by the rupture of a cerebral aneurysm, a weakened area in a blood vessel. The detection and quantification of cerebral aneurysms are essential for the prevention and treatment of aneurysmal rupture and cerebral infarction. Here, we propose a novel aneurysm detection method in a three-dimensional (3D) cerebrovascular model based on convolutional neural networks (CNNs). The multiview method is used to obtain a sequence of 2D images on the cerebral vessel branch model. The pretrained CNN is used with transfer learning to overcome the small training sample problem. The data augmentation strategy with rotation, mirroring and flipping helps improve the performance dramatically, particularly on our small datasets. The hyperparameter of the view number is determined in the task. We have applied the labeling task on 56 3D mesh models with aneurysms (positive) and 65 models without aneurysms (negative). The average accuracy of individual projected images is 87.86%, while that of the model is 93.4% with the best view number. The framework is highly effective with quick training efficiency that can be widely extended to detect other organ anomalies.

1 Introduction

Cerebral aneurysms are localized pathological dilatations of the cerebral arteries. Their rupture causes subarachnoid hemorrhage and is associated with a high morbidity and mortality rate [3]. For the average person, the incidence of aneurysms is 2–3%, and this proportion increases with age [13]. The early detection, growth monitoring and early treatment of aneurysms is the most effective

X. Wang and Z. Wu—Equal contribution to this work.

sequence method for preventing aneurysmal rupture. However, the early detection of aneurysms in the brain vessel network is quite challenging.

Conventional aneurysm detection methods use the machine learning method to classify the aneurysm and vessel segments. Three main methods have been used to identify areas in which aneurysms may occur based on vascular shape, vascular skeleton, and image differences. Algorithms based on vascular morphology depend on the assumption that aneurysms are approximately spherical. Suniaga used Hessian eigenvalues analysis to find spherical objects in 3D images [15]. Lauric constructed a geometric descriptor "writhe number" to distinguish between areas of tubular and nontubular structures [8]. The nontubular structures may be aneurysms. The dot filter [16] and blobness filter [2,4,15] have also been used to detect cluster structures in images based on prior knowledge of aneurysm morphology. The algorithm in [15,16] is based on the skeleton to find the endpoints and branch points of the vascular structure and considers the distance between the endpoints and the branch points as the parameters of the classifier. Several hybrid algorithms have been used to train the classifier after feature extraction, incorporating classification strategies such as feature thresholding [8], rule-based systems [16] or case-based reasoning [6]. Almost all proposed algorithms are intended to work with magnetic resonance angiography (MRA) datasets; one, however, implements a multimodal approach on three-dimensional rotational angiography (3DRA) and computed tomography angiography (CTA) datasets [8]. The conventional methods for aneurysms detection are not generalizable; they extract features using descriptors of a dot filter or a blobness filter, or they extract customized features such as those related to geometry or distance. Since the use of CNNs has been successful in computer vision and image processing, many studies have examined aneurysm detection in medical images, such as MRA or 3DRA using a CNN. Jerman [5] used a Hessian-based filter to enhance spherical and elliptical structures such as aneurysms and attenuate other structures on the angiograms. Next, they boosted the classification performance using a 2D CNN trained on intravascular distance maps computed by casting rays from the preclassified voxels and detecting the first-hit edges of the vascular structures. Nakao [10] employed a voxel-based CNN classifier. The inputs of the network were 2D images generated from volumes of interest of the MRA images by applying a mixed-integer programming algorithm. The network architecture they used was not very deep: 4 convolution layers in one [5] and 2 convolution layers in the other [10]. More adjustable parameters (weights and bias) correspond to greater freedom of adjustment and a better approximation effect.

In medical image analysis, 2D images are widely used as input, but this approach is not well suited for aneurysm detection due to four limitations. First, we are interested in detecting aneurysms from different types of imaging modalities such as CT, MRA or 3DRA. The image resolution and file size may adversely affect the CNN performance. Second, even for the subjects having aneurysms, the percentage of the aneurysm volume data is quite small, which causes an imbalance of positive and negative samples in the learning process. Third, doctors detect aneurysms relying more on anisotropic shape representation than the intensity or texture in image, which results in a starting research point of separation aneurysms directly from the 3D cerebral mesh model. Finally, due to ethics and case selection problems, the availability of large population databases is not assured. Training on a small sample dataset is a common problem for many tasks in medical image analysis, such as segmentation or registration. To manage these limits, we detect aneurysms with a CNN in 3D cerebral mesh models with a pretrained neural network. After the segmentation and reconstruction of the cerebral vessel mesh model, the heterogeneous nature of the image format and resolution can be eliminated. The cerebral vessel network model can be divided into branch models with two or three bifurcations. Relative to the volume data of an image, the imbalance of the training sample of a model can be significantly reduced. We overcome the influence of texture and intensity using the 3D cerebral branch model, which focuses on shape. In several view experiments, the classification accuracy of images is approximately 87%, while the classification accuracy of mesh models is approximately 92%. To our knowledge, we are the first group to apply CNN transfer learning to the aneurysm detection task on mesh models instead of medical images. The main contributions of the paper are as follows: (1)We present a novel aneurysm mesh model detection method based on a CNN. Due to the challenges of direct calculation convolution on the mesh model, we use the projection idea to change the 3D mesh model as a sequence of multiview projection images. (2)We use the transfer learning method and data augmentation of the input image to overcome the small training sample problem. The pretraining was performed using GoogleNet Inception V3 on ImageNet. We use the data augmentation with mirroring, rotation and flipping operations on the input image, which obtains 6 times more training samples than before.

2   Methodology

Problem Formulation. The aneurysm detection task is formulated as a classification problem in this paper. Assume we have a training dataset composed of branches with or without aneurysms in a featured space $T = \{(b_1, l_1), (b_2, l_2), ..., (b_n, l_n)\} \subset B \times L$ where $B = \mathbb{R}^n$ is the feature vector space and $L = \{0, 1\}$ is the label space, where $l_i = 1$ represents that b_i includes an aneurysm (positive). The objective is to predict the label from the feature vector by a classification function. $\hat{l} = f(b)$ The training of $f(b)$ is based on minimization of the error between the predicted value $\hat{l}_i$ and the ground truth l_i, and the parameters in the classification function f are updated so that the classifier can be more effective.

Architecture of the Network. The neural network of the project in called the multiview aneurysm model label network ($MVML$) with combined f^G and f^C. We consider as a classifier the pretrained GoogleNet Inception V3 (f^G) with a modified full-connection layer (f^C). The feature vector is each of the 2D images generated from multiview rendered images of the mesh models, which are used as inputs of the network. The outputs of the network are two probability values,

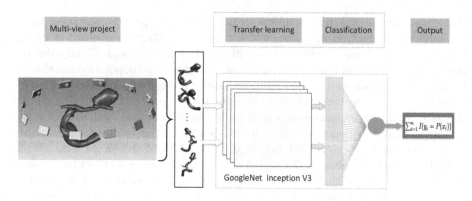

Fig. 1. Training process for the aneurysm detection with multiview CNN.

corresponding to the negative and positive cases. Finally, the accuracy of the mesh model is determined by majority voting according to the image accuracy. Figure 1 shows the framework of aneurysm detection proposed in this paper.

Case Selection. The positive dataset (vessel model with aneurysms) of 56 patients is drawn from a large multicenter database created within the EU-funded project @neurIST [9] based on the 3DRA image. The negative model set (vessel model without aneurysms), derived from the public dataset distributed by the MIDAS Data Server at Kitware Inc. [1], The segments of the mesh model included the similar branches as the positive dataset such as the anterior cerebral artery (ACA) or the internal carotid artery (ICA) bifurcation. No other information was considered during the selection process.

Multiview Images from 3D Aneurysm Models. To obtain the multiview images from each 3D aneurysm model, the model coordination must be determined, and the projected method should be chosen. We first use PCA to determine the coordination of the mesh model. The vascular vertexes of each sample are denoted by $A = \{\left[x_i^j, y_i^j, z_i^j\right], j = 1, 2, 3 \ldots N\}, i = 1, 2, 3 \ldots n_j$, where N denotes the number of samples within the dataset and n_j denotes the number of vascular vertexes in the dataset. The vertex coordinates of each sample are represented by $A_j(a) = \bar{A}_j + \sum_{i=1}^{n_j-1} a_i^j v_i^j$ $\bar{A}$ denotes the average vertex of the model j in the dataset and $a^j = [a_1{}^j, a_2{}^j, a_3{}^j, \ldots, a_{n_j-1}^j]$ denotes the eigenvalues of the covariance matrix in descending order of the model j and $v^j = [v_1{}^j, v_2{}^j, v_3{}^j, \ldots, v_{n-1}^j]$ denotes the corresponding orthogonal eigenvectors. The first three eigenvectors $[v_1{}^j, v_2{}^j, v_3{}^j]$, complemented with the right-hand rule, define the adapted coordinates of the 3D model j. The eigenvector corresponding to the largest eigenvalue is the rotation axis, while the one with the smallest eigenvalue of the first three is the beginning of the view projection. We then use the Phong reflection model [12] to create multiview images of the mesh model. We set up different numbers of viewpoints (virtual cameras) to obtain the mesh model rendering results. All the viewpoints (virtual cameras) are positioned on the ground plane and pointed toward the centroid of the mesh. We render the images from all the viewpoints to obtain a sequence of highly representative images of the mesh model. We shrink the white space around the view to enlarge its effective area. Different shading coefficients or illumination models do not affect our output descriptors due to the invariance of the learned filter to illumination changes as observed in an image-based CNN [7,14]. Thus, the view color is set to gray. We create v rendered views by placing v virtual cameras around the mesh every $360/v$ degrees. The selection of the value for the hyperparameter v is discussed in detail in the experimental description. The Toolbox Graph [11] is used to generate the rendering result of the 3D mesh model.

Data Augmentation with Rotation and Reflection. We enrich the dataset using mirroring, rotation, and flipping operations. Each image of the view is transformed to create 2 additional images by flipping its horizontal and vertical edges, and another 3 additional images are created by rotating with 90, 180, and 270°. Thus, we obtain 6 projected images from each view. The number of datasets is $v \times 6 \times 121$. With these treatments of the images, the difference of rotation axis orientation of the PCA eigenvector is eliminated.

Aneurysm Labeling with Transfer Learning. To address the limitations of the aneurysm model dataset, we use the transfer learning method of the image to realize the learning result. The pretrained CNN based on large annotated image databases (ImageNet) is used for various classification tasks in the images of the different domains. The original network architecture can be maintained, and the network can be initialized with pretrained weights. The representation of each layer can be calculated from the representation of the previous layer. The end-to-end back-propagation algorithm, which combines feature extraction and classification processes, is widely used in CNN training. Generally, the convolution layers are considered as feature extractors, while the fully connected layers are seen as a classifier. The network architecture $MVML$ of the project is composed by f^G and f^C. We accept the pretrained GoogleNet Inception V3 model (f^G) as the feature extractors and the two-layer fully connected neuron network (f^C) as the classification. The latter outputs probabilities of the two classes with each input image view with the Softmax function. The cross-entropy loss function is adopted. $C = -\frac{1}{n} \sum_{i=1}^{n} [y_i \ln \hat{y}_i + (1 - y_i) \ln(1 - \hat{y}_i)]$ When the network training, only the weights of the fully connected neuron network are updated with the pretrained GoogleNet weights frozen. From the resulting decision for each view, we obtained the mesh group decision with majority voting, $E_i = \sum_{i=1}^{m} I(y_i = $

$$P(x_i)) \quad s.t. \quad I(y_i = P(x_i)) = \begin{cases} 1, & \text{if } y_i = P(x_i); \\ 0, & \text{if } y_i \neq P(x_i). \end{cases} \text{ where } m = k \times v \text{ is the}$$

total number of projected images per mesh model and v is the number of views. In our task, k is the multiplying factor of the data augmentation. x_i is the input image, and y_i is the label of the image. $P(x_i)$ is the prediction of the image by the classifier. The final label for the mesh model is the one satisfying $E_i > \frac{m}{2}$. For instance, for $v = 12$ and $k = 6$, an aneurysm mesh model with more than

36 different positive labeled projected images is assigned a final positive label. Model performance is measured by first classifying views of testing mesh models, and the classification results of all views through a majority voting process are used to obtain the final class label for each mesh model.

3   Experiments and Data Analysis

We conduct our research platform based on TensorFlow using an NVIDIA 960 M GPU on an Ubuntu 16.10 Linux OS 64-bit operating system. The initial fully connected classification is randomly set from 0 to 1. A stochastic gradient descent optimizer is employed to train the loss function of cross-entropy. A learning rate of 0.01 is suitable. The epoch step $K = 500$. The mini-batch size $N' = 128$. A five-fold cross-validation is used on the classifier performance. In the following, we test the effectiveness of the classification algorithm, the effect of the data augmentation, and the computational time of the network training.

Optimization of the View Number Hyperparameter. First, we aim to verify the effect of the different view numbers v on the classification results using the accuracy of the mesh model and image data. We collect 3, 6, 9, 12, 15, and 18 views of the mesh model for the experiments. The views of the mesh models used for training the classifier are never used for testing. The overall prediction accuracy of the classifier on the image is evaluated, that is, the ratio of the number of images correctly classified to the total number of images evaluated (Table 1). The classification of each view is only an instrumental task. The real result is the classification of the model. The mesh model label is achieved by a majority voting process based on the predicted probability for every view. The data show that when the number of views is large (such as 18), more images can be created to identify the aneurysms, but image mislabeling will greatly influence the results. For the proposed method, the equal possibility of aneurysms with the voting result of the images without aneurysms reduces the accuracy of the final result. The view number in this research is a nonlinear and unpredictable hyperparameter that greatly influences the result. The small view number of the model cannot offer sufficient images to reveal the aneurysm's shape; however, the large view number creates more branch clip images, resulting in mislabeling. From these results, we selected the number of views $v = 9$ as the optimal one, with a mean accuracy of 93.40%.

Table 1. Classification accuracy of the image and mesh model (%).

View	3	6	9	12	15	18
Image	87.4 ± 2.4	87.6 ± 3.0	87.9 ± 2.8	88.0 ± 2.7	87.9 ± 2.6	87.7 ± 2.8
Model	90.9 ± 1.7	91.7 ± 0.1	93.4 ± 2.0	92.6 ± 1.6	92.6 ± 1.6	92.6 ± 1.6

Effect of the Data Augmentation. To validate the effect of the data augmentation on the images, we test the model with or without data augmentation of the images. For the without-mirroring and rotation data augmentation view, the sizes of the dataset are 363, 726, 1089, 1452, 1815, and 2178. The accuracy of the classifier experiment on the images and the mesh model is shown in Table 2. Thus, the data augmentation appears not to greatly influence the accuracy. Inception V3 can bring out the strong features of the image to clearly illustrate aneurysms. However, the data augmentation has a strong influence on the mesh model. Without data augmentation, the accuracy of the model decreases by an average of 2%. First, the convolution layer of the Inception V3 is local on the image. After the data augmentation, the augmented image can be labeled identically to the original image. Second, the data augmentation brings more training data, which can increase the learning result of the classifier in the fully connected neural network. Third, deep learning with small training data is relatively instable in learning. More data can bring better results. In this case, the image data greatly influence the model accuracy.

Table 2. Average classification accuracy of images and mesh models without data augmentation(%).

View	3	6	9	12	15	18
Image	87.9 ± 3.3	87.9 ± 3.7	88.3 ± 3.4	88.6 ± 2.7	87.9 ± 2.5	87.7 ± 2.9
Model	90.9 ± 3.1	90.1 ± 4.2	90.9 ± 4.1	91.8 ± 2.5	90.9 ± 3.0	91.8 ± 3.6

Computation and Convergence Time. The time-consuming processes that are involved constitute a major challenge encountered in deep learning. We use transfer learning with GoogleNet to limit the training data and decrease the test time. The average change in the total lost function is smaller than 0.01 for 20 steps. We can identify the convergence of the training. The convergence steps of the training process are shown in Tables 3. For data that are not mirrored and rotated, the average numbers of convergence steps of the classifier in different views are approximately 330, 301, 327, 332, 331, and 300. The average numbers of convergence steps for different view classifiers are approximately 384, 361, 367, 348, 363, and 373 for different view classifiers through the mirroring and rotation data.

Table 3. Convergence steps of the training process.

View	3	6	9	12	15	18
Without data augmentation	330	301	327	332	331	300
Data augmentation	384	361	367	348	363	373

4   Conclusions

In this paper, we present a new multiview CNN to identify aneurysms in a 3D cerebrovascular model. No registration or alignment is necessary in the method for any of the models. With the projection of the 3D mesh model, we can obtain the multiview images. The transfer learning method with data augmentation is used in the model. The final mesh model identification is obtained by the voting algorithm. The method is simple to understand and implement. In a future study, we plan to incorporate postprocessing adjustment that is known to slightly improve the identification of some datasets. The development of a more sophisticated automatic adjustment will also necessitate further research.

Acknowledgement. The authors want to thank the anonymous reviewers for their constructive comments. This research was partially supported by the National Key Cooperation between the BRICS of China (No. 2017YFE0100500), National Key R&D Program of China (No. 2017YFB1002604, No. 2017YFB1402105) and Beijing Natural Science Foundation of China (No. 4172033). AFF is supported by the Royal Academy of Engineering Chair in Emerging Technologies Scheme (CiET1819\19), and the OCEAN project (EP/M006328/1) and the MedIAN Network (EP/N026993/1) both funded by the Engineering and Physical Sciences Research Council (EPSRC).

References

1. Aylward, S.R., Bullitt, E.: Initialization, noise, singularities, and scale in height ridge traversal for tubular object centerline extraction. IEEE Trans. Med. Imaging **21**(2), 61–75 (2002)
2. Hentschke, C.M., Beuing, O., Nickl, R., Tönnies, K.D.: Detection of cerebral aneurysms in MRA, CTA and 3D-RA data sets. In: Proceedings of SPIE - The International Society for Optical Engineering, p. 83151I (2012)
3. Hop, J.W., Rinkel, G.J.E., Algra, A., van Gijn, J.: Case-fatality rates and functional outcome after subarachnoid hemorrhage: a systematic review. Stroke **28**(3), 660–664 (1997)
4. Jerman, T., Pernuš, F., Likar, B., Špiclin, Ž.: Computer-aided detection and quantification of intracranial aneurysms. In: Navab, N., Hornegger, J., Wells, W.M., Frangi, A.F. (eds.) MICCAI 2015. LNCS, vol. 9350, pp. 3–10. Springer, Cham (2015). https://doi.org/10.1007/978-3-319-24571-3_1
5. Jerman, T., Pernus, F., Likar, B., Špiclin, Ž.: Aneurysm detection in 3D cerebral angiograms based on intra-vascular distance mapping and convolutional neural networks. In: IEEE International Symposium on Biomedical Imaging, pp. 612–615 (2017)
6. Kobashi, S., Kondo, K., Hata, Y.: Computer-aided diagnosis of intracranial aneurysms in mra images with case-based reasoning. IEICE Trans. Inf. Syst. **89**(1), 340–350 (2006)
7. Krizhevsky, A., Sutskever, I., Hinton, G.E.: ImageNet classification with deep convolutional neural networks. In: Advances in Neural Information Processing Systems, pp. 1097–1105 (2012)
8. Lauric, A., Miller, E., Frisken, S., Malek, A.M.: Automated detection of intracranial aneurysms based on parent vessel 3D analysis. Med. Image Anal. **14**(2), 149–159 (2010)

9. Villa-Uriol, M.C., et al.: @neurIST complex information processing toolchain for the integrated management of cerebral aneurysms. Interface Focus **1**(3), 308–319 (2011)
10. Nakao, T., et al.: Deep neural network-based computer-assisted detection of cerebral aneurysms in MR angiography. J. Magn. Reson. Imaging **47**(4), 948–953 (2018)
11. Peyre, G.: MATLAB central file exchange select, 2 edn. (2009)
12. Phong, B.T.: Illumination for computer generated pictures. Commun. ACM **18**(6), 311–317 (1975)
13. Rinkel, G.J.E., Djibuti, M., Algra, A., Van Gijn, J.: Prevalence and risk of rupture of intracranial aneurysms: a systematic review. Stroke **29**(1), 251–256 (1998)
14. Su, H., Maji, S., Kalogerakis, E., Learned-Miller, E.: Multi-view convolutional neural networks for 3D shape recognition. In: Computer Vision and Pattern Recognition, pp. 945–953 (2015)
15. Suniaga, S., Werner, R., Kemmling, A., Groth, M., Fiehler, J., Forkert, N.D.: Computer-aided detection of aneurysms in 3D time-of-flight MRA datasets. In: Wang, F., Shen, D., Yan, P., Suzuki, K. (eds.) MLMI 2012. LNCS, vol. 7588, pp. 63–69. Springer, Heidelberg (2012). https://doi.org/10.1007/978-3-642-35428-1_8
16. Yang, X., Blezek, D.J., Cheng, L.T.E., Ryan, W.J., Kallmes, D.F., Erickson, B.J.: Computer-aided detection of intracranial aneurysms in MR angiography. Journal Digit. Imaging **24**(1), 86–95 (2011). https://doi.org/10.1007/s10278-009-9254-0

Predicting Clinical Outcome of Stroke Patients with Tractographic Feature

Po-Yu Kao[1](✉) (iD), Jefferson W. Chen[2], and B. S. Manjunath[1](✉)

[1] University of California, Santa Barbara, CA, USA
{poyu_kao,manj}@ucsb.edu
[2] University of California, Irvine, CA, USA

Abstract. The volume of stroke lesion is the gold standard for predicting the clinical outcome of stroke patients. However, the presence of stroke lesion may cause neural disruptions to other brain regions, and these potentially damaged regions may affect the clinical outcome of stroke patients. In this paper, we introduce the tractographic feature to capture these potentially damaged regions and predict the modified Rankin Scale (mRS), which is a widely used outcome measure in stroke clinical trials. The tractographic feature is built from the stroke lesion and average connectome information from a group of normal subjects. The tractographic feature takes into account different functional regions that may be affected by the stroke, thus complementing the commonly used stroke volume features. The proposed tractographic feature is tested on a public stroke benchmark Ischemic Stroke Lesion Segmentation 2017 and achieves higher accuracy than the stroke volume and the state-of-the-art feature on predicting the mRS grades of stroke patients. Also, the tractographic feature yields a lower average absolute error than the commonly used stroke volume feature.

Keywords: modified Rankin Scale (mRS) · Stroke · Clinical outcome prediction · Tractographic feature · Machine learning

1 Introduction

According to the World Health Organization, 15 million people suffer strokes each year, the second leading cause of death (5.8 million) and the third leading cause of disability worldwide [8,13]. Around 87% of strokes are ischemic strokes, which result from an obstruction within a blood vessel in the brain [18]. The corresponding lack of oxygen results in different degrees of disability of people, and the modified Rankin Scale (mRS) is commonly used to measure the degree of disability or dependence in the daily activities of stroke patients [2,5,22].

Several studies [2,12,14,17,21,23] demonstrate significant correlations between stroke volume and mRS grades, with larger lesions predicting more severe disability. However, only a few studies [3,4,15] extracted different features, including first-order features and deep features, other than the volume

© Springer Nature Switzerland AG 2020
A. Crimi and S. Bakas (Eds.): BrainLes 2019, LNCS 11992, pp. 32–43, 2020.
https://doi.org/10.1007/978-3-030-46640-4_4

from stroke lesion to predict the mRS grades of stroke patients. The study of Maier and Handels [15] is most relevant to our work. They extracted 1650 image features and 12 shape characteristics from the stroke volume, the volume surrounding the stroke and the remaining brain volume, and they applied a random forest regressor with 200 trees on these 1662 features to predict the mRS grades of stroke patients. However, the presence of stroke lesion may disrupt other brain regions that may affect the clinical outcome of stroke patients.

The main contribution of this paper is the introduction of a new second-order feature, the tractographic feature, that couples the stroke lesion of a patient with the average connectome information from a group of normal subjects. The tractographic feature describes the potentially damaged brain regions due to the neural disruptions of the stroke lesion. Ideally one would like to use the diffusion images from the stroke patient, but this is not a realistic scenario. For instance, the patient with mental in their body is unsafe for getting an MRI scan. Instead, we use the "normal" subject data from the HCP project with the assumption that the parcellations and the associated tracts computed from that data are a reasonable approximation to extract the connectivity features. These tractographic features coupled with the stroke lesion information are used to predict the mRS grades of stroke patients. The concept of the tractographic feature was first proposed by Kao et al. [9] who used these to predict the overall survival of brain tumor patients. We modify their method to adapt to the size of the lesions and propose a new weighted vector of the tractographic feature. Our experimental results demonstrate that the proposed approach improves upon the state-of-the-art method and the gold standard in predicting the clinical outcome of stroke patients.

2   Materials and Methods

2.1   Dataset

Ischemic Stroke Lesion Segmentation (ISLES) 2017 [10,16] provides 43 subjects in the training dataset. Each subject has two diffusion maps (DWI, ADC), five perfusion maps (CBV, CBF, MTT, TTP, Tmax), one ground-truth lesion mask and clinical parameters. The ground-truth lesion mask is built in the follow-up anatomical sequence (T2w or FLAIR) and the corresponding mRS grade was given on the same day. The clinical parameters include mRS grade ranging from 0 to 4, time-to-mRS (88 to 389 days), TICI scale grade from 0 to 3, time-since-stroke (in minutes), and time-to-treatment (in minutes). Since TICI scale grade, time-since-stroke, and time-to-treatment were missing for some subjects, these three clinical parameters are not used in this work. The dimension and voxel spacing of MR images are different between each subject, but they are the same within each subject. We only focus on the subjects who obtain an mRS grade at 3 months (90 days) following hospital discharge since ascertainment of disability at 3-month post-stroke is an essential component of outcome assessment in stroke patients [5], and the tractographic data may change at a different time. Therefore, only 37 subjects are considered in this paper.

2.2   Tractographic Feature

The tractographic feature describes the potentially damaged region impacted by the presence of the stroke lesion through the average connectome information from 1021 Human Connectome Project (HCP) subjects [20]. For each HCP subject, q-space diffeomorphic reconstruction [24] is used to compute the diffusion orientation distribution function. Figure 1 shows the workflow of building a tractographic feature for a stroke patient.

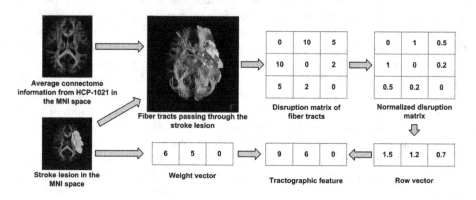

Fig. 1. The workflow for constructing a tractographic feature from a stroke region.

Given the stroke lesion in the subject space, we first map the stroke lesion to the Montreal Neurological Institute (MNI) space [6]. Second, we place one seed within each voxel of the brain region, and a deterministic diffusion fiber tracking method [25] is used to find all possible tracts passing through the stroke volume inside the brain from the average diffusion orientation distribution function of 1021 HCP subjects. Topology-informed pruning [26] is used to remove false-positive tracts. Third, an existing brain parcellation atlas is used to create a disruption matrix D, which describes the degree of disruption between different brain parcellation regions due to the presence of the stroke lesion.

$$D = \begin{bmatrix} d_{11} & d_{12} & \dots & d_{1N} \\ d_{21} & d_{22} & \dots & d_{2N} \\ \vdots & \vdots & \ddots & \vdots \\ d_{N1} & d_{N2} & \dots & d_{NN} \end{bmatrix} \tag{1}$$

d_{ij} notes the number of tracts starting from a region i and ending in a region j, and N is the total number of brain parcellation regions in the existing atlas. Then, this disruption matrix is normalized by its maximum value, i.e., $\hat{D} = D/d_m$ where $\hat{D}$ is the normalized disruption matrix, and d_m is the maximum element of the disruption D. Afterward, we sum up each column in this normalized disruption matrix $\hat{D}$ to form a row vector $\boldsymbol{L} = \sum_{i=1}^{N} \hat{d}_{ij} = [l_1, l_2, \dots, l_N]$.

From the stroke lesion, we build a weight vector $\gamma = [s_1, s_2, \ldots, s_N]$, which is the distribution of the stroke volume in the different brain parcellation regions. s_i is the volume of the stroke lesion in the i-th brain parcellation region. In the end, the row vector L is multiplied by this weight vector γ element-wisely to form the tractographic feature T.

$$T = \gamma \circ L \qquad (2)$$

$\circ$ is the Hadamard-product. This vector T is the proposed tractographic feature extracted from stroke lesion without any diffusion information of a patient. In this paper, the Automated Anatomical Labeling (AAL) [19] template is used to define 116 brain regions so the dimension of the tractographic feature is 116. The reasons for choosing AAL rather than other existing atlases are (i) this atlas contains an optimal number of brain regions that could make each region large enough to compensate possible stroke-induced lesion effect or distortion, and (ii) this atlas contains cortical, subcortical and cerebellar regions, which could be equally important for mRS prediction. The source code is available on GitHub[1].

Parameters of Fiber Tracking

DSI Studio[2] is used to build the fiber tracts for each subject. Table 1 shows the tracking parameters[3] we used in this paper. The type of stroke lesion is set to ROI (–roi=stroke_lesion) that found all possible tracts passing through the stroke lesion.

Parameters of Connectivity Matrix

DSI studio is used to create the connectivity matrix[4] followed by fiber tracking. Automated Anatomical Labeling is chosen to form a 116×116 connectivity matrix. The type of the connectivity matrix is set to end, the value of each element in the connectivity matrix is the count of fiber tracts, and the threshold to remove the noise in the connectivity matrix is set to 0.

2.3   Evaluation Metrics

The employed evaluation metrics are (i) the accuracy, which is the percentage of the predicted mRS scores matching the corresponding ground-truth mRS scores, and (ii) the average absolute error between the predicted mRS scores and the corresponding ground-truth mRS scores.

---

[1] https://github.com/pykao/ISLES2017-mRS-prediction.

[2] https://github.com/frankyeh/DSI-Studio.

[3] parameter_id=F168233E9A99193F32318D24ba3Fba3Fb404b0FA43D21D22cb01ba0 2a01d.

[4] http://dsi-studio.labsolver.org/Manual/command-line-for-dsi-studio.

Table 1. Tracking parameters of building the fiber tracts for stroke patients in this paper. More details of parameters can be found at http://dsi-studio.labsolver.org/Manual/Fiber-Tracking.

Parameter	Value
Termination index	qa
Threshold	0.15958
Angular threshold	90
Step size (mm)	0.50
Smoothing	0.50
Min length (mm)	3.0
Max length (mm)	500.0
Topology-informed pruning (iteration)	1
Seed orientation	All orientations
Seed position	Voxel
Randomize seeding	Off
Check ending	Off
Direction interpolation	Tri-linear
Tracking algorithm	Streamline (Euler)
Terminate if	2,235,858 Tracts
Default otus	0.60

3 Experimental Results

First Experiment: In this experiment, we compare the mRS prediction performance of the tractographic feature with other first-order features extracted from the lesion mask. These first-order features include the volumetric feature, spatial feature, morphological feature and volumetric-spatial feature depicted in Table 2. The framework of the first experiment is shown in Fig. 2.

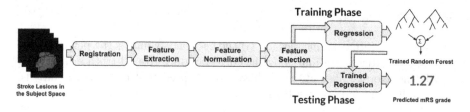

Fig. 2. The framework of the first experiment. In the end, the predicted mRS grade is rounded to an integer.

We first register the stroke lesions from subject space to the MNI space to overcome the differences of the voxel spacing and image dimension between different subjects. The tractographic feature and other first-order features are extracted from these normalized stroke lesions. After feature extraction, we apply a standard feature normalization on the extracted features to ensure that each dimension of the features has the same scale. Then, we remove the dimensions of the features with zero variance between subjects and apply a recursive feature elimination with leave-one-out cross-validation to find the best subset of the feature that yields the lowest average mean absolute error. In the training phase, we train one random forest regressor for each type of feature, i.e., five random forest regressors are trained. Each random forest regressor has 300 trees of which maximum depth is 3. In the testing phase, we use different types of features with the corresponding trained random forest regressors to predict the mRS grades of stroke patients, and the predicted mRS grade is rounded to an integer. We evaluate the mRS prediction performance of different types of features with leave-one-out cross-validation on ISLES 2017 training dataset. The quantitative results are reported in Table 3. From Table 3, the tractographic feature has the highest accuracy and lowest average absolute error on predicting the mRS grades of stroke patients compared to other first-order features.

Table 2. First-order features extracted from the stroke lesion.

Type of feature	Descriptions
Volumetric feature	Volumetric feature is the volume of the lesion in the MNI space, and it only has one dimension
Spatial feature	Spatial feature describes the location of the lesion in the brain. The centroid of the lesion is extracted as the spatial feature for each subject, and the spatial feature has three dimensions
Morphological feature	Morphological feature describes shape information of the lesion. The length of the major axis and minor axis of the lesion, the ratio of the length of the major axis and minor axis of the lesion, the solidity and roundness of the lesion, and the surface of the lesion are extracted as the morphological feature. The morphological feature has six dimensions for each subject
Volumetric-spatial feature	Volumetric-spatial feature describes the distribution of the stroke lesion in different brain parcellation regions from an existing structural atlas. Automated Anatomical Labeling (AAL) [19] is used to build the volumetric-spatial feature so the dimension of the volumetric-spatial feature is 116

Second Experiment: We compare the mRS prediction performance of the tractographic feature with the state-of-the-art feature proposed by Maier and Handels [15]. We implement their feature extraction method on ISLES 2017 dataset. First, 1650 image features and 12 shape features are extracted from the lesion volume and the apparent diffusion coefficients (ADC) maps in the subject space. Thereafter, these two types of features are concatenated to build a 1662-dimension feature. Then, we apply the same feature normalization, feature selection, cross-validation, and random forest regressor as the first experiment to predict the mRS of stroke patients. The quantitative results of the state-of-the-art feature are also shown in Table 3. From Table 3, the tractographic feature also achieves higher accuracy and similar average absolute error ($p = 0.81$) compared to the state-of-the-art feature.

Table 3. The mRS prediction performance of different types of features on ISLES 2017 training dataset with leave-one-out cross-validation. The bold numbers show the best performance. (The average absolute error is reported as mean $\pm$ std.)

Type of feature	Accuracy	Average absolute error
Tractographic feature	**0.622**	0.487 ± 0.683
Volumetric feature	0.514	0.595 ± 0.715
Volumetric-spatial feature	0.568	0.621 ± 0.817
Morphological feature	0.378	0.703 ± 0.609
Spatial feature	0.351	0.919 ± 0.882
Maier and Handels [15]	0.595	$\mathbf{0.460 \pm 0.597}$

4   Discussion and Conclusion

From the first experiment, the tractographic feature has the best mRS prediction accuracy and the lowest average absolute error compared to other first-order features. The main reason is that the tractographic feature integrates volumetric-spatial information of the stroke lesion and the average diffusion information from a group of normal subjects that describes the potentially damaged regions impacted by the stroke lesion. These potentially damaged regions are formatted in the disruption matrix D from Eq. (1), and the weight vector γ from Eq. (2) carries spatial and volumetric information of the stroke lesion to the tractographic feature T. Also, it is worth noting that the volumetric-spatial feature is the same as the weight vector γ of the tractographic feature, and the mRS prediction performance of volumetric-spatial feature is improved by considering the average connectome information from a group of normal subjects.

The second experiment demonstrates that the tractographic feature also has better mRS prediction accuracy than the state-of-the-art feature [15]. It should be noted that their approach requires ADC maps that are not necessarily always available, and using only the lesion shape information degrades the overall performance significantly in their approach. We also note that the tractographic feature is of much lower dimensions (116) compared to the state-of-the-art feature (1662).

In both experiments, we apply the recursive feature selection with cross-validation on different types of features, and this procedure reduces one dimension of feature recursively until finding the best subset of the feature with the lowest mean absolute error. For the tractographic feature, this reduces the dimensionality from 116 to 8. This selected tractographic feature comes from eight AAL regions shown in Fig. 3 (left and right inferior temporal gyrus, right Rolandic operculum, left middle frontal gyrus, orbital part and triangular part of right inferior frontal gyrus, left angular gyrus and left putamen).

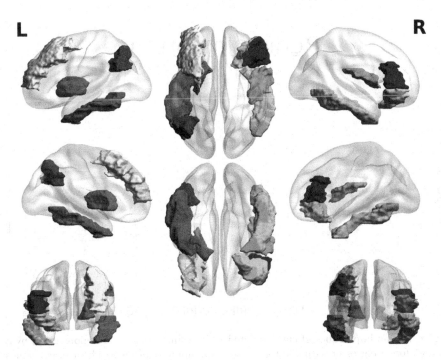

Fig. 3. Selected tractographic feature from eight AAL regions including left (in red) and right (in pink) inferior temporal gyrus red, right Rolandic operculum (in orange), left middle frontal gyrus (in yellow), orbital part (in green) and triangular part (in blue) of right inferior frontal gyrus, left angular gyrus (in purple) and left putamen (in grey) after applying the recursive feature selection with cross-validation on the original tractographic features. These tractographic features are extracted from 37 ISLES 2017 training subjects. **Best viewed in color.**

After feature selection, we use a random forest regressor to predict the mRS grades of stroke patients. The random forest regressor gives the importance to each dimension within a given type of feature shown in Fig. 4.

For the selected tractographic feature, left inferior temporal gyrus yields the highest average importance compared to the other seven regions within 37 ISLES 2017 training subjects on the task of predicting the mRS grades. The reasons left inferior temporal gyrus has the greatest effect on the mRS of stroke patients are (i) this region is important for language processing and speech production [1], and (ii) a large number of fiber tracts, passing through this region, goes across the splenium of the corpus callosum which connects the visual, parietal and auditory cortices [7,11] (See Fig. 5).

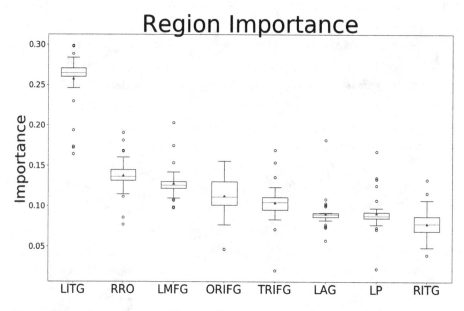

Fig. 4. Region importance of eight selected AAL brain parcellation regions given by a random forest regressor with 300 trees whose maximum depth is 3. The average values are marked in the green triangles. Left inferior temporal gyrus (LITG) yields a higher mean importance (0.26) than right Rolandic operculum (RRO, 0.14), left middle frontal gyrus (LMFG, 0.13), orbital part (ORIFG, 0.11) and triangular part (TRIFG, 0.10) of right inferior frontal gyrus, left angular gyrus (LAG, 0.09), left putamen (LP, 0.09) and right inferior temporal gyrus (RITG, 0.08) within 37 ISLES 2017 training subjects on the task of predicting the mRS grades of stroke patients. **Best viewed in color.**

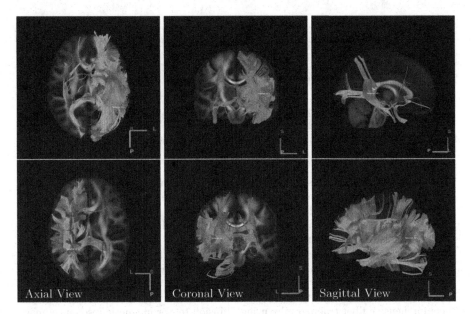

Fig. 5. The fiber tracts passing through the left inferior temporal gyrus from the average connectome information of 1024 HCP subjects. We place a seed in each voxel inside the whole brain to find all possible tracts passing through the left inferior temporal gyrus. **Best viewed in color.**

In conclusion, the paper presents for the first time the use of tractographic features for predicting the clinical outcome of stroke patients. The tractographic feature leads to promising mRS prediction results on ISLES 2017 dataset but needs to be further validated using a larger and representative independent dataset to rule out a potential methodical bias and over-fitting effects. The proposed tractographic feature has the potential to be improved if we build a disruption matrix from each HCP subject given the stroke lesion in MNI space and construct the average disruption matrix from these individual disruption matrices.

Limitation. The proposed tractographic feature cannot be generated if the stroke lesion is not located in the brain parcellation regions.

Acknowledgement. This research was partially supported by a National Institutes of Health (NIH) award # 5R01NS103774-03. We thank Oytun Ulutan for technical support and Dr. Robby Nadler for writing assistance and language editing.

References

1. Antonucci, S.M., et al.: Lexical retrieval and semantic knowledge in patients with left inferior temporal lobe lesions. Aphasiology **22**(3), 281–304 (2008)

2. Banks, J.L., Marotta, C.A.: Outcomes validity and reliability of the modified Rankin scale: implications for stroke clinical trials: a literature review and synthesis. Stroke **38**(3), 1091–1096 (2007)
3. Choi, Y., Kwon, Y., Lee, H., Kim, B.J., Paik, M.C., Won, J.H.: Ensemble of deep convolutional neural networks for prognosis of ischemic stroke. In: Crimi, A., Menze, B., Maier, O., Reyes, M., Winzeck, S., Handels, H. (eds.) BrainLes 2016. LNCS, vol. 10154, pp. 231–243. Springer, Cham (2016). https://doi.org/10.1007/978-3-319-55524-9_22
4. Forkert, N.D., et al.: Multiclass support vector machine-based lesion mapping predicts functional outcome in ischemic stroke patients. PLoS One **10**(6), e0129569 (2015)
5. Goyal, M., et al.: Endovascular thrombectomy after large-vessel ischaemic stroke: a meta-analysis of individual patient data from five randomised trials. Lancet **387**(10029), 1723–1731 (2016)
6. Grabner, G., Janke, A.L., Budge, M.M., Smith, D., Pruessner, J., Collins, D.L.: Symmetric atlasing and model based segmentation: an application to the hippocampus in older adults. In: Larsen, R., Nielsen, M., Sporring, J. (eds.) MICCAI 2006. LNCS, vol. 4191, pp. 58–66. Springer, Heidelberg (2006). https://doi.org/10.1007/11866763_8
7. Hofer, S., Frahm, J.: Topography of the human corpus callosum revisited—comprehensive fiber tractography using diffusion tensor magnetic resonance imaging. Neuroimage **32**(3), 989–994 (2006)
8. Johnson, W., et al.: Stroke: a global response is needed. Bull. World Health Organ. **94**(9), 634 (2016)
9. Kao, P.-Y., Ngo, T., Zhang, A., Chen, J.W., Manjunath, B.S.: Brain tumor segmentation and tractographic feature extraction from structural MR images for overall survival prediction. In: Crimi, A., Bakas, S., Kuijf, H., Keyvan, F., Reyes, M., van Walsum, T. (eds.) BrainLes 2018. LNCS, vol. 11384, pp. 128–141. Springer, Cham (2019). https://doi.org/10.1007/978-3-030-11726-9_12
10. Kistler, M., et al.: The virtual skeleton database: an open access repository for biomedical research and collaboration. J. Med. Internet Res. **15**(11), e245 (2013)
11. Knyazeva, M.G.: Splenium of corpus callosum: patterns of interhemispheric interaction in children and adults. Neural Plast. **2013**, 12 (2013)
12. Lev, M.H., et al.: Utility of perfusion-weighted CT imaging in acute middle cerebral artery stroke treated with intra-arterial thrombolysis: prediction of final infarct volume and clinical outcome. Stroke **32**(9), 2021–2028 (2001)
13. Lindley, R.I.: Stroke, 2nd edn. Oxford University Press, Oxford (2017)
14. Löuvbld, K.O., et al.: Ischemic lesion volumes in acute stroke by diffusion-weighted magnetic resonance imaging correlate with clinical outcome. Ann. Neurol.: Off. J. Am. Neurol. Assoc. Child Neurol. Soc. **42**(2), 164–170 (1997)
15. Maier, O., Handels, H.: Predicting stroke lesion and clinical outcome with random forests. In: Crimi, A., Menze, B., Maier, O., Reyes, M., Winzeck, S., Handels, H. (eds.) BrainLes 2016. LNCS, vol. 10154, pp. 219–230. Springer, Cham (2016). https://doi.org/10.1007/978-3-319-55524-9_21
16. Maier, O., et al.: ISLES 2015 - A public evaluation benchmark for ischemic stroke lesion segmentation from multispectral MRI. Med. Image Anal. **35**, 250–269 (2017)
17. Parsons, M., et al.: Combined 1H MR spectroscopy and diffusion-weighted MRI improves the prediction of stroke outcome. Neurology **55**(4), 498–506 (2000)
18. Summers, D., et al.: Comprehensive overview of nursing and interdisciplinary care of the acute ischemic stroke patient: a scientific statement from the American Heart Association. Stroke **40**(8), 2911–2944 (2009)

19. Tzourio-Mazoyer, N., et al.: Automated anatomical labeling of activations in SPM using a macroscopic anatomical parcellation of the MNI MRI single-subject brain. Neuroimage **15**(1), 273–289 (2002)
20. Van Essen, D.C., et al.: The WU-Minn human connectome project: an overview. Neuroimage **80**, 62–79 (2013)
21. Van Everdingen, K., et al.: Diffusion-weighted magnetic resonance imaging in acute stroke. Stroke **29**(9), 1783–1790 (1998)
22. Van Swieten, J., et al.: Interobserver agreement for the assessment of handicap in stroke patients. Stroke **19**(5), 604–607 (1988)
23. Vogt, G., et al.: Initial lesion volume is an independent predictor of clinical stroke outcome at day 90: an analysis of the virtual international stroke trials archive (VISTA) database. Stroke **43**(5), 1266–1272 (2012)
24. Yeh, F.C., Tseng, W.Y.I.: NTU-90: a high angular resolution brain atlas constructed by q-space diffeomorphic reconstruction. Neuroimage **58**(1), 91–99 (2011)
25. Yeh, F.C., et al.: Deterministic diffusion fiber tracking improved by quantitative anisotropy. PLoS One **8**(11), e80713 (2013)
26. Yeh, F.-C., et al.: Automatic removal of false connections in diffusion MRI tractography using topology-informed pruning (TIP). Neurotherapeutics **16**(1), 52–58 (2018). https://doi.org/10.1007/s13311-018-0663-y

Towards Population-Based Histologic Stain Normalization of Glioblastoma

Caleb M. Grenko[1,2], Angela N. Viaene[3], MacLean P. Nasrallah[4],
Michael D. Feldman[4], Hamed Akbari[1,5], and Spyridon Bakas[1,4,5(✉)]

[1] Center for Biomedical Image Computing and Analytics (CBICA),
University of Pennsylvania, Philadelphia, PA, USA
sbakas@upenn.edu

[2] Center for Interdisciplinary Studies, Davidson College, Davidson, NC, USA

[3] Department of Pathology and Laboratory Medicine, Children's Hospital
of Philadelphia, University of Pennsylvania, Philadelphia, PA, USA

[4] Department of Pathology and Laboratory Medicine, Perelman School of Medicine,
University of Pennsylvania, Philadelphia, PA, USA

[5] Department of Radiology, Perelman School of Medicine,
University of Pennsylvania, Philadelphia, PA, USA

Abstract. Glioblastoma (*'GBM'*) is the most aggressive type of primary malignant adult brain tumor, with very heterogeneous radiographic, histologic, and molecular profiles. A growing body of advanced computational analyses are conducted towards further understanding the biology and variation in glioblastoma. To address the intrinsic heterogeneity among different computational studies, reference standards have been established to facilitate both radiographic and molecular analyses, e.g., anatomical atlas for image registration and housekeeping genes, respectively. However, there is an apparent lack of reference standards in the domain of digital pathology, where each independent study uses an arbitrarily chosen slide from their evaluation dataset for normalization purposes. In this study, we introduce a novel stain normalization approach based on a composite reference slide comprised of information from a large population of anatomically annotated hematoxylin and eosin (*'H&E'*) whole-slide images from the Ivy Glioblastoma Atlas Project (*'IvyGAP'*). Two board-certified neuropathologists manually reviewed and selected annotations in 509 slides, according to the World Health Organization definitions. We computed summary statistics from each of these approved annotations and weighted them based on their percent contribution to overall slide (*'PCOS'*), to form a global histogram and stain vectors. Quantitative evaluation of pre- and post-normalization stain density statistics for each annotated region with PCOS $> 0.05\%$ yielded a significant (largest $p = 0.001$, two-sided Wilcoxon rank sum test) reduction of its intensity variation for both *'H'* & *'E'*. Subject to further large-scale evaluation, our findings support the proposed approach as a potentially robust population-based reference for stain normalization.

Keywords: Histology · Digital pathology · Computational pathology · Stain normalization · Pre-processing · Glioblastoma · Brain tumor

© Springer Nature Switzerland AG 2020
A. Crimi and S. Bakas (Eds.): BrainLes 2019, LNCS 11992, pp. 44–56, 2020.
https://doi.org/10.1007/978-3-030-46640-4_5

1 Introduction

Glioblastoma (*'GBM'*) is the most aggressive, and common, type of primary malignant adult brain tumor. GBMs are usually *de novo*, meaning they frequently appear without any precursor lesions. If left untreated, the tumor is quickly fatal, and even with treatment, median survival is about 16 months [1,2]. If a GBM is suspected, multi-parametric magnetic resonance imaging (*'mpMRI'*) will be done to follow up, and presumptive diagnosis can typically be given. Ideally, surgical gross total resection is performed. When extensive surgery is not possible, often needle or excisional biopsies are performed to confirm the diagnosis. GBMs, due to their serious and sudden nature, lack of effective treatment options, as well as their reported heterogeneity [3], have been the subject of research in the realm of personalized medicine and diagnostics. However, investigating their precise characterization requires large amounts of data. Fortunately, publicly available datasets with abundant information are becoming much more available.

With the advent of data collection and storage, not only are large datasets becoming available for public use, but they are also becoming more detailed in multiple scales, i.e., macro- and micro-scopic. Large comprehensive datasets publicly available in various repositories, such as The Cancer Imaging Archive (TCIA - www.cancerimagingarchive.net) [4] have shown promise on expediting discovery. One of the exemplary data collections of glioblastoma, is the TCGA-GBM [5], which since its initiation has included longitudinal radiographic scans of GBM patients, with corresponding detailed molecular characterization hosted in the National Cancer Institute's Genomic Data Commons (gdc.cancer.gov). This dataset enabled impactful landmark studies on the discovery on integrated genomic analyses of gliomas [6,7]. TCIA has also made possible the release of 'Analysis Results' from individual research groups, with the intention of avoiding study replication and allowing reproducibility analyses, but also expediting further discoveries. Examples of these 'Analysis Results' describe the public release of expert annotations [8–11], as well as exploratory radiogenomic and outcome prediction analyses [12–15]. The further inclusion of available histology whole-slide images (*'WSIs'*) corresponding to the existing radiographic scans of the TCGA-GBM collection contributes to the advent of integrated diagnostic analyses, which in turn raises the need for normalization. Specifically, such analyses attempt to identify integrated tumor phenotypes [16,17] and are primarily based on extracting visual descriptors, such as the tumor location [18,19], and intensity values [20], as well as subtle sub-visual imaging features [21–28].

Intensity normalization is considered an essential step for performing such computational analysis of medical images. Specifically, for digital pathology analyses, stain normalization is an essential pre-processing step directly affecting subsequent computational analyses. Following the acquisition of tissue up until the digitization and storage of WSIs, nearly every step introduces variation into the final appearance of a slide. Prior to staining, tissue is fixed for variable amounts of time. Slide staining is a chemical process, and thus is highly prone to not only the solution preparation, but also the environmental conditions.

While preparing a specimen, the final appearance can be determined by factors such as: stain duration, manufacturer, pH balance, temperature, section thickness, fixative, and numerous other biological, chemical, or environmental conditions. Additionally, the advent of digital pathology has incurred even more variation on the final appearance of WSIs, including significant differences in the process of digitization that varies between scanners (vary by manufacturers and models within a given company).

Various approaches have been developed to overcome these variations in slide appearance. Techniques such as Red-Green-Blue (*'RGB'*) histogram transfers [29] and Macenko *et al.* [30] use the general approach of converting an image to an appropriate colorspace, and using a single example slide as a target for modifying the colors and intensities of a source image. Recently, techniques such as Reinhard [31], Vahadane [32], and Khan [33] have been developed to separate the image into optical density (*'OD'*) stain vectors (S), as well as corresponding densities (W) of each stain per pixel. This process (known as 'stain deconvolution') has been one of the more successful and popular techniques in recent years. Additionally, a number of generative deep learning techniques, such as StainGAN [34] and StaNoSA [35], have also been developed for stain normalization. While such techniques [34,35] have been shown to outperform many transfer-based approaches [30–33], they also have multiple downsides. First, these techniques are generative, which means that rather than modifying existing information, they attempt to generate their own information based on distributive models. These generative techniques apply a "blackbox" to input data, making it difficult to discern if the model is biased, and hence may influence all downstream processing without notice. For example, if a StainGAN model had not seen an uncommon structure during training, it would be unable to accurately model the staining of that structure, and fail in producing an accurate result. For a much more thorough review of stain normalization algorithms, see [36]. Our approach, in comparison to StainGAN or other generative methods, attempts to expand upon prior transformative stain transfer techniques. Our motivation is that an approach as the one we propose will obviate the "black box" presented by generative methods, and prevent a potential entry for insidious bias in the normalization of slides, while still maintaining a robust and accurate representation of a slide batch.

While the technical aspects of stain transfer algorithms have progressed significantly, their application remains fairly naive. In demonstration, most studies arbitrarily pick either a single WSIs, or even a single patch within a WSI, as a normalization step prior to further analysis. In this paper, we sought to build upon current stain normalization techniques by using a publicly available dataset in an effort to form a composite reference slide for stain transfer, and avoid the use of arbitrarily chosen slides from independent studies. By doing this, we have developed a standardized target for stain normalization, thus allowing the creation of more robust, accurate, and reproducible digital pathology techniques through the employment of a universal pre-processing step.

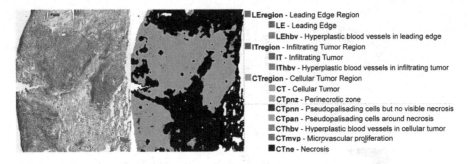

Fig. 1. An example slide and corresponding label map for IvyGAP is shown on the left, and on the right is the list of anatomical features annotated in the IvyGAP dataset, as well as their corresponding color in the label map.

2   Materials and Methods

2.1   Data

The Ivy Glioblastoma Atlas Project (*'IvyGAP'*) [37,38] describes a comprehensive *radio-patho-genomic* dataset of GBM patients systematically analyzed, towards developing innovative diagnostic methods for brain cancer patients [37].

IvyGAP is a collaborative project between the *Ben and Catherine Ivy Foundation*, the *Allen Institute for Brain Science*, and the *Ben and Catherine Ivy Center for Advanced Brain Tumor Treatment*. The radiographic scans of IvyGAP are made available on TCIA (wiki.cancerimagingarchive.net/display/Public/Ivy+GAP). *In situ* hybridization (*'ISH'*), RNA sequencing data, and digitized histology slides, along with corresponding anatomic annotations are available through the Allen Institute (glioblastoma.alleninstitute.org). Furthermore, the detailed clinical, genomic, and expression array data, designed to elucidate the pathways involved in GBM development and progression, are available through the Swedish Institute (ivygap.swedish.org).

The histologic data contains approximately 11,000 digitized and annotated frozen tissue sections from 42 tumors (41 GBM patients) [37] in the form of hematoxylin and eosin (*'H&E'*) stained slides along with accompanying ISH tests and molecular characterization. Tissue acquisition, processing, and staining occurred at different sites and times by different people following specific protocols [37]. Notably to this study, this resource contains a large number of H&E-stained GBM slides, each with a corresponding set of annotations corresponding to structural components of the tumor (Fig. 1).

2.2   Data Selection/Tissue Review

As also mentioned in the landmark manuscript describing the complete IvyGAP data [37], we note that the annotation labels (Fig. 1) a) do not comply with the current World Health Organization (*'WHO'*) classification, and b) have as

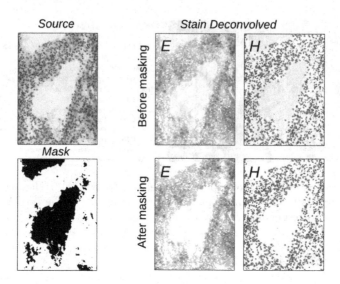

Fig. 2. An example of stain deconvolution on a patch from an IvyGAP WSI. The top and bottom row illustrate the effect of deconvolution before and after applying background masking, where is a substantial artifact of background intensity.

low as 60% accuracy due to their semi-supervised segmentation approach (as noted by IvyGAP's Table S11 [37]). Therefore, for the purpose of this study, and to ensure consistency and compliance with the clinical evaluation WHO criteria during standard practice, 509 IvyGAP annotated histological images were reviewed by two board-certified neuropathologists (A.N.V. and M.P.N.). For each image, the structural features/regions that were correctly identified and labeled by IvyGAP's semi-automated annotation application according to their published criteria [37] were marked for inclusion in this study, and all others were excluded from the analysis.

2.3   Stain Deconvolution

Each slide was paired with a corresponding label map of anatomical features, and then on a per-annotation basis, each region was extracted and the stains were deconvolved. A example deconvolution is illustrated in Fig. 2. Our method of stain deconvolution was based off the work of Ruifrok and Johnston [39], whose work has become the basis for many popular normalization methods. More directly, our method stems from Vahadane *et al.* [32], but modified so the density transformation is not a linear mapping, but a two-channel stain density histogram matching.

First, a source image is flattened to a list of pixels, and represented by the matrix I. It can be stated that:

$$I \in R^{m \times n} \tag{1}$$

where m is the number of color channels in an RGB image ($r = 3$), and n is the number of pixels in the flattened source image. This source image can be deconvolved into the color basis for each stain, S, and the density map, W, with each matrix element representing how much one of the stains contributes to the overall color of one of the pixels. In matrix form, let:

$$W \in R^{r \times n} \tag{2}$$

where r is the number of stains in the slide (in this case, we consider $r = 2$ for '$H\&E$'), and n is the number of pixels in the image, Also let:

$$S \in R^{m \times r} \tag{3}$$

where m is once more the number of color channels in an RGB image, and r is the number of stains present in the slide. Additionally, I_0 is a matrix which represents the background of light shining through the slide, which in the case is an RGB value of (255, 255, 255). Putting it all together, we get:

$$I = I_0^{-WH}. \tag{4}$$

To accomplish this, we used an open source sparse modeling library for python, namely SPAMS [40]. However, it should be noted that more robust libraries for sparse modeling exist, such as SciPy. First, we used dictionary learning to find a sparse representation of an input reach, resulting in a 2×3 OD representation of the stain vectors, S. Then, using these stain vectors, we used SPAMS' lasso regression function to deconvolve the tissue into the density map W, which is a matrix showing the per-pixel contribution of each of the stains to each pixel's final color.

The actual process of anatomical region extraction begins by pairing an '$H\&E$' with its corresponding label (Fig. 2). Then, each tissue region corresponding to each anatomical label was extracted. The pixels were converted to the CIELAB color space, and pixels over $0.8 L$ were thresholded out as background intensity. The remaining pixels were stain-deconvolved, and the stain vectors were saved. The ratio of red to blue was found for each stain vector, and each stain's identity was then inferred from it being more pink (eosin) or blue (hematoxylin) when compared with the other stain vector. The stain densities were converted into a sparse histogram, and also saved. Additional region statistics (mean, standard deviation, median, and interquartile range) were also saved for each anatomical region for validation.

2.4 Composite Histograms

The global histogram composition of our approach is based on the assessment of each independent region (Fig. 3(a)). To create each composite histograms, first, a slide and associated label map were loaded. Then, the label map was downsampled to one tenth of its original size, and a list of colors present in

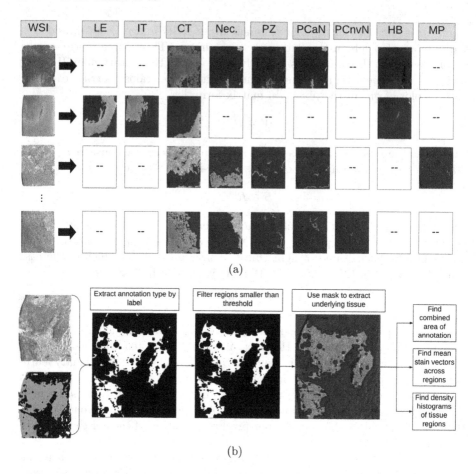

(a)

(b)

Fig. 3. (a) Examples of decomposing WSIs into various annotated anatomical regions. The labels correspond to the following respective components: leading edge, infiltrating tumor, cellular tumor, necrosis, perinecrotic zone, pseudopalisading cells around necrosis, pseudopalisading but no visible necrosis, hyperplastic blood, and microvascular proliferation. (b) Overview of the process used to separate and analyze regions from WSIs. First, a slide and a corresponding annotation map were loaded. Then, on a per-annotation basis, a pixel mask was created, and regions smaller than a threshold were removed. Then, each underlying region of tissue was extracted, and broken down into stain vectors and density maps via SNMF encoding [32]. Finally, a histogram was computed for the annotated tissue region. Repeat for each annotation present in the slide.

the label map were found. Next, each present annotation was converted into a binary mask, and all connected components for that annotation were found. Components with an area smaller than 15,000 (0.5% of WSI size) pixels were discarded. Next, the remaining components were used as a mask to extract the underlying tissue, and flatten the region into a list of pixels. Then, stain

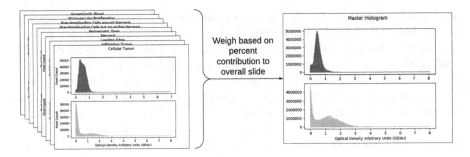

Fig. 4. Weighing all summary histograms based on total area, and merging to create a final summary histogram.

vectors were estimated for the region, and a density map was found. Two sparse histograms were then created from the associated density values for each stain (Fig. 3(b)). This process was repeated for each annotation within an image, and then across all images. Histograms and stain vectors were kept specific to each annotation type. Each annotation type had an associated cumulative area, master histogram, and list of stain vectors.

2.5 Image Transformation

Statistics across regions were summed across all slides. To account for differences in the area representing the whole slide, each annotated anatomical structure's overall histograms were weighed according to their percent contribution to overall slide (*'PCOS'*), then merged to create a master histogram (Fig. 4). Additionally, mean stain vectors from each annotation region were computed, then weighted according to their annotation's PCOS, and finally combined to give a master set of stain vectors.

With the target histogram and stain vectors computed, we are able to transform a source slide in a number of ways, but we choose to use a technique derived from [32]. To do this, we first converted the slide to CIELAB color space, thresholded out background pixels, and transformed the remaining pixels using nonnegative matrix factorization (*'NMF'*) as proposed in [41] and extended in [32]. Then, a cumulative density function (*'CDF'*) was found for each stain in both the source image and the target image, and a two-channel histogram matching function transformed each of the source's stain density maps to approximate the distribution of the target. Finally, the stain densities and corresponding stain vectors were reconvolved, and the transformed source image closely resembles the composite of all target images in both stain color and density distributions.

3 Results

We validate our approach by quantitatively evaluating the reduction in variability of each annotated region in WSIs of our dataset. To do this, we computed the standard deviation of each anatomical annotation independently across all selected IvyGAP slides by extracting the annotated anatomical structure per slide, filtering background pixels, deconvolving, and finding the distibutions of standard deviation of each stain's density within that tissue type. We then used the master stain vectors and histograms to batch normalize each of these slides. Following stain normalization, we recomputed the standard deviation of each transformed slide by using the same process as with the pre-convolved slides.

Results of each distribution of standard deviations for each stain are shown in Fig. 5. In other words, the boxes of Fig. 5 denote the spread of the spreads. Comparing pre- and post-transformation slides, through the distributions of standard deviations, shows a significant (largest $p = 0.001$, two-sided Wilcoxon rank sum test) decrease in standard deviation across tissue types, for all regions contributing more than 0.05% PCOS.

We identified the master stain vectors for hematoxylin ('H') and eosin ('E') as $RGB_H = [141, 116, 203]$ and $RGB_E = [148, 117, 180]$, respectively.

4   Discussion and Conclusion

We found that it is feasible to create composite statistics of a batch of images, to create a robust and biologically significant representation of the target GBM slide for future pre-processing. The multi-site nature of the dataset used for validating the proposed approach further emphasizes its potential for generalizability.

The technique proposed here, when compared with deep learning approaches [34, 35], obviates a "black box" entirely by nature of being a transformative technique, and not a generative one. Through the law of large numbers, we attempt to approximate the general distribution of stain densities for GBM slides from a large batch, and use it to transform slides to match said specifications. Thus, no new information in synthesized, and previously unseen structures can be transformed without issue.

Using the very specific set of slides in this study, we identified specific master stain vectors for 'H' and 'E', provided in the "Results" section above. The colors of these master stain vectors seem fairly close to each other in an RGB space, owing this partly to the "flattened color appearance" of slides fixed in frozen tissue sections, as the ones provided in the IvyGAP dataset. We expect to obtain master stain vectors of more distinct colors with formalin-fixed paraffin-embedded (FFPE) tissue slides.

While this approach has been shown to be feasible, there is still much room for improvement. For instance, the current approach is limited by the slide transformation algorithm that it implements. Stain transfer algorithms are heavily prone to artifacting [33]. While this algorithm avoids issues such as artificially staining background density, it still struggles in certain areas. Notably, NMF approaches rely on the assumption that the number of stains in a slide is already known, and that every pixel can be directly reconstructed from a certain combination of the stains. While with '$H\&E$', we can assuredly say there are only two stains, it neglects other elements such as red blood cells that introduce a third element of

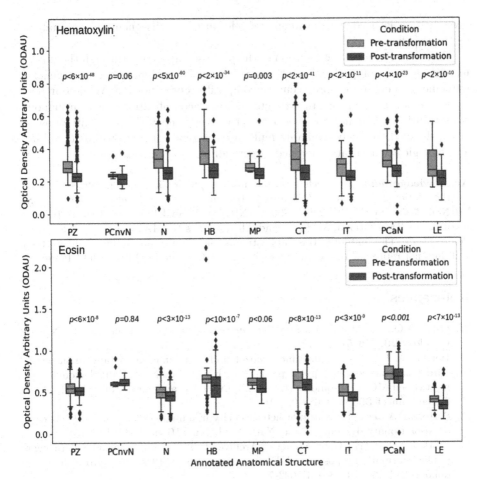

Fig. 5. The overall reduction in intra-cohort variability of stain density by annotated anatomical structure. p value based on two-sided Wilcoxon rank sum test. Abbreviations for annotated regions are as follows: 'PZ' = Perinecrotic Zone, 'PCnvN' = Pseudopalisading Cells with no visible Necrosis, 'N' = Necrosis, 'HB' = Hyperplastic Blood, 'MP' = Microvascular Proliferation, 'CT' = Cellular Tumor, 'IT' = Infiltrating Tumor, 'PCaN' = Psuedopalisading Cells around Necrosis, 'LE' = Leading Edge

color into the pixel. Thus, an area for expansion is in the ability of stain transfer algorithms to cope with other variations in color.

The current application of this approach has been on a dataset containing frozen tissue, which is not representative of most slides in anatomical pathology cases or research. However, it offers the methodology to apply on a larger, more representative set of curated slides to yield a more accurate target for clinically relevant stain normalization. Furthermore, even though we note the benefit of the proposed approach by the overall reduction in the intra-cohort variability of stain density across multiple annotated anatomical structures (Fig. 5), further investigation is needed to evaluate its relevance in subsequent image analysis methods [42].

Future work includes expansion of the proposed approach through the implementation and comparison of other stain transfer techniques, and a general refinement to create a more accurate composite representation. While our proposed method is still prone to artifacts and other complications seen throughout stain transfer techniques, we believe that this study shows the feasibility of using large, detailed, publicly available multi-institutional datasets to create robust and biologically accurate reference targets for stain normalization.

Acknowledgement. Research reported in this publication was partly supported by the National Institutes of Health (NIH) under award numbers NIH/NINDS: R01NS042645, NIH/NCI:U24CA189523, NIH/NCATS:UL1TR001878, and by the Institute for Translational Medicine and Therapeutics (ITMAT) of the University of Pennsylvania. The content of this publication is solely the responsibility of the authors and does not represent the official views of the NIH, or the ITMAT of the UPenn.

References

1. Ostrom, Q., et al.: Females have the survival advantage in glioblastoma. Neuro-Oncology **20**, 576–577 (2018)
2. Herrlinger, U., et al.: Lomustine-temozolomide combination therapy versus standard temozolomide therapy in patients with newly diagnosed glioblastoma with methylated MGMT promoter (CeTeG/NOA-09): a randomised, open-label, phase 3 trial. Lancet **393**, 678–688 (2019)
3. Sottoriva, A., et al.: Intratumor heterogeneity in human glioblastoma reflects cancer evolutionary dynamics. Proc. Natl. Acad. Sci. **110**, 4009–4014 (2013)
4. Clark, K., et al.: The Cancer Imaging Archive (TCIA): maintaining and operating a public information repository. J. Digit. Imaging **26**(6), 1045–1057 (2013). https://doi.org/10.1007/s10278-013-9622-7
5. Scarpace, L., et al.: Radiology data from the cancer genome atlas glioblastoma [TCGA-GBM] collection. Cancer Imaging Arch. **11**(4) (2016)
6. Verhaak, R.G.W., et al.: Integrated genomic analysis identifies clinically relevant subtypes of glioblastoma characterized by abnormalities in PDGFRA, IDH1, EGFR, and NF1. Cancer Cell **17**(1), 98–110 (2010)
7. Cancer Genome Atlas Research Network: Comprehensive, integrative genomic analysis of diffuse lower-grade gliomas. N. Engl. J. Med. **372**(26), 2481–2498 (2015)
8. Beers, A., et al.: DICOM-SEG conversions for TCGA-LGG and TCGA-GBM segmentation datasets. Cancer Imaging Arch. (2018)
9. Bakas, S., et al.: Advancing the cancer genome atlas glioma MRI collections with expert segmentation labels and radiomic features. Nat. Sci. Data **4**, 170117 (2017)
10. Bakas, S., et al.: Segmentation labels and radiomic features for the pre-operative scans of the TCGA-GBM collection. Cancer Imaging Arch. **286** (2017)
11. Bakas, S., et al.: Segmentation labels and radiomic features for the pre-operative scans of the TCGA-LGG collection. Cancer Imaging Arch. (2017)
12. Gevaert, O.: Glioblastoma multiforme: exploratory radiogenomic analysis by using quantitative image features. Radiology **273**(1), 168–174 (2014)

13. Gutman, D.A., et al.: MR imaging predictors of molecular profile and survival: multi-institutional study of the TCGA glioblastoma data set. Radiology **267**(2), 560–569 (2013)
14. Binder, Z., et al.: Epidermal growth factor receptor extracellular domain mutations in glioblastoma present opportunities for clinical imaging and therapeutic development. Cancer Cell **34**, 163–177 (2018)
15. Jain, R.: Outcome prediction in patients with glioblastoma by using imaging, clinical, and genomic biomarkers: focus on the nonenhancing component of the tumor. Radiology **272**(2), 484–93 (2014)
16. Aerts, H.J.W.L.: The potential of radiomic-based phenotyping in precision medicine: a review. JAMA Oncol. **2**(12), 1636–1642 (2016)
17. Davatzikos, C., et al.: Cancer imaging phenomics toolkit: quantitative imaging analytics for precision diagnostics and predictive modeling of clinical outcome. J. Med. Imaging **5**(1), 011018 (2018)
18. Bilello, M., et al.: Population-based MRI atlases of spatial distribution are specific to patient and tumor characteristics in glioblastoma. NeuroImage: Clin. **12**, 34–40 (2016)
19. Akbari, H., et al.: In vivo evaluation of EGFRvIII mutation in primary glioblastoma patients via complex multiparametric MRI signature. Neuro-Oncology **20**(8), 1068–1079 (2018)
20. Bakas, S., et al.: In vivo detection of EGFRvIII in glioblastoma via perfusion magnetic resonance imaging signature consistent with deep peritumoral infiltration: the φ-index. Clin. Cancer Res. **23**, 4724–4734 (2017)
21. Zwanenburg, A., et al.: Image biomarker standardisation initiative, arXiv:1612.07003 (2016)
22. Lambin, P., et al.: Radiomics: extracting more information from medical images using advanced feature analysis. Eur. J. Cancer **48**(4), 441–446 (2012)
23. Haralick, R.M., et al.: Textural features for image classification. IEEE Trans. Syst. Man Cybern. **3**, 610–621 (1973)
24. Galloway, M.M.: Texture analysis using grey level run lengths. Comput. Graph. Image Process. **4**, 172–179 (1975)
25. Chu, A., et al.: Use of gray value distribution of run lengths for texture analysis. Pattern Recogn. Lett. **11**, 415–419 (1990)
26. Dasarathy, B.V., Holder, E.B.: Image characterizations based on joint gray level—run length distributions. Pattern Recogn. Lett. **12**, 497–502 (1991)
27. Tang, X.: Texture information in run-length matrices. IEEE Trans. Image Process. **7**, 1602–1609 (1998)
28. Amadasun, M., King, R.: Textural features corresponding to textural properties. IEEE Trans. Syst. Man Cybern. **19**, 1264–1274 (1989)
29. Jain, A.: Fundamentals of Digital Image Processing. Prentice Hall, Englewood Cliffs (1989)
30. Macenko, M., et al.: A method for normalizing histology slides for quantitative analysis. In: 2009 IEEE International Symposium on Biomedical Imaging: From Nano to Macro, pp. 1107–1110 (2009)
31. Reinhard, E., et al.: Color transfer between images. IEEE Comput. Graphics Appl. **21**(5), 34–41 (2001)
32. Vahadane, A., et al.: Structure-preserving color normalization and sparse stain separation for histological images. IEEE Trans. Med. Imaging **35**, 1962–1971 (2016)
33. Khan, A., et al.: A non-linear mapping approach to stain normalisation in digital histopathology images using image-specific colour deconvolution. IEEE Trans. Biomed. Eng. **61**(6), 1729–1738 (2014)

34. Shaban, M.T., et al.: StainGAN: stain style transfer for digital histological images. In: 2019 IEEE 16th International Symposium on Biomedical Imaging (ISBI), pp. 953–956 (2019)

35. Janowczyk, A., et al.: Stain normalization using sparse AutoEncoders (StaNoSA). Comput. Med. Imaging Graph. **50–61**, 2017 (2017)

36. Bianconi, F., Kather, J.N., Reyes-Aldasoro, C.C.: Evaluation of colour pre-processing on patch-based classification of H&E-stained images. In: Reyes-Aldasoro, C.C., Janowczyk, A., Veta, M., Bankhead, P., Sirinukunwattana, K. (eds.) ECDP 2019. LNCS, vol. 11435, pp. 56–64. Springer, Cham (2019). https://doi.org/10.1007/978-3-030-23937-4_7

37. Puchalski, R., et al.: An anatomic transcriptional atlas of human glioblastoma. Science **360**, 660–663 (2018)

38. Shah, N., et al.: Data from Ivy GAP. Cancer Imaging Arch. (2016)

39. Ruifrok, A., Johnston, D.: Quantification of histochemical staining by color deconvolution. Anal. Quant. Cytol. Histol. **23**(4), 291–299 (2001)

40. Mairal, J., et al.: Online learning for matrix factorization and sparse coding. J. Mach. Learn. Res. **11**, 19–60 (2010)

41. Rabinovich, A., et al.: Unsupervised color decomposition of histologically stained tissue samples. In: Advances in Neural Information Processing Systems, vol. 16, pp. 667–674 (2004)

42. Li, X., et al.: A complete color normalization approach to histopathology images using color cues computed from saturation-weighted statistics. IEEE Trans. Biomed. Eng. **62**(7), 1862–1873 (2015)

Skull-Stripping of Glioblastoma MRI Scans Using 3D Deep Learning

Siddhesh P. Thakur[1,2], Jimit Doshi[1,3], Sarthak Pati[1,3], Sung Min Ha[1,3], Chiharu Sako[1,3], Sanjay Talbar[2], Uday Kulkarni[2], Christos Davatzikos[1,3], Guray Erus[1,3], and Spyridon Bakas[1,3,4(✉)]

[1] Center for Biomedical Image Computing and Analytics (CBICA), University of Pennsylvania, Philadelphia, PA, USA
sbakas@upenn.edu
[2] Shri Guru Gobind Singhji Institute of Engineering and Technology (SGGS), Nanded, Maharashtra, India
[3] Department of Radiology, Perelman School of Medicine, University of Pennsylvania, Philadelphia, PA, USA
[4] Department of Pathology and Laboratory Medicine, Perelman School of Medicine, University of Pennsylvania, Philadelphia, PA, USA

Abstract. Skull-stripping is an essential pre-processing step in computational neuro-imaging directly impacting subsequent analyses. Existing skull-stripping methods have primarily targeted non-pathologically-affected brains. Accordingly, they may perform suboptimally when applied on brain Magnetic Resonance Imaging (MRI) scans that have clearly discernible pathologies, such as brain tumors. Furthermore, existing methods focus on using only T1-weighted MRI scans, even though multi-parametric MRI (mpMRI) scans are routinely acquired for patients with suspected brain tumors. Here we present a performance evaluation of publicly available implementations of established 3D Deep Learning architectures for semantic segmentation (namely DeepMedic, 3D U-Net, FCN), with a particular focus on identifying a skull-stripping approach that performs well on brain tumor scans, and also has a low computational footprint. We have identified a retrospective dataset of 1,796 mpMRI brain tumor scans, with corresponding manually-inspected and verified gold-standard brain tissue segmentations, acquired during standard clinical practice under varying acquisition protocols at the Hospital of the University of Pennsylvania. Our quantitative evaluation identified DeepMedic as the best performing method ($Dice = 97.9$, $Hausdorff_{95} = 2.68$). We release this pre-trained model through the Cancer Imaging Phenomics Toolkit (CaPTk) platform.

Keywords: Skull-stripping · Brain extraction · Glioblastoma · GBM · Brain tumor · Deep learning · DeepMedic · U-Net · FCN · CaPTk

1 Introduction

Glioblastoma (GBM) is the most aggressive type of brain tumors, with a grim prognosis in spite of current treatment protocols [1,2]. Recent clinical advance-

© Springer Nature Switzerland AG 2020
A. Crimi and S. Bakas (Eds.): BrainLes 2019, LNCS 11992, pp. 57–68, 2020.
https://doi.org/10.1007/978-3-030-46640-4_6

ments in the treatment of GBMs have not increased the overall survival rate of patients with this disease by any substantial amount. The recurrence of GBM is virtually guaranteed and its management is often indefinite and highly case-dependent. Any assistance that can be gleaned from the computational imaging and machine learning communities could go a long way towards making better treatment plans for patients suffering from GBMs [3–11]. One of the first steps towards the goal of a good treatment plan is to ensure that the physician is observing only the areas that are of immediate interest, i.e., the brain and the tumor tissues, which would ensure better visualization and quantitative analyses.

Skull-stripping is the process of removing the skull and non-brain tissues from brain magnetic resonance imaging (MRI) scans. It is an indispensable pre-processing operation in neuro-imaging analyses that directly affects the efficacy of subsequent analyses. The effects of skull-stripping on subsequent analyses have been reported in the literature, including studies on brain tumor segmentation [12–14] and neuro-degeneration [15]. Manual removal of the non-brain tissues is a very involved and grueling process [16], which often results in inter- and intra-rater discrepancies affecting reproducibility in large scale studies.

In recent years, with theoretical advances in the field and with the pro-liferation of inexpensive computing power, including consumer-grade graphical processing units [17], there has been an explosion of deep learning (DL) algo-rithms that use heavily parallelized learning techniques for solving major seman-tic segmentation problems in computer vision. These methods have the added advantage of being easy to implement by virtue of the multitude of mature tools available, most notable of these being TensorFlow [18] and PyTorch [19]. Importantly, DL based segmentation techniques, which were initially adopted from generic applications in computer vision, have promoted the development of novel methods and architectures that were specifically designed for segmenting 3-dimensional (3D) MRI images [20–23]. DL, specifically convolutional neural net-works, have been applied for segmentation problems in neuroimaging (including skull-stripping), obtaining promising results [16]. Unfortunately, most of these DL algorithms either require a long time to train or have unrealistic run-time inference requirements.

In this paper, we evaluate the performance of 3 established and validated DL architectures for semantic segmentation, which have out-of-the-box publicly-available implementations. Our evaluation is focusing on skull-stripping of scans that have clearly discernible pathologies, such as scans from subjects diagnosed with GBM. We also perform extensive comparisons using models trained on var-ious combinations of different MRI modalities, to evaluate the benefit of utilizing multi-parametric MRI (mpMRI) data that are typically acquired in routine clin-ical practice for patients with suspected brain tumors on the final segmentation.

2   Materials and Methods

2.1   Data

We retrospectively collected 1,796 mpMRI brain tumor scans, from 449 glioblastoma patients, acquired during standard clinical practice under varying acquisition protocols at the Hospital of the University of Pennsylvania. Corresponding brain tissue annotations were manually-approved by an expert and used as the gold-standard labels to quantitatively evaluate the performance of the algorithms considered in this study.

In this study, we have chosen to take advantage of the richness of the mpMRI protocol that is routinely acquired in the cases of subjects with suspected tumors. Specifically, four structural modalities are included at baseline pre-operative time-point: native (T1) and post-contrast T1-weighted (T1Gd), native T2-weighted (T2), and T2-weighted Fluid Attenuated Inversion Recovery (FLAIR) MRI scans (Fig. 1). To conduct our quantitative performance evaluation we split the available data, based on an 80/20 ratio, in a training and testing subset of 1,432 and 364 mpMRI brain tumor scans, from 358 and 91 patients, respectively.

2.2   Pre-processing

To guarantee the homogeneity of the dataset, we applied the same pre-processing pipeline across all the mpMRI scans. Specifically, all the raw DICOM scans obtained from the scanner were initially converted to the NIfTI [24] file format and then followed the protocol for pre-processing, as defined in the International Brain Tumor Segmentation (BraTS) challenge [12–14,25,26]. Specifically, each patient's T1Gd scan was rigidly registered to a common anatomical atlas of $240 \times 240 \times 155$ image size and resampled to its isotropic resolution of 1 mm^3 [27]. The remaining scans of each patient (namely, T1, T2, FLAIR) were then rigidly co-registered to the same patient's resampled T1Gd scan. All the registrations were done using "Greedy" (github.com/pyushkevich/greedy) [28], which is a CPU-based C++ implementation of the greedy diffeomorphic registration algorithm [29]. "Greedy" is integrated into the ITK-SNAP (itksnap.org) segmentation software [30,31], as well as the Cancer Imaging Phenomics Toolkit (CaPTk - www.cbica.upenn.edu/captk) [32,33]. After registration, all scans were downsampled from a resolution of $240 \times 240 \times 155$ to a resolution of $128 \times 128 \times 128$, with anisotropic spacing of $1.875 \times 1.875 \times 1.25$ mm^3 with proper padding measures to ensure the anisotropic spacing is attained. Finally, the intensities found on each scan below the 2nd percentile and above the 95th percentile were capped, to ensure suppression of spurious intensity changes due to the scanner acquisition parameters.

2.3   Network Topologies

For our comparative performance evaluation, we focused on the most well-established DL network topologies for 3D semantic segmentation. The selection

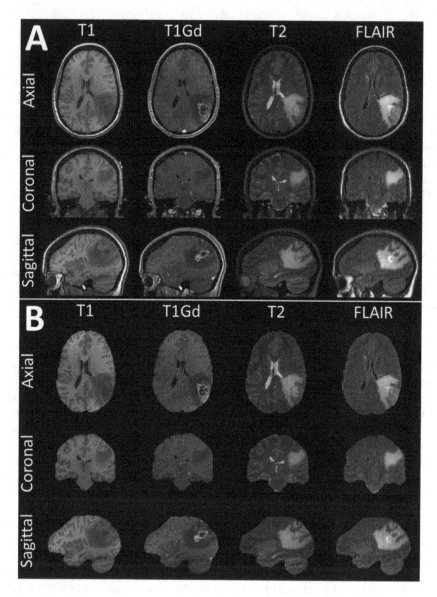

Fig. 1. Example mpMRI brain tumor scans from a single subject. The original scans including the non-brain-tissues are illustrated in A, whereas the same scans after applying the manually-inspected and verified gold-standard brain tissue segmentations are illustrated in B.

was done after taking into consideration their wide application in related literature, their state-of-the-art performance on other segmentation tasks, as established by various challenges [12,13,34,35], as well as their out-of-the-box applicability of their publicly available implementation, in low resource settings. The specific architectures included in this evaluation comprise the a) DeepMedic [20,21], b) 3D U-Net [22], and c) Fully Convolutional Neural network (FCN) [23].

DeepMedic [20,21] is a novel architecture, which came into the foreground after winning the 2015 ISchemic LEsion Segmentation (ISLES) challenge [34]. DeepMedic is essentially a 3D convolutional neural network with a depth of 11-layers, along with a double pathway to provide sufficient context and detail in resolution, simultaneously. In our study, we have applied DeepMedic using its default parameters, as provided in its GitHub repository github.com/deepmedic/deepmedic. As a post-processing step, we also include a hole filling algorithm.

As a second method, we have applied a 3D U-Net [22], an architecture that is widely used in neuroimaging. We used 3D U-Net with an input image size of 128 × 128 × 128 voxels. Taking into consideration our requirement for a low computational footprint, we reduced the initial number of "base" filters from 64 (as was originally proposed) to 16.

The third method selected for our comparisons was a 3D version of an FCN [23]. Similarly to the 3D U-Net, we used an input image size of 128 × 128 × 128 voxels. For both 3D U-Net and FCN, we used 'Leaky ReLU' instead of 'ReLU' for back-propagation with leakiness defined as $\alpha = 0.01$. Furthermore, we used instance normalization instead of batch normalization due to batch size being equal to 1, due to the high memory consumption.

2.4   Experimental Design

Current state of the art methods typically use only the T1 modality for skull-stripping [36–41]. Here, we followed a different approach, by performing a set of experiments using various input image modality combinations for training DL models. Our main goal was to investigate potential contribution of different modalities, which are obtained as part of routine mpMRI acquisitions in patients with suspected brain tumors, beyond using T1 alone for skull-stripping. Accordingly, we first trained and inferred each topology on each individual modality separately to measure segmentation performance using different independent modalities, resulting in 4 models for each topology ("T1", "T1Gd", "T2", "Flair"). Additionally, we trained and inferred models on a combination of modalities; namely, using a) both T1 and T2 modalities ("Multi-2"), and b) all 4 structural modalities together ("Multi-4"). The first combination was chosen as it has been shown that addition of the T2 modality improves the skull-stripping performance [14] and can also be used in cases where contrast medium is not used and hence the T1Gd modality is not available, i.e., in brain scans without any tumors. The second combination approach (i.e., "Multi-4") was chosen to evaluate a model that uses all available scans. Finally, we utilized an ensembling approach (i.e., "Ens-4") where the majority voting of the 4 models trained and inferred on individual modalities was used to produce the final label for skull-stripping.

We ensured that the learning parameters stayed consistent across each experiment. Each of the applied topologies needed different time for convergence, based on their individual parameters. For Deep-Medic, we trained with default parameters (as provided at the original github repository - 0.7.1 [commit dbdc1f1]) and it trained for 44 h. 3D U-Net and FCN were trained with Adam optimizer with a learning rate of 0.01 over 25 epochs. The number of epochs was determined according to the amount of improvement observed. Each of them trained for 6 h.

The average inference time for DeepMedic including the pre-processing and post-processing for a single brain tumor scan was 10.72 s, while for 3D U-Net and FCN was 1.06 s. These times were estimated based on the average time taken to infer on 300 patients. The hardware we used to train and infer were NVIDIA P100 GPUs with 12 GB VRAM utilizing only a single CPU core with 32 GB of RAM from nodes of the CBICA's high performance computing (HPC) cluster.

2.5   Evaluation Metrics

Following the literature on semantic segmentation we use the following metrics to quantitatively evaluate the performance of the trained methods.

Dice Similarity Coefficient. The Dice Similarity Coefficient (*Dice*) is typically used to evaluate and report on the performance of semantic segmentation. Dice measures the extent of spatial overlap between the predicted masks (*PM*) and the provided ground truth (*GT*), and is mathematically defined as:

$$\text{Dice} = \frac{2|GT \cap PM|}{|GT| + |PM|} * 100 \tag{1}$$

where it would range between 0-100, with 0 describing no overlap and 100 perfect agreement.

Hausdorff$_{95}$. Evaluating volumetric segmentations with spatial overlap agreement metrics alone can be insensitive to differences in the slotted edges. For our stated problem of brain extraction, changes in edges might lead to minuscule differences in spatial overlap, but major differences in areas close to the brain boundaries resulting in inclusion of skull or exclusion of a tumor region. To robustly evaluate such differences, we used the 95th percentile of the *Hausdorff*$_{95}$ distance to measure the maximal contour distance d, on a radial assessment, between the PM and GT masks.

$$Hausdorff_{95} = percentile\left(d_{PM,GT} \cup d_{GT,PM}, 95^{th}\right) \tag{2}$$

3   Results

The median and inter-quartile range for *Dice* and *Hausdorff*$_{95}$ scores for each of the constructed models using 3 topologies and 7 input image combinations, are shown in Tables 1 and 2 and Fig. 2 and 3. DeepMedic showed a consistent superior performance to 3D U-Net and FCN, for all input combinations.

Table 1. Median and inter-quartile range for *Dice* scores of all trained models and input image combinations. DM:DeepMedic, 3dU:3D U-Net.

	T1	T2	T1Gd	Flair	Multi-4	Multi-2	Ens-4
DM	98.09 ± 1.18	97.88 ± 1.07	97.86 ± 1.08	97.88 ± 1.03	98.19 ± 1.08	98.13 ± 1.08	97.94 ± 1.10
3dU	94.77 ± 2.30	96.01 ± 1.58	97.15 ± 1.66	96.08 ± 1.92	98.20 ± 1.19	98.13 ± 1.02	98.05 ± 0.76
FCN	97.65 ± 0.74	96.34 ± 1.32	97.16 ± 0.99	96.74 ± 1.11	93.34 ± 1.13	97.82 ± 1.18	97.46 ± 1.07

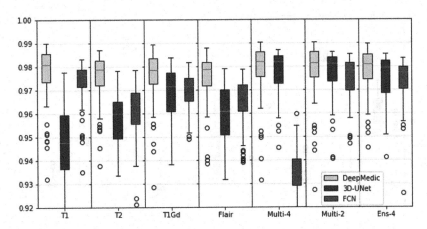

Fig. 2. Median and inter-quartile range for *Dice* scores of all trained models and input image combinations.

Overall, best performance was obtained for the *DeepMedic-Multi-4* model (*Dice* = 97.9, *Hausdorff*$_{95}$ = 2.68). However, for the model trained using DeepMedic and only the T1 modality obtained comparable (statistically insignificant, $p > 0.05$ - Wilcoxon signed-rank test) performance (*Dice* = 97.8, *Hausdorff*$_{95}$ = 3.01). This result reaffirms the use of T1 in current state of the art methods for skull-stripping.

Performance of 3D U-Net was consistently lower when the network was trained on single modalities. However, the *3D U-Net-Multi-4* model obtained performance comparable to DeepMedic. Despite previous literature reporting a clear benefit of the ensemble approach [12,13], in our validations we found that the ensemble of models trained and inferred on individual modalities did not offer a noticeable improvement.

Illustrative examples of the segmentations for the best performing model (*DeepMedic-Multi-4*) are shown in Fig. 4. We showcase the best and the worst segmentation results, selected based on the *Dice* scores.

Table 2. Median and inter-quartile range for $Hausdorff_{95}$ scores for all trained models and input image combinations. DM:DeepMedic, 3dU:3D U-Net.

	T1	T2	T1Gd	Flair	Multi-4	Multi-2	Ens-4
DM	2.24 ± 1.41	2.24 ± 1.00	2.24 ± 1.27	2.24 ± 1.00	2.00 ± 1.41	2.24 ± 1.41	2.24 ± 1.00
3dU	11.45 ± 2.98	19.21 ± 2.85	3.00 ± 2.01	5.20 ± 5.85	1.73 ± 1.41	2.00 ± 1.50	2.00 ± 0.61
FCN	2.24 ± 1.00	3.61 ± 1.18	2.83 ± 0.93	3.00 ± 1.55	5.92 ± 0.93	2.24 ± 1.00	2.24 ± 0.76

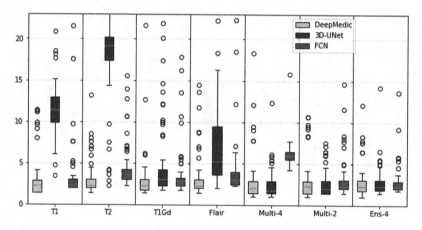

Fig. 3. Median and inter-quartile range for $Hausdorff_{95}$ scores of all trained models and input image combinations.

4   Discussion

We compared three of the most widely used DL architectures for semantic segmentation in the specific problem of skull-stripping of images with brain tumors. Importantly, we trained models using different combinations of input image modalities that are typically acquired as part of routine clinical evaluations of patients with suspected brain tumors, to investigate contribution of these different modalities to overall segmentation performance.

DeepMedic consistently outperformed the other 2 methods with all input combinations, suggesting that it is more robust. In contrast, 3D U-Net and FCN had highly variable performance with different image combinations. With the addition of mpMRI input data, the 3D U-Net models ("Multi-4" and "Multi-2") performed comparably with DeepMedic.

We have made the pre-trained DeepMedic model, including all pre-processing and post-processing steps, available for inference to others through our cancer image processing toolkit, namely the CaPTk [32,33], which provides readily deployable advanced computational algorithms to facilitate clinical research.

In future work, we intend to extend application of these topologies to multi-institutional data along with other topologies for comparison.

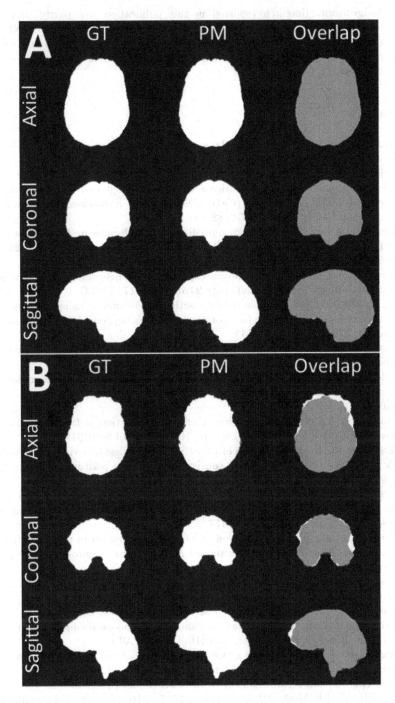

Fig. 4. Visual example of the best (A) and the worst (B) output results.

Acknowledgement. Research reported in this publication was partly supported by the National Institutes of Health (NIH) under award numbers NIH/NINDS: R01NS042645, NIH/NCI:U24CA189523, NIH/NCI:U01CA242871. The content of this publication is solely the responsibility of the authors and does not represent the official views of the NIH.

References

1. Ostrom, Q., et al.: Females have the survival advantage in glioblastoma. Neuro-Oncology **20**, 576–577 (2018)
2. Herrlinger, U., et al.: Lomustine-temozolomide combination therapy versus standard temozolomide therapy in patients with newly diagnosed glioblastoma with methylated MGMT promoter (CeTeG/NOA 2009): a randomised, open-label, phase 3 trial. Lancet **393**, 678–688 (2019)
3. Gutman, D.A., et al.: MR imaging predictors of molecular profile and survival: multi-institutional study of the TCGA glioblastoma data set. Radiology **267**(2), 560–569 (2013)
4. Gevaert, O.: Glioblastoma multiforme: exploratory radiogenomic analysis by using quantitative image features. Radiology **273**(1), 168–174 (2014)
5. Jain, R.: Outcome prediction in patients with glioblastoma by using imaging, clinical, and genomic biomarkers: focus on the nonenhancing component of the tumor. Radiology **272**(2), 484–93 (2014)
6. Aerts, H.J.W.L.: The potential of radiomic-based phenotyping in precision medicine: a review. JAMA Oncol. **2**(12), 1636–1642 (2016)
7. Bilello, M., et al.: Population-based MRI atlases of spatial distribution are specific to patient and tumor characteristics in glioblastoma. NeuroImage: Clin. **12**, 34–40 (2016)
8. Zwanenburg, A., et al.: Image biomarker standardisation initiative. Radiology, arXiv:1612.07003 (2016). https://doi.org/10.1148/radiol.2020191145
9. Bakas, S., et al.: In vivo detection of EGFRvIII in glioblastoma via perfusion magnetic resonance imaging signature consistent with deep peritumoral infiltration: the phi-index. Clin. Cancer Res. **23**, 4724–4734 (2017)
10. Binder, Z., et al.: Epidermal growth factor receptor extracellular domain mutations in glioblastoma present opportunities for clinical imaging and therapeutic development. Cancer Cell **34**, 163–177 (2018)
11. Akbari, H., et al.: In vivo evaluation of EGFRvIII mutation in primary glioblastoma patients via complex multiparametric MRI signature. Neuro-Oncology **20**(8), 1068–1079 (2018)
12. Menze, B.H., et al.: The multimodal Brain Tumor Image Segmentation Benchmark (BRATS). IEEE Trans. Med. Imaging **34**(10), 1993–2024 (2015)
13. Bakas, S., et al.: Identifying the best machine learning algorithms for brain tumor segmentation, progression assessment, and overall survival prediction in the BRATS challenge. arXiv e-prints. arxiv:1811.02629 (2018)
14. Bakas, S., et al.: Advancing the cancer genome atlas glioma MRI collections with expert segmentation labels and radiomic features. Nat. Sci. Data **4**, 170117 (2017)
15. Gitler, A.D., et al.: Neurodegenerative disease: models, mechanisms, and a new hope. Dis. Models Mech. **10**(5), 499–502 (2017). https://www.ncbi.nlm.nih.gov/pmc/articles/PMC5451177/

16. Souza, R., et al.: An open, multi-vendor, multi-field-strength brain MR dataset and analysis of publicly available skull stripping methods agreement. NeuroImage **170**, 1053–8119 (2018)
17. Lin, H.W., et al.: Why does deep and cheap learning work so well? J. Stat. Phys. **168**, 1223–1247 (2017)
18. Abadi, M., et al.: TensorFlow: large-scale machine learning on heterogeneous systems, Software. tensorflow.org (2015)
19. Paszke, A., et al.: Automatic differentiation in PyTorch. In: NIPS Autodiff Workshop (2017)
20. Kamnitsas, K., et al.: Multi-scale 3D convolutional neural networks for lesion segmentation in brain MRI. Ischemic Stroke Lesion Segment. **13**, 46 (2015)
21. Kamnitsas, K., et al.: Efficient multi-scale 3D CNN with fully connected CRF for accurate brain lesion segmentation. Med. Image Anal. **36**, 61–78 (2017)
22. Çiçek, Ö., Abdulkadir, A., Lienkamp, S.S., Brox, T., Ronneberger, O.: 3D U-Net: learning dense volumetric segmentation from sparse annotation. In: Ourselin, S., Joskowicz, L., Sabuncu, M.R., Unal, G., Wells, W. (eds.) MICCAI 2016. LNCS, vol. 9901, pp. 424–432. Springer, Cham (2016). https://doi.org/10.1007/978-3-319-46723-8_49
23. Long, J. et al.: Fully convolutional networks for semantic segmentation. In: Proceedings of the IEEE Conference on Computer Vision and Pattern Recognition, pp. 3431–3440 (2015)
24. Cox, R., et al.: A (Sort of) new image data format standard: NIfTI-1: WE 150. Neuroimage **22** (2004)
25. Bakas, S., et al.: Segmentation labels and radiomic features for the pre-operative scans of the TCGA-GBM collection. Cancer Imaging Arch. **286** (2017)
26. Bakas, S., et al.: Segmentation labels and radiomic features for the pre-operative scans of the TCGA-LGG collection. Cancer Imaging Arch. (2017)
27. Rohlfing, T., et al.: The SRI24 multichannel atlas of normal adult human brain structure. Hum. Brain Mapp. **31**, 798–819 (2010)
28. Yushkevich, P.A., et al.: Fast automatic segmentation of hippocampal subfields and medial temporal lobe subregions in 3 Tesla and 7 Tesla T2-weighted MRI. Alzheimer's Dement. J. Alzheimer's Assoc. **12**(7), P126–P127 (2016)
29. Joshi, S., et al.: Unbiased diffeomorphic atlas construction for computational anatomy. NeuroImage **23**, S151–S160 (2004)
30. Yushkevich, P.A., et al.: User-guided 3D active contour segmentation of anatomical structures: significantly improved efficiency and reliability. NeuroImage **31**(3), 1116–1128 (2006)
31. Yushkevich, P.A., et al.: User-guided segmentation of multi-modality medical imaging datasets with ITK-SNAP. Neuroinformatics **17**(1), 83–102 (2018). https://doi.org/10.1007/s12021-018-9385-x
32. Davatzikos, C., et al.: Cancer imaging phenomics toolkit: quantitative imaging analytics for precision diagnostics and predictive modeling of clinical outcome. J. Med. Imaging **5**(1), 011018 (2018)
33. Rathore, S., et al. Brain cancer imaging phenomics toolkit (brain-CaPTk): an interactive platform for quantitative analysis of glioblastoma. In: Brainlesion : Glioma, Multiple Sclerosis, Stroke and Traumatic Brain Injuries: Third International BrainLes Workshop Held in Conjunction with MICCAI 2017, Quebec City, QC, Canada, vol. 10670, pp. 133–145 (2018)
34. Maier, O., et al.: ISLES 2015 - a public evaluation benchmark for ischemic stroke lesion segmentation from multispectral MRI. Med. Image Anal. **35**, 250–269 (2017)

35. Simpson, A.L., et al.: A large annotated medical image dataset for the development and evaluation of segmentation algorithms, arXiv e-prints, arXiv:1902.09063 (2019)

36. Shattuck, D.W., et al.: Magnetic resonance image tissue classification using a partial volume model. NeuroImage **13**(5), 856–876 (2001)

37. Smith, S.M.: Fast robust automated brain extraction. Hum. Brain Mapp. **17**(3), 143–155 (2002)

38. Iglesias, J.E., et al.: Robust brain extraction across datasets and comparison with publicly available methods. IEEE Trans. Med. Imaging **30**(9), 1617–1634 (2011)

39. Eskildsen, S.F.: BEaST: brain extraction based on nonlocal segmentation technique. NeuroImage **59**(3), 2362–2373 (2012)

40. Doshi, J., et al.: Multi-atlas skull-stripping. Acad. Radiol. **20**(12), 1566–1576 (2013)

41. Doshi, J., et al.: MUSE: MUlti-atlas region segmentation utilizing ensembles of registration algorithms and parameters, and locally optimal atlas selection. Neuroimage **127**, 186–195 (2016)

Estimation of the Principal Ischaemic Stroke Growth Directions for Predicting Tissue Outcomes

Christian Lucas[1][✉], Linda F. Aulmann[2], André Kemmling[3],
Amir Madany Mamlouk[4], and Mattias P. Heinrich[1]

[1] Institute of Medical Informatics, University of Lübeck, Lübeck, Germany
lucas@imi.uni-luebeck.de
[2] Department of Neuroradiology, University Hospital UKSH, Lübeck, Germany
[3] Department of Neuroradiology, Westpfalz Hospital, Kaiserslautern, Germany
[4] Institute for Neuro- and Bioinformatics, University of Lübeck, Lübeck, Germany

Abstract. Convolutional neural networks (CNN) have been widely used for the medical image analysis of brain lesions. The estimates of traditional segmentation networks for the prediction of the follow-up tissue outcome in strokes are, however, not yet accurate enough or capable of properly modeling the growth mechanisms of ischaemic stroke.

In our previous shape space interpolation approach, the prediction of the follow-up lesion shape has been bounded using core and penumbra segmentation estimates as priors. One of the challenges is to define well-suited growth constraints, as the transition from one to another shape may still result in a very unrealistic spatial evolution of the stroke.

In this work, we address this shortcoming by explicitly incorporating vector fields for the spatial growth of the infarcted area. Since the anatomy of the cerebrovascular system defines the blood flow along brain arteries, we hypothesise that we can reasonably regularise the direction and strength of growth using a lesion deformation model. We show that a Principal Component Analysis (PCA) model computed from the diffeomorphic displacements between a core lesion approximation and the entire tissue-at-risk can be used to estimate follow-up lesions (0.74 F1 score) for a well-defined growth problem with accurate input data better than with the shape model (0.62 F1 score) by predicting the PCA coefficients through a CNN.

Keywords: Stroke · Growth · Prediction · PCA · CNN

1 Introduction

Ischaemic stroke is the blockage of brain arteries by thrombotic or embolic clots leading to a reduced blood flow. The lack of oxygen can cause brain cell death and makes it a potentially deadly medical condition with high mortality rates in the industrialised world. The treatments mainly aim to reperfuse the tissue

© Springer Nature Switzerland AG 2020
A. Crimi and S. Bakas (Eds.): BrainLes 2019, LNCS 11992, pp. 69–79, 2020.
https://doi.org/10.1007/978-3-030-46640-4_7

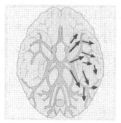 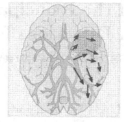 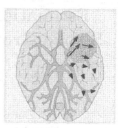

(a) Normal blood flow   (b) Blood flow with AIS   (c) Stroke growth

Fig. 1. Illustrating the motivation: oxygenated blood supply is represented by vectors at sample positions in the left hemisphere. Interpreting vector directions and magnitudes as blood flow, the vectors follow the main arteries in the brain (a). Acute ischaemic stroke (AIS) clots (yellow) reduce the flow and oxygen supply (b). The less oxygen, the faster the stroke lesion growth (c). (Color figure online)

and the outcome is highly dependent on time from stroke onset until successful reperfusion [1]. While necrotic tissue in the core of the stroke cannot be recovered, the hypo-perfused tissue-at-risk in the surrounding penumbra can often be restored. CT perfusion (CTP) imaging in the acute phase of the stroke can provide parameter maps, such as TTD (time to drain) or CBV (cerebral blood volume), to segment core and penumbra. The bigger the potential spatial growth of the lesion into the tissue-at-risk (Fig. 1), the higher could be the clinical benefit of an immediate treatment saving most of the recoverable brain area.

Today, the estimation of the spatial growth of the stroke is strongly depending on the subjective experience of the doctor. The lack of longitudinal data ("time is brain", i.e. patients are treated as quick as possible after the emergency imaging) is a major obstacle for proposing a data-driven algorithmic model that accurately describes the relation from imaging or clinical parameters to future outcome. Regularisation is crucial, because the space of outcomes at different times of treatment is often under-sampled by the available ground truth data.

1.1   Related Work

In recent years, machine learning methods have become increasingly popular in the field of stroke prediction research [2,3]. Several deep learning approaches for stroke tissue outcome prediction – often exploiting standard segmentation networks for prediction – have been investigated and compared [4]. Recently presented methods [5,6] learn the tissue outcome prediction from weakly supervised data over time (single follow-up imaging) and use clinical meta-data as predictors to a convolutional neural network, e.g. by upsampling scalar to spatial data in order to combine it with the image data. However, the performance of the deep learning based follow-up predictions is still too low for clinical usage.

While the task of segmenting the CTP perfusion lesion (all tissue-at-risk incl. necrosis, "core+penumbra") is rather straight-forward, the true necrotic core is usually estimated in the acute phase by comparing prolonged perfusion times with significantly decreased blood flow and volume, which can be quite difficult to assess. Thus, also automatic methods struggle to segment the core on CTP. At the ISLES challenge 2018, only the winning [7] out of 50 algorithms in the leaderboard achieved the maximum Dice score of 0.51 (www.isles-challenge.org).

Principal component analysis (PCA) serves well to obtain representative low-dimensional encodings that keep the main information. It has been combined with neural learning in various ways, e.g. to cope with the respiratory motion of lungs [8], to integrate prior shape knowledge when predicting segmentations of the heart [9], or to predict deformations for landmark regression tasks [10]. Inspired by spatial transformer networks [11], those methods do not need to directly regress and compare the coefficients for the PCA embeddings but a final resampled mask result that is matched with the ground truth mask.

Some algorithms for the prediction of disease progression incorporate prior knowledge with the help of autoencoders from retrospective outcome data during training. The inter- or extrapolated disease progression along a trajectory in the latent space can be set to be linear [12] or non-linear as in [13]. The learnt decoders continuously morph the shape of a segmentation mask or the change in gray values of a medical image to simulate the growth of the disease.

Using the shape distributions in our previous work [12] as a surrogate for a stroke growth model relies on several assumptions, which are difficult to choose and result in a highly under-sampled problem. In particular, it is not ensured how the progression between the observed ground truth labels continues. This is completely allocated to the optimisation process under some constraints defined before (e.g. monotony or trajectory criteria). The learnt shape sample distributions serve as a surrogate for a non-existing growth supervision.

In order to better guide the learning of the spatial growth process itself over time (rather than morphing between shape "snapshots"), we aim to regularise the change of the stroke lesion segmentation in both temporal directions (forward/backward) by using non-linear deformations reconstructed from a PCA on diffeomorphic displacements. The underlying hypothesis implies that the follow-up lesion mask should be located somewhere on the deformation trajectory of the displacements from core to core+penumbra.

1.2   Contribution

Our contribution is based on different aspects of the method's novel perspective on stroke growth modelling:

(1) Indirect *prior knowledge of the brain*'s common blood vessel anatomy (Fig. 1) is incorporated through the registration of stroke lesion masks.
(2) The feasability of using only the *displacements from core+penumbra to core for tissue outcome* (follow-up mask) prediction is demonstrated.
(3) *Valid evolution of the lesion shape* is ensured through the dense resampling of displacements from biophysically plausible diffeomorphic registrations.
(4) *Generalisation of the spatial evolution modes* is ensured by using only principal components (PCA) of the training data's growth vectors.
(5) *Non-linearity of stroke growth* is learned through a CNN estimating the coefficients of the principal directions of the lesion deformation.

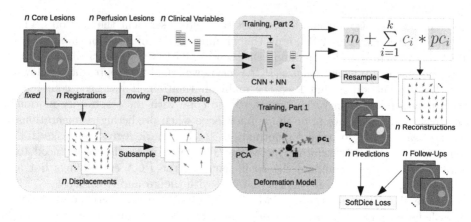

Fig. 2. Training method overview: n pairs of core and perfusion lesion are being registered in a preprocessing step. The resulting displacements will be downsampled before extracting their principal components pc_i $(1 \leq i \leq k)$ and mean m. A CNN estimates the coefficients c_i to reconstruct displacements generating a follow-up prediction. While part 1 is only computed once, part 2 and after are computed for each of the n samples.

2   Materials and Methods

The idea for stroke tissue outcome prediction is to mimic the stroke shape progression (spatial growth) as a registration problem of its segmentations at various times (core at $t = 0$, follow-up at $0 < t < 1$, core+penumbra at $t = 1$). We propose a two-part training (Fig. 2) to estimate principal directions of stroke growth by using a surrogate of principal displacement modes between core and core+penumbra from an independent registration procedure.

These modes are modelled using statistical machine learning, namely a PCA (Part 1), of displacement vector fields (Sect. 2.1) that warp the core+penumbra onto the core segmentation. Instead of a high number of deformation parameters (regular vector field), we have simplified the task and regularised it through the robust PCA model along a small number of $k \leq n$ main deformation directions (n total training cases).

The CNN (Part 2) learns the PCA coefficients c_i for a suitable weighting of each component $i \in \{1..k\}$ to predict the displacements for the follow-up lesion. The follow-up mask is being resampled from these displacements on the core+penumbra mask through a *GridSampler* module implemented in the PyTorch deep learning package [14].

Inserting the coefficients c_i in Eq. (1) together with both mean m and the linear combination of the principal directions pc_i of the PCA gives us a reconstructed displacement field d that warps core+penumbra onto the follow-up segmentation. Using invertible diffeomorphic ground truth displacements as supervision, we can think of the growth from core to follow-up (forward in time,

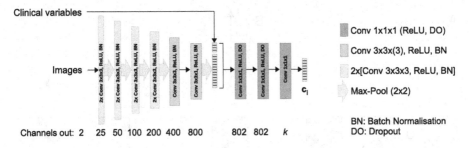

Fig. 3. The CNN architecture of Fig. 2, Part 2, consists of two sides: Before concatenation, the images are convolved and downsampled, while increasing the number of channels (bottom). After concatenation with clinical time data, the channels are reweighted and reduced to the dimensionality of the PCA space. (Best viewed on screen)

d^{-1}) also in the reverse direction from core+penumbra to follow-up (backward in time, d). This enables us to use the core+penumbra segmentation with more shape information as a *moving* image. During training, a *SoftDice* loss as in [12] measures the difference to the actual ground truth follow-up lesion mask. In order to update the parameter weights of the CNN, the loss can easily be back-propagated through the derivative of Eq. (1) before passing backward through the CNN.

$$d = m + \sum_{i=1}^{k} c_i * pc_i \tag{1}$$

The CNN receives core and core+penumbra masks as input, repeatedly doubles the feature channels from 25 to 800, while reducing the spatial size down to $1 \times 1 \times 1$ (Fig. 3, before concatenation). Time meta-data, i.e. both time periods *onset-to-imaging* and *imaging-to-treatment*, is concatenated to this vector and processed through two 802-channel $1 \times 1 \times 1$ convolutional layers followed by dropouts to regress the output vector c_i that contains the PCA model coefficients (Fig. 3, after concatenation).

2.1 Data

We used the same dataset of 29 patients consisting of CBV segmentations, TTD segmentations, and follow-up CT segmentations after successful thrombectomy as in [12]. The TTD segmentation outlines the core+penumbra, and the CBV segmentation mask serves as an approximation of the core segmentation. However, this does not always meet the clinical hypothesis of a monotonously growing segmentation shape, because the follow-up lesion in CT can be sometimes smaller than the acute CBV lesion in the original dataset.

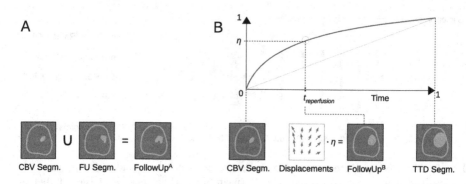

Fig. 4. Dataset *Modified A* (left) uses the union of CBV and follow-up segmentations as target, while for *Modified B* (right) an intermediate deformation from core to core+penumbra according to a non-linear growth rate is used as target.

Thus, we replaced the training target in dataset *Modified A* by the union of the original follow-up (*FU Segm.*) and the core segmentation (*CBV Segm.*) as shown in Fig. 4 (left). In dataset *Modified B* we first registered core and core+penumbra and applied the resulting displacements by a weighting factor η to adjust the vector magnitudes according to the time-to-reperfusion t (Fig. 4, right). This simulates a non-linearly growing artificial lesion with a growth rate of $\eta = t^{0.5}$ for a time $t \in \{0..1\}$ between the initial necrosis ($t = 0$) and the worst case ($t = 1$). Using dataset B we are able to demonstrate the general applicability of the method for a well-specified growth problem based on the clinical observations of higher growth at the beginning [1].

2.2   Experimental Setup

For the preprocessing step in Fig. 2, a diffeomorphic B-Spline transformation model optimised through stochastic gradient descent [15] was used in order to generate the displacement ground truths from core+penumbra to core. The choice of registration method is up to the user and can be tuned independently from the rest of our method. The displacement vector fields are trilinearly downsampled from $256 \times 256 \times 28 \times 3$ to $32 \times 32 \times 28 \times 3$ voxels for reduced computational demand.

Since our aim is to learn the main displacement along the cerebral vessel system, we do not work on a canonical (centered) basis but keep the location in the common brain space of the images (within a single hemisphere, Fig. 5a). Similar to [12] we ran the same 5-fold validation. The first two components of a learnt displacement model roughly show the main variation in height and width, which explains about 2/3 of the variance in the data (Fig. 5b). To improve robustness of the PCA, we also applied affine augmentation directly on the displacement fields (9 random transformations per patient) for which we also had to increase k to cover a similar variance than without the augmentation.

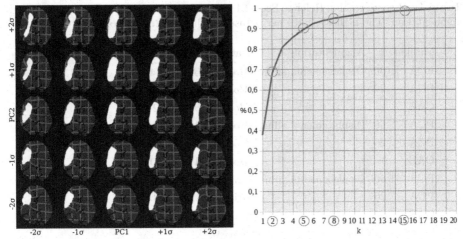

(a) The first two principal components   (b) Accumulated variance for increasing k

Fig. 5. Displacement model from the training data of [12]. (a) First two principal modes deforming a mean shape of core+penumbra. (b) Accumulated percentage of variance explained by the first k modes: ~90% for k = 5, ~95% for k = 8, and ~99% for k = 15.

2.3   Previous Approach

The proposed method that predicts PCA displacement coefficients has been compared to our previous approach using shape interpolations within a learnt shape space [12]. It incorporates prior knowledge about stroke growth in the form of a learnt stroke tissue shape space. By enforcing follow-up segmentation to lay on the trajectory between core and penumbra representations within the shape space, we could continuously evolve the shape to simulate growth of ischaemic strokes restricted to a lower (core) and upper bound (core+penumbra).

The training of the CAE is conducted in a two-step manner: First, reconstructions of segmentation masks are learned. Second, the decoded interpolation of the low-dimensional representations of core and core+penumbra is forced to match with the follow-up segmentation mask while ensuring monotonous growth of the reconstructed lesion. Apart from the results reported in [12] we added the results of the CAE on the two modified datasets in the evaluation of this paper.

3   Results

Find the overlap metric results in Table 1: The "F1 Oracle" score quantifies the capability of the PCA to enable a deformation of core+penumbra that overlaps with the follow-up, while "F1 Prediction" quantifies the capability of the CNN to find the right coefficients in the PCA model for an unknown follow-up given the input of core mask, core+penumbra mask, and time meta-data.

Table 1. F1 score results: Refer to Sect. 2.1 for a description of the datasets and affine augmentation. The highest scores per dataset are highlighted in **bold**.

Method	k	Augmentation	Dataset	F1 Prediction	F1 Oracle
PCA	60	Displacements affine	*Modified B*	**0.74**	**0.92**
PCA	20	–	*Modified B*	0.71	0.88
CAE	–	–	*Modified B*	0.62	0.71
PCA	60	Displacements affine	*Modified A*	0.52	**0.84**
PCA	20	–	*Modified A*	0.48	0.73
CAE	–	–	*Modified A*	**0.60**	0.71
PCA	60	Displacements affine	Original [12]	0.37	**0.84**
PCA	20	–	Original [12]	0.38	0.70
CAE	–	–	Original [12]	**0.46**	0.53

Considering the diffeomorphic registration overlap of the core+penumbra warped onto the core lesion from the preprocessing step, we have an initial average F1 score of 0.84 for the supervision (not listed in Table 1). The registration on the core is thereby not perfect, but diffeomorphic, so that the subsequent PCA receives rather plausible displacement fields as training input.

If we take the principal components of each PCA ($k = 20$) for the five training folds (with $n = 23$ samples) with an *oracle* knowing the right coefficients for all k components (by gradient descent optimisation on the ground truth), we achieve an average F1 score of 0.70 on the original data or even more on the two modified datasets.

4 Discussion

These oracle results indicate that the principal components are capable of representing meaningful displacements from core+penumbra to follow-up lesions, although the samples are drawn from displacements of core+penumbra to core. This supports our hypothesis that the final lesion on the follow-up image roughly represents an intermediate lesion when the core deforms to the core+penumbra lesion.

In particular, the follow-up lesions can – in theory – be much better represented by the PCA space than by the learnt shape space model of [12]. Once learnt, the CAE shape space, which encodes the low-dimensional representations of a core-penumbra pair, offers only a single degree of freedom to find the right follow-up shape (linear trajectory position between both representations in the latent space). With the PCA model, however, we still have k degrees of freedom to find the right deformation for getting the follow-up shape in a space with dimensions that explain a known percentage of the variance in the original data.

Even though the prediction F1 scores of the PCA model with coefficients from the CNN are lower on the original and *Modified A* data than for the shape space

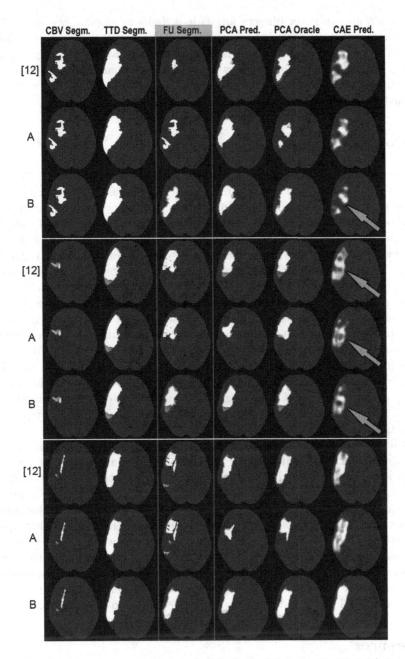

Fig. 6. Results of each dataset (Original of [12], Modified **A**, Modified **B**) for three sample subjects (top, middle, bottom) are shown. The first three columns show the ground truth masks, while the last three columns show the predictions of the follow-up mask (*FU Segm.*). Note: Implausible interpolations (green arrows) can still occur with the CAE interpolations but not with the PCA of displacements. (Color figure online)

reconstructions, we never observe any fade-in/fade-out appearance or implausible shape change over time of progression. This can still occur when training a shape space with the CAE and is difficult to fully avoid (see Fig. 6, green arrows). The more plausible a follow-up shape is compared to the given core approximation, the better are the predictions of the PCA+CNN (F1 scores on *Modified B*: 0.74 for PCA, 0.62 for CAE).

5 Conclusion

Here we present the first results on a stroke lesion deformation model that can serve as the basis for the prediction of ischaemic stroke tissue outcome. It is capable of learning the main direction of spatial growth over time from the core to the perfusion lesion (core+penumbra) based on a small number of samples. We show that these deformation modes allow to reconstruct displacements that deform the perfusion lesion shape close to the follow-up lesion shape (F1 $\geq$ 0.70).

We did not strive for the highest overlap accuracy but look for consistent growth (in terms of spatial regularisation) over time by a model of main deformation modes that can be more plausible to the doctor than an interpolation based on the weak supervision of a single follow-up shape sample per patient. It seems quite clear from the results that displacement fields offer richer supervision than the segmentation masks used for training the shape CAE. While the theoretical upper bound (oracle) of overlap with the follow-up lesion is higher than with the compared shape-interpolation method, we cannot yet achieve high overlap on data with CBV lesion segmentations used as core approximation in general, and in particular by learning the coefficients of the principal components with a CNN to predict the follow-up lesion.

Although being work-in-progress that requires further investigations, e.g. to find the optimal hyper-parameter k for the k-dimensional PCA basis, we see this as a promising research direction to quantify potential stroke growth not just by a global volume measure but locally and spatially within the tissue to assist the doctor in treatment planning. Since it was shown that displacements between core and perfusion lesion are suitable to generate actual follow-ups if reconstructed from low-dimensional representations, we would like to integrate the displacement fields into a CAE architecture in the future directly to learn a displacement model end-to-end along with the CNN to eventually improve the prediction of the unseen follow-up segmentations. Further auxiliary supervision (e.g. computing loss on c_i directly) might be necessary to achieve this.

References

1. Broocks, G., Rajput, F., Hanning, U., et al.: Highest lesion growth rates in patients with hyperacute stroke. Stroke **50**(1), 189–192 (2019)
2. Lee, E.-J., Kim, Y.-H., Kim, N., et al.: Deep into the brain: artificial intelligence in stroke imaging. J. Stroke **19**(3), 277–285 (2017)

3. Kamal, H., Lopez, V., Sheth, S.A.: Machine learning in acute ischemic stroke neuroimaging. Front. Neurol. **9**, 945 (2018)
4. Winzeck, S., Hakim, A., McKinley, R., et al.: ISLES 2016 and 2017-benchmarking ischemic stroke lesion outcome prediction based on multispectral MRI. Front. Neurol. **9**, 679 (2018)
5. Robben, D., Boers, A., Marquering, H., et al.: Prediction of final infarct volume from native CT perfusion and treatment parameters using deep learning. arXiv:1812.02496 (2018)
6. Pinto, J., Mckinley, R., Alves, V., et al.: Stroke lesion outcome prediction based on MRI imaging combined with clinical information. Front. Neurol. **9**, 1060 (2018)
7. Song, T., Huang, N.: Integrated extractor, generator and segmentor for ischemic stroke lesion segmentation. In: Crimi, A., Bakas, S., Kuijf, H., Keyvan, F., Reyes, M., van Walsum, T. (eds.) BrainLes 2018. LNCS, vol. 11383, pp. 310–318. Springer, Cham (2019). https://doi.org/10.1007/978-3-030-11723-8_31
8. Foote, M.D., Zimmerman, B., Sawant, A., et al.: Real-time patient-specific lung radiotherapy targeting using deep learning. In: International Conference on Medical Imaging with Deep Learning (MIDL) (2018)
9. Milletari, F., Rothberg, A., Jia, J., et al.: Integrating statistical prior knowledge into convolutional neural networks. In: International Conference on Medical Image Computing and Computer Assisted Intervention (MICCAI) (2017)
10. Yu, X., Zhou, F., Chandraker, M.: Deep deformation network for object landmark localization. In: Leibe, B., Matas, J., Sebe, N., Welling, M. (eds.) ECCV 2016. LNCS, vol. 9909, pp. 52–70. Springer, Cham (2016). https://doi.org/10.1007/978-3-319-46454-1_4
11. Jaderberg, M., Simonyan, K., Zisserman, A., et al.: Spatial transformer networks. In: Advances in Neural Information Processing Systems (NIPS), vol. 28 (2015)
12. Lucas, C., Kemmling, A., Bouteldja, N., et al.: Learning to predict ischemic stroke growth on acute CT perfusion data by interpolating low-dimensional shape representations. Front. Neurol. **9**, 989 (2018)
13. Couronne, R., Louis, M., Durrleman, S.: Learning disease progression models with longitudinal data and missing values. In: IEEE International Symposium on Biomedical Imaging (ISBI) (2019)
14. Paszke, A., Gross, S., Chintala, S., et al.: Automatic differentiation in PyTorch. In: Advances in Neural Information Processing Systems (NIPS) - Autodiff Workshop (2017)
15. Sandkühler, R., Jud, C., Andermatt, S., et al.: AirLab: autograd image registration laboratory. arXiv:1806.09907 (2018)

Global and Local Multi-scale Feature Fusion Enhancement for Brain Tumor Segmentation and Pancreas Segmentation

Huan Wang, Guotai Wang[✉], Zijian Liu, and Shaoting Zhang

University of Electronic Science and Technology of China, Chengdu, China
guotai.wang@uestc.edu.cn

Abstract. The fully convolutional networks (FCNs) have been widely applied in numerous medical image segmentation tasks. However, tissue regions usually have large variations of shape and scale, so the ability of neural networks to learn multi-scale features is important to the segmentation performance. In this paper, we improve the network for multi-scale feature fusion, in the medical image segmentation by introducing two feature fusion modules: i) global attention multi-scale feature fusion module (GMF); ii) local dense multi-scale feature fusion module (LMF). GMF aims to use global context information to guide the recalibration of low-level features from both spatial and channel aspects, so as to enhance the utilization of effective multi-scale features and suppress the noise of low-level features. LMF adopts bottom-up top-down structure to capture context information, to generate semantic features, and to fuse feature information at different scales. LMF can integrate local dense multi-scale context features layer by layer in the network, thus improving the ability of network to encode interdependent relationships among boundary pixels. Based on the above two modules, we propose a novel medical image segmentation framework (GLF-Net). We evaluated the proposed network and modules on challenging brain tumor segmentation and pancreas segmentation datasets, and very competitive performance has been achieved.

1 Introduction

The segmentation of target tissue is one of basic problems in medical image analysis. However, it would be time-consuming to label a large amount of medical images manually. In this case, reliable and automatic segmentation technology has a potential to improve the efficiency in clinical practice, promoting quantitative assessment in pathology and detection in illness progress [1]. With the development of fully convolutional network (FCN), it has achieved good performance in numerous medical image segmentation tasks [2, 3].

Nevertheless, tissue areas in medical images often present complex morphology and multi-scale variation, so different multi-scale context features are required to encode local and global information for accurate classification of pixels. For instance, segmentation of large structures requires more global and wider receptive field, while segmentation of small structures needs to concentrate on local high-resolution information. Especially, to locate vague boundary in tissue areas precisely, local dense multi-scale features are

© Springer Nature Switzerland AG 2020
A. Crimi and S. Bakas (Eds.): BrainLes 2019, LNCS 11992, pp. 80–88, 2020.
https://doi.org/10.1007/978-3-030-46640-4_8

required to encode interdependent relationships among boundary pixels. In addition, for a specific-scale object, because of the large difference in resolution and semantic level between high- and low-level features, these features have different importance for discrimination, some of which may indicate false information [4]. Additionally, low-level features have complex background noise. Simply integrating high- and low-level features by feature concatenation is unable to utilize the multi-scale information thoroughly. Therefore, it is necessary to select the discriminative and effective features.

Based on motivation of these two aspects, and with the inspiration of literature [5], we introduce attention mechanism based on the skip connection and propose a global attention multi-scale feature fusion module (GMF). This module utilizes the global context information of high-level features, guising improvement and recalibration in spatial and channel information of low-level features respectively, aiming to strengthen relative features and suppress irrelative noise. Before the high- and low-level features fusion, this module would perform discriminative selection and optimization for low-level features, urging lower-level features to provide more effective multi-scale information for higher-level features. In addition, we introduce a local dense multi-scale feature fusion module (LMF) to learn the dense multi-scale features. The LMF uses the advantage of bottom-up top-down structure, obtaining larger receptive field and higher-level semantic features by down-sampling, and fusing multi-scale context information by up-sampling and skip connection. It could be regarded as a fundamental module used in network, and with the cascade of LMF in multiple layers, the network could obtain multi-scale information under arbitrary feature resolutions. As far as we know, it is the first time to propose local dense multi-scale feature fusion module based on mini bottom-up top-down structure in the neural network, and it is also the first time to apply it to medical image segmentation.

We integrate the proposed modules (GMF & LMF) into a typical encoder-decoder network for medical image segmentation, to demonstrate that these are two generic components to boost performance, so as to propose a novel medical image segmentation framework (GLF-Net). On challenging brain tumor segmentation and pancreas segmentation tasks, GLF-Net and the proposed modules have been extensively evaluated. The conclusion indicates that GLF-Net is perfectly adaptive to these two different segmentation tasks, and the modules also improve the performance considerably.

Related Work. Researchers mainly try to improve the ability of networks to learn multi-scale features from the following three aspects: (i) multi-scale feature extraction based on image pyramid [6]; (ii) multi-scale feature fusion layer by layer based on encoder-decoder structure and skip connection [2]; (iii) multi-scale features extraction based on dilated convolution and dilated spatial pyramid module [7]. Our method improves the performance of feature learning on the basis of the second aspect. Besides, attention mechanism offers a feasible scheme for the adaptive recalibration of feature maps. Both of [5, 8] adopt different attention mechanism to calibrate features and improve the network performance.

2 Method

Convolution layer uses a set of feature maps as inputs, constructs information features by fusing information from space and channel in local receptive field, and generates a new set of feature maps as outputs. Here, we define a group of convolutional transformations $F_{tr} : X \to X', X \in \mathbb{R}^{H \times W \times C}, X' \in \mathbb{R}^{H' \times W' \times C'}$, where H and W are the spatial height and width, with C and C' being the number of input and output channels, respectively. Every convolution operation $F_{tr}(\cdot)$ encodes the spatial information in adjacent locations in the input feature X, and then outputs X' which is able to encode more abundant spatial information. Therefore, features of FCN in different layers encode information at different scales. Many researchers proposed different schemes (such as skip connection [3], ASPP [7], etc.) to use multi-scale information in FCN. However, most of methods based on feature fusion [2] only consider simple fusion of high- and low-level features in the decoder, to enhance the reconstruction in high-resolution details. In this study, we focus on using the rich multi-scale information among different feature maps in FCN to achieve accurate and robust segmentation. We emphasize the learning of local multi-scale features in encoder and enhance the efficiency of high- and low-level feature fusion in decoder respectively. Next, we would introduce the methods proposed in this paper in detail.

2.1 Global Attention Multi-scale Feature Fusion Module (GMF)

We introduce global attention multi-scale feature fusion module, by combining skip connection with attention mechanism. It uses the high-level features' perception abilities to global information to guide the low-level features optimization, and offers more effective multi-scale information to high-level features. GMF emphasizes that high-level features could guide the low-level features' optimization effectively to enhance the semantic consistence of high- and low-level features, and improve the fuse efficiency of multi-scale features.

Here, we use the global information of high-level features to recalibrate local features in space and channel respectively. We assume that the input high- and low-level features $U = [u_1, u_2, u_3, \cdots, u_C], L = [l_1, l_2, l_3, \cdots, l_C]$ have c channels and the i-th channel is denoted as $u_i \in \mathbb{R}^{H \times W}, l_i \in \mathbb{R}^{H \times W}$ respectively. In the branch of spatial feature recalibration, we use a feature transformation to obtain the spatial projected map $s = F_{s-sq}(U)$ of high-level feature U, where the $F_{s-sq}(\cdot)$ indicates the 1×1 convolutional operation and the output channel number is 1. Then, using a sigmoid function to obtain spatial weight map $\tilde{s} = \sigma(s), \tilde{s} \in \mathbb{R}^{H \times W}$. In the branch of channel feature recalibration, we use a global average pooling to compress the global spatial information of u_i into a channel descriptor, and generate channel-wise statistics vector $z = F_{c-sq}(U)$. The vector is transformed to $z' = F_{ex}(z)$ by two 1×1 convolution layers, and the final channel weight vector $(\tilde{z} = \sigma(z'), \tilde{z} \in \mathbb{R}^{1 \times 1 \times C})$ is gained by a sigmoid function. The weight vectors $\tilde{s}$ and $\tilde{z}$ encode spatial and channel information in the high-level feature U respectively. We encode the global information into low-level features by element-wise multiplication, so as to improve the encoding quality of low-level features, and to suppress irrelative background noise. Subsequently, the final multi-scale feature is obtained by adding the corresponding elements of the recalibrated low-level feature and the high-level feature U. It could be defined as: $\tilde{L}_s = \tilde{s} \otimes L + U, \tilde{L}_c = \tilde{z} \otimes L + U$, where $\otimes$ denotes element-wise multiplication. Finally, we concatenate $\tilde{L}_s, \tilde{L}_c$ and send the result to a 3×3 convolution layer to fuse their corresponding feature information, and reduce the channel dimensions. The architecture of GMF is illustrated in Fig. 1.

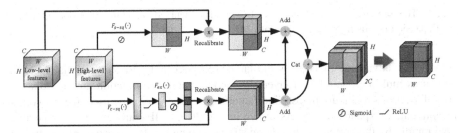

Fig. 1. The schematic illustration of the global attention multi-scale feature fusion module.

2.2 Local Dense Multi-scale Feature Fusion Module (LMF)

Generally, tissue areas in medical images have vague contours. Precise location of tissue contours requires local dense multi-scale features to encode the interdependent relationships among boundary pixels. However, every feature map in FCN only encodes corresponding scale information, which is unable to meet the network's need to encode this relationship. In this case, we introduce a local dense multi-scale feature fusion module (LMF), which could obtain context information under arbitrary feature resolutions, generate semantic features and fuse feature information at different scales.

In particular, LMF uses bottom-up top-down structure to learn and obtain context information to generate semantic features, and to fuse multi-scale features in adjacent layers by skip connections between bottom-up and top-down. The whole module is shown in Fig. 2. LMF uses two 3×3 convolution layers and a down-sampling to expand the receptive field rapidly and obtain context information from the input feature maps. Then, it expands global information and generates semantic features by two 3×3 convolution layers and an up-sampling. In LMF, two skip connections are used to fuse features in adjacent scales. It is worth noting that LMF could change the quantity of these components according to different tasks for getting better performance. Networks based on bottom-up top-down structure [2] are widely applied in medical image segmentation field. Unlike those methods, LMF aims to enhance the encoding ability to learn local dense multi-scale features by obtaining and fusing adjacent multi-scale features, rather than to solve specific segmentation tasks.

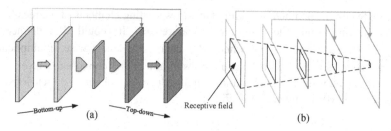

Fig. 2. The schematic illustration of LMF. (a) The detailed structure of proposed LMF. (b) Expansion details in receptive field of LMF.

Another advantage of LMF is the larger receptive field. In conventional FCN, it only uses multiple convolution layers to encode feature information among adjacent spatial locations in feature maps. Since small convolution kernel is used, the network's receptive field expands slowly. As shown in Fig. 2(b), LMF could expand its receptive field rapidly by the bottom-up top-down structure, which obtains spatial information at different scales respectively. By fusing the obtaining information, LMF could learn denser multi-scale features while provide larger respective field. Therefore, the network is able to learn dense multi-scale features from low to high layer by layer with simple repetition of LMFs.

2.3   Multi-scale Feature Fusion Framework

The proposed multi-scale feature fusion enhancement modules could be integrated into existing segmentation frameworks, improving their ability to learn multi-scale features by substituting standard convolution layers and skip connection. To demonstrate the effectiveness of these modules, we choose the Unet [2] as the backbone structure, which is the most widely used in medical image segmentation, leading to a new medical image segmentation network (GLF-Net). As shown in Fig. 3, in the Unet decoder, we use GMF to enhance the fusion of high- and low-level features. Then, we substitute all standard convolutions in the encoder with LMF to improve the encoder's ability to learn local dense multi-scale information. In this paper, each convolution module in GLF-Net consists of a 3×3 convolution layer, a group normalization layer [9] and a rectified linear unit (ReLU) layer. Finally, we get the segmentation probability map by Sigmoid function.

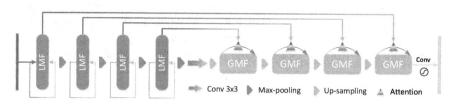

Fig. 3. Illustration of the architecture of our proposed GLF-Net.

3 Experiments

3.1 Data and Experiments Setups

We first investigated automatic segmentation of the whole tumor from Fluid Attenuated Inversion Recovery (FLAIR) images. We randomly chose 175 cases of FLAIR images from 2018 Brain Tumor Segmentation Challenge (BraTS) [10–14] training set. Different from previous works focus on multi-label and multimodality segmentation, we only used FLAIR images and only segmented the whole tumor in 2D slices due to the limitation of memory. We used 100 volumes with 8000 slices for training, 25 volumes with 2000 slices for validation and 50 volumes with 4000 slices for testing. Then we studied the automatic segmentation of pancreas in CT images, provided by Memorial Sloan Kettering Cancer Center form MSD challenge[1]. This dataset contains 280 cases of patients in total. We performed data splitting at patient level and used images from 180, 20, 80 patients for training, validation and testing, respectively. Similarly, we only used 2D slices to train the network, and we obtained 6886 training images in total, 800 validation images and 3200 testing images. Noting that in both datasets, we manually discarded some slices that contain only background.

Our GLF-Net was implemented using Pytorch on a Linux system with an Nvidia 1080Ti GPU. During training, we used the dice loss and the Adam optimizer with a learning rate of 1×10^{-4}, with a learning rate reduction of 0.1 times after every 15 epochs. In each experiment, we saved the model that performed best on the validation set during training as the final test model. Data augmentation including random cropping and flipping were used to improve the robustness of the model. As for BraTS dataset, we first re-scaled all images to 224×224 pixels and normalized the pixel values of the images to the range of 0 to 1. As for the pancreas dataset, we first normalized all images to 0 to 1 and resized the images to 256×256. In these two experiments, the batch size used for training was 10 and 5, respectively.

For verifying the effectiveness of the proposed modules and network, we conducted ablation studies on two datasets respectively, and compared GLF-Net with Unet-24 [2], Res-Unet-24. Res-Unet-24 was a modified Unet where each convolution block was replaced by the bottleneck building block used in the ResNet [15]. The number of basic channels of Unet-24 and Res-Unet-24 is 24 to ensure that the number of parameters is close to that of GLF-Net. Dice coefficient and Jaccard index were used to quantitative evaluation of the segmentation performance.

3.2 Results and Discussion

Table 1 shows the evaluation results of different variants of the proposed method (only GMF, only LMF and GLF-Net) on brain tumor dataset and pancreas dataset respectively. It can be seen that the GLF-Net method, which only includes GMF, has better

---

[1] http://medicaldecathlon.com/.

performance on these two datasets though its parameters are only a half of Unet-24. This phenomenon indicates that GMF could use multi-scale information of low-level features more efficiently to improve the segmentation results of network. In addition, LMF gets better performance than GMF. Though LMF augments parameters, its segmentation results improve a lot compared with Unet-24. This proves the effectiveness of proposed LMF. The GLF-Net, integrating GMF and LMF, gets the best performance among all the compared methods. Compared with Unet-24, the Jaccard index of GLF-Net increases by 3.8% and 7.4% on pancreas and brain tumor datasets respectively. The results show that compared with ordinary organ segmentation, GLF-Net could obtain better performance on lesion segmentation tasks with complicated multi-scale variation (such as brain tumor segmentation).

Table 1. Quantitative evaluation of different networks on brain tumor and pancreas datasets.

Method	Brain Tumor dataset		Pancreas dataset		
	Dice	Jaccard	Dice	Jaccard	Parameters
Unet-24	0.841 ± 0.125	0.737 ± 0.153	0.764 ± 0.103	0.628 ± 0.121	4.3×10^6
Res-Unet-24	0.862 ± 0.066	0.766 ± 0.099	0.774 ± 0.079	0.640 ± 0.102	4.5×10^6
GLF-Net (only GMF)	0.875 ± 0.065	0.786 ± 0.097	0.773 ± 0.095	0.637 ± 0.111	1.9×10^6
GLF-Net (only LMF)	0.884 ± 0.061	0.796 ± 0.093	0.788 ± 0.093	0.658 ± 0.109	4.4×10^6
GLF-Net	$\mathbf{0.893 \pm 0.055}$	$\mathbf{0.811 \pm 0.086}$	$\mathbf{0.795 \pm 0.069}$	$\mathbf{0.666 \pm 0.092}$	4.5×10^6

The qualitative segmentation results of three examples with different appearances from brain tumor and pancreas datasets are shown in Fig. 4. For example A, though the tumor area is obvious, there are two independent segmentation regions which influence network's judgment, leading Unet-24 makes wrong prediction. However, GMF improves one of the segmentation performances, and LMF almost completely avoids the error prediction of these two regions. For example B, the background is very close to the tumor area, and Unet-24 is unable to distinguish boundary information of tumor areas, thus, causing segmentation results with larger errors. GMF and LMF improve segmentation results in different areas successively. This indicates that both of GMF and LMF are able to enhance the ability of network to learn multi-scale features effectively, and get better performance. Moreover, LMF has a more precise segmentation result than GMF, which indicates that LMF could encode the interdependent relationship among boundary pixels more effectively, so as to locate boundaries precisely. Example C also proves the validity of GMF and LMF. Besides, we visualize spatial weight maps in GMF, and it is obvious that different spatial weight maps encode context information on diverse scales in target areas. By using this kind of global information to improve encoding of low-level features, it is able to enhance target-related feature information in low-level features.

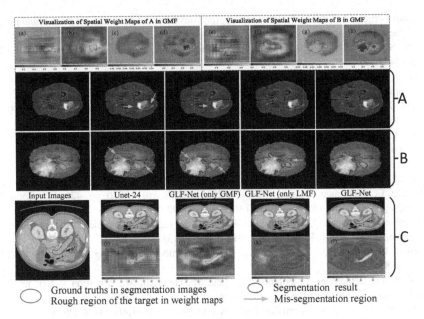

Fig. 4. The qualitative segmentation results of three examples (A, B, C) from brain tumor and pancreas datasets. Each example contains different network segmentation results and the visualization of GMF spatial weight maps. From left to right (a-b, e-h, i-l), feature resolution goes from low to high.

4 Conclusion

This paper propose two modules used in multi-scale feature fusion enhancement for better medical image segmentation performance. Before the fusion of high-level and low-level features, GMF uses attention mechanism to select the optimal low-level features information to improve the high- and low-level feature fusion efficiency. LMF aims to obtain more abundant local dense multi-scale features with a bottom-up top-down structure, to improve the tissue contours segmentation precision of network. Based on these two modules, we propose a novel medical image segmentation network. We evaluated the proposed methods on brain tumor and pancreas datasets, and got very competitive results. This indicates that the proposed methods have effectiveness and wide adaptability. Besides, as a general solution, future work aims to apply the methods to 3D segmentation or other segmentation tasks.

References

1. Qin, Y., et al.: Autofocus layer for semantic segmentation. In: Frangi, A.F., Schnabel, J.A., Davatzikos, C., Alberola-López, C., Fichtinger, G. (eds.) MICCAI 2018. LNCS, vol. 11072, pp. 603–611. Springer, Cham (2018). https://doi.org/10.1007/978-3-030-00931-1_69

2. Ronneberger, O., Fischer, P., Brox, T.: U-Net: convolutional networks for biomedical image segmentation. In: Navab, N., Hornegger, J., Wells, W.M., Frangi, A.F. (eds.) MICCAI 2015. LNCS, vol. 9351, pp. 234–241. Springer, Cham (2015). https://doi.org/10.1007/978-3-319-24574-4_28

3. Wang, H., Gu, R., Li, Z.: Automated segmentation of intervertebral disc using fully dilated separable deep neural networks. In: Zheng, G., Belavy, D., Cai, Y., Li, S. (eds.) CSI 2018. LNCS, vol. 11397, pp. 66–76. Springer, Cham (2019). https://doi.org/10.1007/978-3-030-13736-6_6

4. Yu, C., Wang, J., Peng, C., Gao, C., Yu, G., Sang, N.: Learning a discriminative feature network for semantic segmentation. In: CVPR, pp. 1857–1866 (2018)

5. Roy, A.G., Navab, N., Wachinger, C.: Concurrent spatial and channel 'squeeze & excitation' in fully convolutional networks. In: Frangi, A.F., Schnabel, J.A., Davatzikos, C., Alberola-López, C., Fichtinger, G. (eds.) MICCAI 2018. LNCS, vol. 11070, pp. 421–429. Springer, Cham (2018). https://doi.org/10.1007/978-3-030-00928-1_48

6. Farabet, C., Couprie, C., Najman, L., LeCun, Y.: Learning hierarchical features for scene labeling. IEEE Trans. PAMI 35(8), 1915–1929 (2013)

7. Zhao, H., Shi, J., Qi, X., Wang, X., Jia, J.: Pyramid scene parsing network. In: CVPR, pp. 2881–2890 (2017)

8. Oktay, O., Schlemper, J., Folgoc, L.L., Lee, M., Heinrich, M.: Attention U-Net: learning where to look for the pancreas. arXiv preprint arXiv:1804.03999 (2018)

9. Wu, Y., He, K.: Group normalization. arXiv preprint arXiv:1803.08494 (2018)

10. Menze, B.H., Jakab, A., Bauer, S., et al.: The multimodal brain tumor image segmentation benchmark (BRATS). IEEE TMI 34(10), 1993–2024 (2015)

11. Bakas, S., Akbari, H., Sotiras, A., Bilello, M., Rozycki, M., Kirby, J.S., et al.: Advancing The Cancer Genome Atlas glioma MRI collections with expert segmentation labels and radiomic features. Nature Sci. Data 4, 170117 (2017)

12. Bakas, S., et al.: Identifying the best machine learning algorithms for brain tumor segmentation, progression assessment, and overall survival prediction in the BRATS Challenge. arXiv preprint arXiv:1811.02629 (2018)

13. Bakas, S., et al.: Segmentation labels and radiomic features for the pre-operative scans of the TCGA-GBM collection. The Cancer Imaging Archive (2017)

14. Bakas, S., et al.: Segmentation labels and radiomic features for the pre-operative scans of the TCGA-LGG collection. The Cancer Imaging Archive (2017)

15. He, K., Zhang, X., Ren, S., Sun, J.: Deep residual learning for image recognition. In: CVPR, pp. 770–778 (2016)

Optimization with Soft Dice Can Lead to a Volumetric Bias

Jeroen Bertels[1]([⊠])(ID), David Robben[1,2], Dirk Vandermeulen[1], and Paul Suetens[1]

[1] Processing Speech and Images, ESAT, KU Leuven, Leuven, Belgium
jeroen.bertels@kuleuven.be
[2] icometrix, Leuven, Belgium

Abstract. Segmentation is a fundamental task in medical image analysis. The clinical interest is often to measure the volume of a structure. To evaluate and compare segmentation methods, the similarity between a segmentation and a predefined ground truth is measured using metrics such as the Dice score. Recent segmentation methods based on convolutional neural networks use a differentiable surrogate of the Dice score, such as soft Dice, explicitly as the loss function during the learning phase. Even though this approach leads to improved Dice scores, we find that, both theoretically and empirically on four medical tasks, it can introduce a volumetric bias for tasks with high inherent uncertainty. As such, this may limit the method's clinical applicability.

Keywords: Segmentation · Cross-entropy · Soft Dice · Volume

1 Introduction

Automatic segmentation of structures is a fundamental task in medical image analysis. Segmentations either serve as an intermediate step in a more elaborate pipeline or as an end goal by itself. The clinical interest often lies in the volume of a certain structure (e.g. the volume of a tumor, the volume of a stroke lesion), which can be derived from its segmentation [11]. The segmentation task can also carry inherent uncertainty (e.g. noise, lack of contrast, artifacts, incomplete information).

To evaluate and compare the quality of a segmentation, the similarity between the true segmentation (i.e. the segmentation derived from an expert's delineation of the structure) and the predicted segmentation must be measured. For this purpose, multiple metrics exist. Among others, overlap measures (e.g. Dice score, Jaccard index) and surface distances (e.g. Haussdorf distance, average surface distance) are commonly used [13].

The focus on one particular metric, the Dice score, has led to the adoption of a differentiable surrogate loss, the so-called soft Dice [9,15,16], to train convolutional neural networks (CNNs). Many state-of-the-art methods clearly outperform the established cross-entropy losses using soft Dice as loss function [7,12].

J. Bertels and D. Robben—Contributed equally to this work.

© Springer Nature Switzerland AG 2020
A. Crimi and S. Bakas (Eds.): BrainLes 2019, LNCS 11992, pp. 89–97, 2020.
https://doi.org/10.1007/978-3-030-46640-4_9

In this work, we investigate the effect on volume estimation when optimizing a CNN w.r.t. cross-entropy or soft Dice, and relate this to the inherent uncertainty in a task. First, we look into this volumetric bias theoretically, with some numerical examples. We find that the use of soft Dice leads to a systematic under- or overestimation of the predicted volume of a structure, which is dependent on the inherent uncertainty that is present in the task. Second, we empirically validate these results on four medical tasks: two tasks with relatively low inherent uncertainty (i.e. the segmentation of third molars from dental radiographs [8], BRATS 2018 [4–6,14]) and two tasks with relatively high inherent uncertainty (i.e. ISLES 2017 [2,18], ISLES 2018 [3]).

2   Theoretical Analysis

Let us formalize an image into I voxels, each voxel corresponding to a true class label c_i with $i = 0 \dots I - 1$, forming the true class label map $C = [c_i]^I$. Typical in medical image analysis, is the uncertainty of the true class label map C (e.g. due to intra- and inter-rater variability; see Sect. 2.2). Under the assumption of binary image segmentation with $c_i \in \{0, 1\}$, a probabilistic label map can be constructed as $Y = [y_i]^I$, where each $y_i = P(c_i = 1)$ is the probability of y_i belonging to the structure of interest. Similarly, we have the maps of voxel-wise label predictions $\hat{C} = [\hat{c}_i]^I$ and probabilities $\hat{Y} = [\hat{y}_i]^I$. In this setting, the class label map $\hat{C}$ is constructed from the map of predictions $\hat{Y}$ according to the highest likelihood.

The Dice score $\mathcal{D}$ is defined on the label maps as:

$$\mathcal{D}(C, \hat{C}) = \frac{2|C \cap \hat{C}|}{|C| + |\hat{C}|} \tag{1}$$

The volumes $\mathcal{V}(C)$ of the true structure and $\mathcal{V}(\hat{C})$ of the predicted structure are then, with v the volume of a single voxel:

$$\mathcal{V}(C) = v \sum_{i=0}^{I-1} c_i, \ \ \mathcal{V}(\hat{C}) = v \sum_{i=0}^{I-1} \hat{c}_i \tag{2}$$

In case the label map is probabilistic, we need to work out the expectations:

$$\mathcal{V}(Y) = v\mathbf{E}[\sum_{i=0}^{I-1} y_i], \ \ \mathcal{V}(\hat{Y}) = v\mathbf{E}[\sum_{i=0}^{I-1} \hat{y}_i] \tag{3}$$

2.1   Risk Minimization

In the setting of supervised and gradient-based training of CNNs [10] we are performing empirical risk minimization. Assume the CNN, with a certain topology, is parametrized by $\theta \in \Theta$ and represents the functions $\mathcal{H} = \{\mathfrak{h}_\theta\}^{|\Theta|}$. Further assume we have access to the entire joint probability distribution $P(\mathbf{x}, y)$ at both training and testing time, with $\mathbf{x}$ the information (for CNNs this is typically a centered image patch around the location of y) of the network that is used to make a prediction $\hat{y} = \mathfrak{h}_\theta(\mathbf{x})$ for y. For these conditions, the general risk minimization principle is applicable and states that in order to optimize the performance for a certain non-negative and real-valued loss $\mathcal{L}$ (e.g. the metric or its surrogate loss) at test time, we can optimize the same loss during the learning phase [17]. The risk $\mathcal{R}_\mathcal{L}(\mathfrak{h}_\theta)$ associated with the loss $\mathcal{L}$ and parametrization θ of the CNN, without regularization, is defined as the expectation of the loss function:

$$\mathcal{R}_\mathcal{L}(\mathfrak{h}_\theta) = \mathbf{E}[\mathcal{L}(\mathfrak{h}_\theta(\mathbf{x}), y)] \tag{4}$$

For years, minimizing the negative log-likelihood has been the gold standard in terms of risk minimization. For this purpose, and due to its elegant mathematical properties, the voxel-wise cross-entropy loss ($\mathcal{CE}$) is used:

$$\mathcal{CE}(\hat{Y}, Y) = \sum_{i=0}^{I-1} [\mathcal{CE}(\hat{y}_i, y_i)] = -\sum_{i=0}^{I-1} [y_i \log \hat{y}_i] \tag{5}$$

More recently, the soft Dice loss ($\mathcal{SD}$) is used in the optimization of CNNs to directly optimize the Dice score at test time [9,15,16]. Rewriting Eq. 1 to its non-negative and real-valued surrogate loss function as in [9]:

$$\mathcal{SD}(\hat{Y}, Y) = 1 - \frac{2\sum_{i=0}^{I-1} \hat{y}_i y_i}{\sum_{i=0}^{I-1} \hat{y}_i + \sum_{i=0}^{I-1} y_i} \tag{6}$$

2.2 Uncertainty

There is considerable uncertainty in the segmentation of medical images. Images might lack contrast, contain artifacts, be noisy or incomplete regarding the necessary information (e.g. in ISLES 2017 we need to predict the infarction after treatment from images taken before, which is straightforwardly introducing *inherent* uncertainty). Even at the level of the true segmentation, uncertainty exists due to intra- and inter-rater variability. We will investigate what happens with the estimated volume $\mathcal{V}$ of a certain structure in an image under the assumption of having perfect segmentation algorithms (i.e. the prediction is the one that minimizes the empirical risk).

Assuming independent voxels, or that we can simplify Eq. 3 into J independent regions with true uncertainty p_j and predicted uncertainty $\hat{p}_j$, and corresponding volumes $s_j = vn_j$, with n_j the number of voxels belonging to region $j = 0 \ldots J - 1$ (having each voxel as an independent region when $n_j = 1$), we get:

$$\mathcal{V}(Y) = \sum_{j=0}^{J-1} (s_j p_j), \quad \mathcal{V}(\hat{Y}) = \sum_{j=0}^{J-1} (s_j \hat{p}_j) \tag{7}$$

We analyze for $\mathcal{CE}$ the predicted uncertainty that minimizes the risk $\mathcal{R}_{\mathcal{CE}}(\mathfrak{h}_\theta)$:

$$\arg\min_{\hat{Y}}[\mathcal{R}_{\mathcal{CE}}(\mathfrak{h}_\theta)] = \arg\min_{\hat{Y}}[\mathbf{E}[\mathcal{CE}(\hat{Y}, Y)]] \tag{8}$$

We need to find for each independent region j:

$$\arg\min_{\hat{p}_j}[s_j\mathcal{CE}(\hat{p}_j, p_j)] = \arg\min_{\hat{p}_j}[-p_j\log\hat{p}_j - (1-p_j)\log(1-\hat{p}_j)] \tag{9}$$

This function is continuous and its first derivative monotonously increasing in the interval $]0, 1[$. First order conditions w.r.t. $\hat{p}_j$ give the optimal value for the predicted uncertainty $\hat{p}_j = p_j$. With the predicted uncertainty being the true uncertainty, $\mathcal{CE}$ becomes an unbiased volume estimator.

We analyze for SD the predicted uncertainty that minimizes the risk $\mathcal{R}_{SD}(\mathfrak{h}_\theta)$:

$$\arg\min_{\hat{Y}}[\mathcal{R}_{SD}(\mathfrak{h}_\theta)] = \arg\min_{\hat{Y}}[\mathbf{E}[SD(\hat{Y}, Y)]] \tag{10}$$

We need to find for each independent region j:

$$\arg\min_{\hat{Y}}[\mathbf{E}[SD(\hat{Y}, Y)]] = \arg\min_{\hat{p}_j}[\mathbf{E}[1 - \frac{2\sum_{j=0}^{J-1} s_j\hat{p}_j p_j}{\sum_{j=0}^{J-1} s_j\hat{p}_j + \sum_{j=0}^{J-1} s_j p_j}]] \tag{11}$$

This minimization is more complex and we analyze its behavior by inspecting the values of SD numerically. We will consider the scenarios with only a single region or with multiple independent regions with inherent uncertainty in the image. For each scenario we will vary the inherent uncertainty and the total uncertain volume.

Single Region of Uncertainty. Imagine the segmentation of an image with $K = 3$ independent regions, α, β and γ, as depicted in Fig. 1 (**A0**). Region α is certainly not part of the structure ($p_\alpha = 0$, i.e. background), region β belongs to the structure with probability p_β and region γ is certainly part of the structure ($p_\gamma = 1$). Let their volumes be $s_\alpha = 100$, s_β, $s_\gamma = 1$, respectively, with $\mu = \frac{s_\beta}{s_\gamma} = s_\beta$ the volume ratio of uncertain to certain part of the structure. Assuming a perfect algorithm, the optimal predictions under the empirical risk from Eq. 11 are:

$$\arg\max_{\hat{p}_\alpha, \hat{p}_\beta, \hat{p}_\gamma}[\mathbf{E}[\frac{2(s_\beta\hat{p}_\beta p_\beta + s_\gamma\hat{p}_\gamma)}{s_\alpha\hat{p}_\alpha + s_\beta\hat{p}_\beta + s_\gamma\hat{p}_\gamma + s_\beta p_\beta + s_\gamma}]] \tag{12}$$

It is trivial to show that $\hat{p}_\alpha = 0 = p_\alpha$ and $\hat{p}_\gamma = 1 = p_\gamma$ are solutions for this equation. The behavior of $\hat{p}_\beta$ w.r.t. p_β and μ can be observed qualitatively in Fig. 1 (**A1-A4**). Indeed, only for $p_\beta = \{0, 1\}$ the predicted uncertainty $\hat{p}_\beta$ is exact. The location of the local minimum in $\hat{p}_\beta = [0, 1]$ switches from 0 to 1 when $p_\beta = 0.5$. Therefore, when p_β decreases or increases from 0.5 (different opacity in **A1-A3**), respectively under- or overestimation will occur (**A4**). The resulting volumetric bias will be highest when the inherent uncertainty $p_\beta = 0.5$ and decreases towards the points of complete certainty, being always 0 or 1. The effect of the volume ratio μ (colors) is two-fold. With μ increasing, the optimal loss value increases (**A1-A3**) and the volumetric bias increases (**A4**; solid lines). However, the error on the estimated uncertainty is not influenced by μ (**A4**; dashed lines).

Multiple Regions of Uncertainty. In a similar way we can imagine the segmentation of a structure with $K = N + 2$ independent regions, for which we further divided the region β into N equally large independent sub-regions β_n with $n = 0 \ldots N - 1$. Let us further assume they have the same inherent uncertainty $p_{\beta_n} = p_\beta$ and volume ratio $\mu_{\beta_n} = \frac{\mu_\beta}{N}$ (in order to keep the total uncertain volume the same). If we limit the analysis to a qualitative observation of Fig. 1 with $N = 4$ (**B0-B4**) and $N = 16$ (**C0-C4**), we notice three things. First, the uncertainty p_β for which under- or overestimation will happen decreases (**A4**,

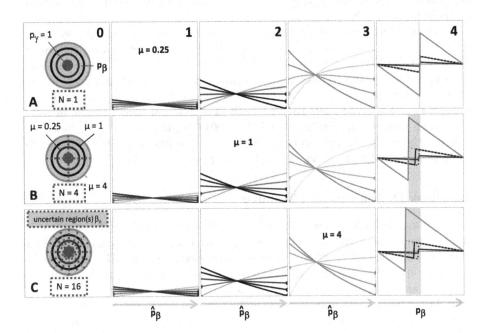

Fig. 1. The effects of optimizing w.r.t. $\mathcal{SD}$ for volume ratios: $\mu = 0.25$ (blue), $\mu = 1$ (black) and $\mu = 4$ (red). ROWS **A-C**: Situations with respectively $N = \{1, 4, 16\}$ independent regions with uncertainty p_β. COLUMN 0: Schematic representation of the situation. COLUMNS **1-3**: $\mathcal{SD} = [0, 1]$ (y-axis) for $p_\beta = \{0, 0.25, 0.5, 0.75, 1\}$ (respectively with increasing opacity) and $\hat{p} = [0, 1]$ (x-axis). COLUMN 4: Influence of $p_\beta = [0, 1]$ (x-axis) on volumetric bias (solid lines) or on the error in predicted uncertainty (dashed lines). With the light red area we want to highlight that easier overestimation of the predicted volume occurs due to a higher volume ratio μ or an increasing number of independent regions N. (Color figure online)

B4, C4). Second, this effect is proportional with μ and the maximal error on the predicted uncertainty becomes higher (**B0-B4, C0-C4**). Third, there is a trend towards easier volumetric overestimation and with the maximal error being more pronounced when the number of regions increases (**A4, B4, C4**).

3 Empirical Analysis

In this section we will investigate whether the aforementioned characteristics can be observed under real circumstances. In a practical scenario, the joint probability distribution $P(\mathbf{x}, y)$ is unknown and presents itself as a training set. The risk $\mathcal{R}_\mathcal{L}$ (Eq. 4) becomes empirical, where the expectation of the loss function becomes the mean of the losses across the training set. Furthermore, the loss $\mathcal{L}$ absorbs the explicit (e.g. weight decay, L2) or implicit (e.g. early stopping, dropout) regularization, which is often present in some aspect of the optimization of CNNs. Finally, the classifier is no longer perfect and additionally to the inherent uncertainty in the task we now have inherent uncertainty introduced by the classifier itself.

To investigate how these factors impact our theoretical findings, we train three models with increasing complexity: LR (logistic regression on the input features), ConvNet (simpler version of the next) and U-Net. We use five-fold cross-validation on the training images from two tasks with relatively low inherent uncertainty (i.e. lower-left third molar segmentation from panoramic dental radiographs (MOLARS) [8], BRATS 2018 [4]) and from two tasks with relatively high inherent uncertainty (i.e. ISLES 2017 [2], ISLES 2018 [3]). Next, we describe the experimental setup, followed by a dissemination of the predicted volume errors $\Delta\mathcal{V}(\hat{Y}, Y) = \mathcal{V}(\hat{Y}) - \mathcal{V}(Y)$ by $\mathcal{CE}$ and $\mathcal{SD}$ trained models.

3.1 Task Description and Training

We (re-)formulate a binary segmentation task for each dataset having one (multimodal) input, and giving one binary segmentation map as output (for BRATS 2018 we limit the task to whole tumor segmentation). For the 3D public benchmarks we use all of the provided images, resampled to an isotropic voxel-size of 2 mm, as input (for both ISLES challenges we omit perfusion images). In MOLARS (2D dataset from [8]), we first extract a 448×448 ROI around the geometrical center of the lower-left third molar from the panoramic dental radiograph. We further downsample the ROI by a factor of two. The output is the segmentation of the third molar, as provided by the experts. All images are normalized according to the dataset's mean and standard deviation.

For our U-Net model we start from the successful No New-Net implementation during last year's BRATS challenge [12]. We adapt it with three $3 \times 3(\times 3)$ average pooling layers with corresponding linear up-sampling layers and strip the instance normalization layers. Each level has two $3 \times 3(\times 3)$ convolutional layers before and after the pooling and up-sampling layer, respectively, with [[10, 20], [20, 10]], [[20, 40], [40, 20]], [[40, 80], [80, 40]] and [40, 20] filters. For the ConvNet model, we remove the final two levels. The LR model uses the inputs directly for classification, thus performing logistic regression on the input features.

The images are augmented intensively during training and inputs are central image crops of $162 \times 162 \times 108$ (in MOLARS 243×243). We train the models w.r.t. $\mathcal{CE}$ or $\mathcal{SD}$ with ADAM, without any explicit regularization, and with the initial learning rate set at 10^{-3} (for LR model at 1). We lower the learning rate by a factor of five when the validation loss did not improve over the last 75 epochs and stop training with no improvement over the last 150 epochs.

3.2   Results and Discussion

In Table 1 the results are shown for each dataset (i.e. MOLARS, BRATS 2018, ISLES 2017, ISLES 2018), for each model (i.e. LR, ConvNet, U-Net) and for each loss (i.e. $\mathcal{CE}$, $\mathcal{SD}$) after five-fold cross-validation. We performed a pairwise non-parametric significance test (bootstrapping) with a p-value of 0.05 to assess inferiority or superiority between pairs of optimization methods.

Table 1. Empirical results for cross-entropy ($\mathcal{CE}$), soft Dice score ($1 - \mathcal{SD}$) and volume error ($\Delta \mathcal{V}$; in 10^2 *pixels* or *ml*) metrics for models optimized w.r.t. $\mathcal{CE}$ and $\mathcal{SD}$ losses. Significant volumetric underestimations in *italic* and overestimations in **bold**.

Dataset ↓	Model →	LR		ConvNet		U-Net	
	Training loss →	$\mathcal{CE}$	$\mathcal{SD}$	$\mathcal{CE}$	$\mathcal{SD}$	$\mathcal{CE}$	$\mathcal{SD}$
	Metric ↓						
MOLARS (2D)	$\mathcal{CE}(\hat{Y}, Y)$	0.240	5.534	0.194	1.456	0.024	0.103
	$1 - \mathcal{SD}(\hat{Y}, Y)$	0.068	0.153	0.150	0.270	0.865	0.931
	$\Delta \mathcal{V}(\hat{Y}, Y)$ (10^2 *pixels*)	−0.069	**302.3**	−0.276	**87.09**	0.092	−0.187
BRATS 2018 (3D)	$\mathcal{CE}(\hat{Y}, Y)$	0.039	0.173	0.030	0.069	0.012	0.027
	$1 - \mathcal{SD}(\hat{Y}, Y)$	0.080	0.355	0.196	0.715	0.585	0.820
	$\Delta \mathcal{V}(\hat{Y}, Y)$ (*ml*)	−2.841	**276.4**	3.936	**19.93**	*−6.778*	−1.905
ISLES 2017 (3D)	$\mathcal{CE}(\hat{Y}, Y)$	0.025	0.155	0.018	0.069	0.014	0.066
	$1 - \mathcal{SD}(\hat{Y}, Y)$	0.099	0.255	0.114	0.321	0.188	0.340
	$\Delta \mathcal{V}(\hat{Y}, Y)$ (*ml*)	15.71	**82.42**	−4.227	**23.83**	−2.875	13.44
ISLES 2018 (3D)	$\mathcal{CE}(\hat{Y}, Y)$	0.055	0.225	0.044	0.139	0.029	0.128
	$1 - \mathcal{SD}(\hat{Y}, Y)$	0.136	0.329	0.200	0.449	0.362	0.518
	$\Delta \mathcal{V}(\hat{Y}, Y)$ (*ml*)	0.773	**34.03**	−0.374	**12.44**	−0.878	**5.442**

Optimizing the $\mathcal{CE}$ loss reaches significantly higher log-likelihoods under all circumstances, while soft Dice *scores* (i.e. $1 - \mathcal{SD}$) are significantly higher for $\mathcal{SD}$ optimized models. Looking at the volume errors $\Delta \mathcal{V}(\hat{Y}, Y)$, the expected outcomes are, more or less, confirmed. For the LR and ConvNet models, $\mathcal{CE}$ optimized models are unbiased w.r.t. volume estimation. For these models, $\mathcal{SD}$ optimization leads to significant overestimation due to the remaining uncertainty, partly being introduced by the models themselves.

The transition to the more complex U-Net model brings forward two interesting observations. First, for the two tasks with relatively low inherent uncertainty (i.e. MOLARS, BRATS 2018), the model is able to reduce the uncertainty to such an extent it can avoid significant bias on the estimated volumes. The significant underestimation for $\mathcal{CE}$ in BRATS 2018 can be due to the optimization difficulties that arise in circumstances with high class-imbalance. Second, although the model now has the ability to extend its view wide enough and propagate the information in a complex manner, the inherent uncertainty that is present in both of the ISLES tasks, brings again forward the discussed bias. In ISLES 2017, having to predict the infarction after treatment straightforwardly introduces uncertainty. In ISLES 2018, the task was to detect the acute lesion, as observed on MR DWI, from CT perfusion-derived parameter maps. It is still unknown to what extent these parameter maps contain the necessary information to predict the lesion.

The $\mathcal{CE}$ optimized U-Net models result in Dice scores (Eq. 1) of 0.924, 0.763, 0.177 and 0.454 for MOLARS, BRATS 2018, ISLES 2017 and ISLES 2018, respectively. The Dice scores obtained with their $\mathcal{SD}$ optimized counterparts are significantly higher, respectively 0.932, 0.826, 0.343 and 0.527. This is in line with recent theory and practice from [7] and justifies $\mathcal{SD}$ optimization when the segmentation quality is measured in terms of Dice score.

4   Conclusion

It is clear that, in cases with high inherent uncertainty, the estimated volumes with soft Dice-optimized models are biased, while cross-entropy-optimized models predict unbiased volume estimates. For tasks with low inherent uncertainty, one can still favor soft Dice optimization due to a higher Dice score.

We want to highlight the importance of choosing an appropriate loss function w.r.t. the goal. In a clinical setting where volume estimates are important and for tasks with high or unknown inherent uncertainty, optimization with cross-entropy can be preferred.

Acknowledgements. J.B. is part of NEXIS [1], a project that has received funding from the European Union's Horizon 2020 Research and Innovations Programme (Grant Agreement #780026). D.R. is supported by an innovation mandate of Flanders Innovation and Entrepreneurship (VLAIO).

References

1. NEXIS - Next gEneration X-ray Imaging System. https://www.nexis-project.eu
2. Ischemic Stroke Lesion Segmentation (ISLES) challenge (2017). http://www.isles-challenge.org/ISLES2017/
3. Ischemic Stroke Lesion Segmentation (ISLES) challenge (2018). http://www.isles-challenge.org/ISLES2017/
4. Multimodal Brain Tumor Segmentation (BRATS) challenge (2018). https://www.med.upenn.edu/sbia/brats2018.html
5. Bakas, S., et al.: Advancing The Cancer Genome Atlas glioma MRI collections with expert segmentation labels and radiomic features. Sci. Data **4**(March), 1–13 (2017). https://doi.org/10.1038/sdata.2017.117

6. Bakas, S., et al.: Identifying the Best Machine Learning Algorithms for Brain Tumor Segmentation, Progression Assessment, and Overall Survival Prediction in the BRATS Challenge (2018). http://arxiv.org/abs/1811.02629
7. Bertels, J., et al.: Optimizing the Dice score and Jaccard index for medical image segmentation: theory and practice. In: Medical Image Computing and Computer-Assisted Intervention (2019)
8. De Tobel, J., Radesh, P., Vandermeulen, D., Thevissen, P.W.: An automated technique to stage lower third molar development on panoramic radiographs for age estimation: a pilot study. J. Forensic Odonto-Stomatol. 35(2), 49–60 (2017)
9. Drozdzal, M., Vorontsov, E., Chartrand, G., Kadoury, S., Pal, C.: Deep Learning and Data Labeling for Medical Applications. LNCS, vol. 10008, pp. 179–187. Springer, Heidelberg (2016). https://doi.org/10.1007/978-3-319-46976-8
10. Goodfellow, I., Bengio, Y., Courville, A.: Deep Learning. MIT Press, Cambridge (2016)
11. Goyal, M., et al.: Endovascular thrombectomy after large-vessel ischaemic stroke: a meta-analysis of individual patient data from five randomised trials. Lancet 387(10029), 1723–1731 (2016). https://doi.org/10.1016/S0140-6736(16)00163-X
12. Isensee, F., Kickingereder, P., Wick, W., Bendszus, M., Maier-Hein, K.H.: No new-net. In: Crimi, A., Bakas, S., Kuijf, H., Keyvan, F., Reyes, M., van Walsum, T. (eds.) BrainLes 2018. LNCS, vol. 11384, pp. 234–244. Springer, Cham (2019). https://doi.org/10.1007/978-3-030-11726-9_21. http://arxiv.org/abs/1809.10483
13. Kamnitsas, K., Ledig, C., Newcombe, V.F.J.: Efficient multi-scale 3D CNN with fully connected CRF for accurate brain lesion segmentation. Med. Image Anal. 36, 61–78 (2017)
14. Menze, B.H., et al.: The multimodal Brain Tumor Image Segmentation Benchmark (BRATS). IEEE Trans. Med. Imaging (2015). https://doi.org/10.1109/TMI.2014.2377694
15. Milletari, F., Navab, N., Ahmadi, S.A.: V-Net: fully convolutional neural networks for volumetric medical image segmentation. In: International Conference on 3D Vision, vol. 4, pp. 1–11 (2016). https://doi.org/10.1109/3DV.2016.79. http://arxiv.org/abs/1606.04797
16. Sudre, C.H., Li, W., Vercauteren, T., Ourselin, S., Jorge Cardoso, M.: Generalised dice overlap as a deep learning loss function for highly unbalanced segmentations. In: Cardoso, M.J., et al. (eds.) DLMIA/ML-CDS -2017. LNCS, vol. 10553, pp. 240–248. Springer, Cham (2017). https://doi.org/10.1007/978-3-319-67558-9_28
17. Vapnik, V.N.: The Nature of Statistical Learning Theory. Springer, Heidelberg (1995). https://doi.org/10.1007/978-1-4757-3264-1
18. Winzeck, S., et al.: ISLES 2016 and 2017-benchmarking ischemic stroke lesion outcome prediction based on multispectral MRI. Front. Neurol. 9(SEP) (2018). https://doi.org/10.3389/fneur.2018.00679

Improved Inter-scanner MS Lesion Segmentation by Adversarial Training on Longitudinal Data

Mattias Billast[1(✉)], Maria Ines Meyer[2,3], Diana M. Sima[2], and David Robben[2,4]

[1] Department of Mathematical Engineering, KU Leuven, Leuven, Belgium
mattias.billast@gmail.com
[2] icometrix, Leuven, Belgium
{ines.meyer,diana.sima,david.robben}@icometrix.com
[3] Department of Health Technology, Technical University of Denmark, Lyngby, Denmark
[4] Medical Image Computing (ESAT/PSI), KU Leuven, Leuven, Belgium

Abstract. The evaluation of white matter lesion progression is an important biomarker in the follow-up of MS patients and plays a crucial role when deciding the course of treatment. Current automated lesion segmentation algorithms are susceptible to variability in image characteristics related to MRI scanner or protocol differences. We propose a model that improves the consistency of MS lesion segmentations in inter-scanner studies. First, we train a CNN base model to approximate the performance of **icobrain**, an FDA-approved clinically available lesion segmentation software. A discriminator model is then trained to predict if two lesion segmentations are based on scans acquired using the same scanner type or not, achieving a 78% accuracy in this task. Finally, the base model and the discriminator are trained adversarially on multi-scanner longitudinal data to improve the inter-scanner consistency of the base model. The performance of the models is evaluated on an unseen dataset containing manual delineations. The inter-scanner variability is evaluated on test-retest data, where the adversarial network produces improved results over the base model and the FDA-approved solution.

Keywords: Deep learning · Inter-scanner · Lesion segmentation · Adversarial training · Longitudinal data · Multiple sclerosis

1 Introduction

Multiple sclerosis (MS) is an autoimmune disorder characterized by a demyelination process which results in neuroaxonal degeneration and the appearance of

This project received funding from the European Union's Horizon 2020 research and innovation program under the Marie Sklodowska-Curie grant agreement No 765148. The computational resources and services used in this work were provided by the VSC (Flemish Supercomputer Center), funded by the Research Foundation – Flanders (FWO) and the Flemish Government – department EWI.

A. Crimi and S. Bakas (Eds.): BrainLes 2019, LNCS 11992, pp. 98–107, 2020.
https://doi.org/10.1007/978-3-030-46640-4_10

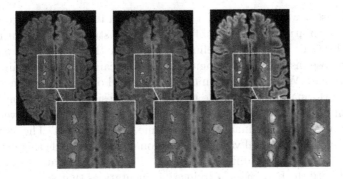

Fig. 1. MRI scans from one patient in three 3T scanners (left to right: Philips Achieva, Siemens Skyra and GE Discovery MR750w). Automated lesion segmentations in green. (Color figure online)

lesions in the brain. The most prevalent type of lesions appear hyperintense on T2-weighted (T2w) magnetic resonance (MR) images and their quantification is an important biomarker for the diagnosis and follow-up of the disease [3].

Over the years methods for automated lesion segmentation have been developed. Several approaches model the distribution of intensities of healthy brain tissue and define outliers to these distributions as lesions [7,15]. Others are either atlas-based [11] or data-driven (supervised) [2,13] classifiers. For a detailed overview of recent methods refer to [3].

Lesion segmentation is particularly interesting for patient follow-up, where data from two or more time-points is available for one patient. Some approaches try to improve segmentation consistency by analysing intensity differences over time [6]. Although these methods achieve good performance in controlled settings, they remain sensitive to changes in image characteristics related to scanner type and protocol. In a test-retest multi-scanner study [1], scanner type was observed to have an effect on MS lesion volume. These findings are supported by [12], where scanner-related biases were found even when using a harmonized protocol across scanners from the same vendor. Figure 1 illustrates such an effect.

Few works have addressed the inter-scanner variability issue in the context of lesion segmentation. Recent approaches attempt to increase the generalization of CNN-based methods to unseen MR scanner types through domain adaptation [8] or transfer learning [4,14] techniques. Nevertheless, these methods share the common downside that they require a training step to adapt to new unseen domains (scanners types and protocols). The consistency of the delineations in longitudinal settings is also not considered. A solution to incorporate consistency information into this type of data-driven solutions would be to train them on a dataset containing intra- and inter-scanner repetitions for the same patient, acquired within a short periods of time. However, in practice this type of *test-retest* dataset is almost impossible to acquire at a large scale, due to time and cost considerations.

In the present work we present a novel approach to improve the consistency of lesion segmentation in the case of multi-scanner studies, by capturing inter-scanner differences from lesion delineations. Given the shortage of test-retest data we propose instead to use longitudinal inter-scanner data to train a cross-sectional method. We start by training a base model on a multi-scanner dataset to achieve performance comparable to an existing lesion segmentation software [7]. We then design a discriminator to identify if two segmentations were generated from images that originate from the same scanner or not. The assumption is that the natural temporal variation in lesion shape can be distinguished from the variation caused by the different scanners. These networks are then combined and trained until the base model produces segmentations that are similar enough to fool the discriminator. We hypothesize that through this training scheme the model will become invariant to scanner differences, thus imposing consistency on the baseline CNN. Finally we evaluate the accuracy on a dataset with manual lesion segmentations and the reproducibility on a multi-scanner test-retest dataset.

2 Methods

We start by building a lesion segmentation *base model* based on a deep convolutional neural network (CNN) architecture [9] that approximates the performance of icobrain, an FDA-approved segmentation software. This method is an Expectation-Maximization (EM) model that uses the distribution of healthy brain tissue to detect lesions as outliers while also using prior knowledge of the location and appearance of lesions [7]. We refer to it as *EM-model*.

Base Model. The base model is based on the DeepMedic architecture [9]. Generally, it is composed of multiple pathways which process different scales of the original image simultaneously. This is achieved by downsampling the original image at different rates before dividing it into input patches, which allows the model to combine the high resolution of the original image and the broader context of a downsampled image to make a more accurate prediction. In our implementation we used three pathways, for which the input volumes were downsampled with factors $(1, 1, 1)$, $(3, 3, 1)$ and $(5, 5, 3)$ and divided into patches of size $(35, 35, 19)$, $(25, 25, 19)$ and $(23, 23, 13)$, respectively. Each pathway is comprised of ten convolutional layers, each followed by a PReLu activation, after which the feature maps from the second and third pathways are upsampled to the same dimensions as the first pathway and concatenated. This is followed by dropout, two fully connected layers and a sigmoid function, returning a $(15, 15, 9)$ probability map. The first five layers have 32 filters and kernel size $(3,3,1)$ and the last five layers 48 filters with kernel size $(3,3,3)$. The values of the output probability map that are above a certain threshold are classified as lesions. The threshold used throughout this article is 0.4. The architecture is represented in Fig. 2. The loss function of the base model is given by

$$L_B = Y log(B(X)) + (1 - Y)log(1 - B(X)), \tag{1}$$

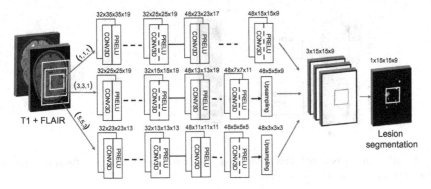

Fig. 2. Architecture of the base model that describes the patch sizes of the different pathways and the overall structure.

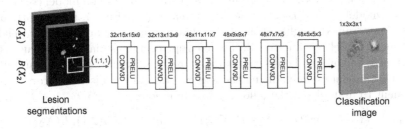

Fig. 3. Architecture of the discriminator that describes the in- and output sizes of the patches and the overall structure.

where X is the concatenation of the T1- and FLAIR MR images, Y is the corresponding lesion segmentation label and B() the output of the base model.

Discriminator. The discriminator is reduced to one pathway with six convolutional layers, since additional pathways with subsampling resulted in a marginal increase in performance. The two first layers have 32 filters of kernel size (3,3,1) and the following layers 48 filters with kernel size (3,3,3). As input it takes two label patches of size $(15, 15, 9)$ and generates a voxel-wise prediction that the two labels are derived from images acquired using the same scanner. The architecture is represented in Fig. 3.

The loss function of the discriminator is given by:

$$L_D = Y log(D(B(X_1), B(X_2))) + (1 - Y)log(1 - D(B(X_1), B(X_2))), \quad (2)$$

where Y is the ground truth indicator variable (0 or 1) indicating whether two time points were acquired on the same scanner or not, X_1 and X_2 are images at different time points and B() and D() are respectively the output of the base model and the discriminator.

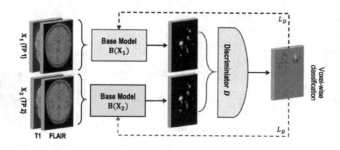

Fig. 4. Adversarial network that combines the base model and the discriminator to reduce the inter-scanner variability.

Adversarial Model. After training, the discriminator was combined adversarially with the base model, as introduced in [5]. The adversarial model consists of two base model blocks (B) and one discriminator (D) (Fig. 4). In our particular case the pre-trained weights of the discriminator are frozen and only the weights of the base model are fine-tuned. The concept of adversarial training uses the pre-trained weights of the discriminator to reduce the inter-scanner variability of the base model by maximizing the loss function of the discriminator. This is equivalent to minimizing the following loss function:

$$L_{Adv} = (1 - Y)log(D(B(X_1), B(X_2))) + Ylog(1 - D(B(X_1), B(X_2))), \quad (3)$$

The loss function of the adversarial network then consists of two terms: one associated with the lesion segmentation labels, and one related to the output image of the discriminator:

$$L = 2 * L_B + L_{Adv} \quad (4)$$

The purpose of L_{Adv} is to ensure that the base model is updated such that the discriminator can no longer distinguish between segmentations that are based on same- or different-scanner studies. We hypothesize that the base model learns to map scans from different scanners to a consistent lesion segmentation.

Model Training. Both the base model and the discriminator were trained using the binary cross entropy objective function and optimized using mini-batch gradient descent with Nesterov momentum $\beta = 0.9$. Initial learning rates were $\alpha = 0.016$ for the base model and $\alpha = 4e-3$ for the discriminator, and were decreased at regular intervals until convergence. For the adversarial network initial learning rate was $\alpha = 2e-3$. All models were trained using an NVIDIA P100. The networks are implemented using the Keras and DeepVoxNet [10] frameworks.

3   Data and Preprocessing

Four different datasets were available: two for training and two for testing the performance of the models. Since for three of the datasets manual delineations were not available, automated segmentations were acquired using the EM-method described in the previous section. All automated delineations were validated by a human expert. Each study in the datasets contains T1w and FLAIR MR images from MS patients.

Cross-Sectional Dataset. 208 independent studies from several centers. The base model is trained on this dataset.

Longitudinal Dataset. 576 multi-center, multi-scanner studies with approved quality MR scans, containing multiple studies from 215 unique patients at different timepoints. For training the adversarial model and the discriminator only studies with less than 2 years interval were used to minimize the effect of the natural evolution of lesions over time and capture the differences between scanners. This resulted in approximately 80% being used since most patients have a follow-up scan every 6 months to one year. The discriminator and adversarial model are trained on this dataset.

Manual Segmentations. 20 studies with manual lesion delineations by experts.

Test-Retest Dataset. 10 MS patients. Each patient was scanned twice in three 3T scanners: *Philips Achieva, Siemens Skyra* and *GE Discovery MR450w* [7].

All the data was registered to Montreal Neurological Institute (MNI) space and intensities were normalized to zero mean and unit standard deviation. Ten studies from each training dataset were randomly selected to use as validation during the training process. The data was additionally augmented by randomly flipping individual samples around the x-axis.

4   Results

The models were evaluated on the manual segmentations and the test-retest datasets described in Sect. 3 and compared to the EM-model. The main results are summarized in Table 1. For the manual segmentations dataset results are described in terms of Dice score, Precision and Recall. For the test-retest dataset we are mainly interested in evaluating the reproducibility in the inter-scanner cases. Since there is no ground truth, we report the metrics between different time points for the same patient. Aside from the total lesion volume (LV) in mm^3 we additionally quantify the absolute differences in lesion volume ($|\Delta LV|$) in mm^3. The results in this table were calculated with a lesion threshold value of 0.4. Figure 5 depicts the distribution of ($|\Delta LV|$) for both inter-scanner and intra-scanner cases of the test-retest dataset.

Base Model. For the manual segmentation dataset, results are comparable to the EM-model. In the test-retest validation, the inter-scanner $|\Delta LV|$ is larger for the base model, which indicates that the model is sensitive to inter-scanner variability.

Discriminator. The discriminator is validated on a balanced sample of the test-retest dataset, so that there is the same number of inter- and intra-scanner

Table 1. Mean performance metrics for the different models on two test sets: manual segmentations and test-retest. For the latter only inter-scanner studies are considered. |LV| represents absolute differences between individual lesion volumes and is given in mm^3.

Model	Manual			Test/Retest	
	Dice	Precision	Recall	|ΔLV|	LV
EM	0.71 ± 0.07	0.85	0.61	2077 ± 2054	8307
Base	0.72 ± 0.10	0.80	0.65	4557 ± 3530	9894
Adversarial	0.68 ± 0.11	0.83	0.59	1331 ± 1020	8584

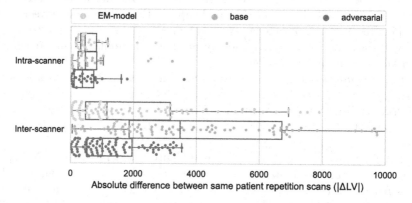

Fig. 5. Absolute intra- and inter-scanner difference in lesion volume, calculated on the test-retest dataset with three different models.

examples. It achieves an accuracy of 78% by looking at the average probability value on the lesion voxels only.

Adversarial Model. On the manual segmentations dataset, again referring to Table 1, the adversarial model achieves a slightly lower but still competitive performance when compared to the EM-model.

Regarding the test-retest dataset, the adversarial model produces lower inter-scanner |ΔLV| when compared to the base model (Wilcoxon Signed-Rank Test, $p = 3.26e - 15$) and to the EM-model, (Wilcoxon Signed-Rank Test, $p = 0.02$). This indicates that the adversarial model produces segmentations that are less sensitive to inter-scanner variation than both the base model and the EM-model.

The mean |ΔLV| values and standard deviation for the EM-model are almost twice as large as the adversarial model. Taking into account the boxplots in Fig. 5, this is partly explained by the fact that the distribution has a positive skew and additionally by three significant outliers, which artificially increase the mean values.

This is evidence that the EM-model has larger variability and lower reproducibility than the adversarial model, while the average predicted lesion volume

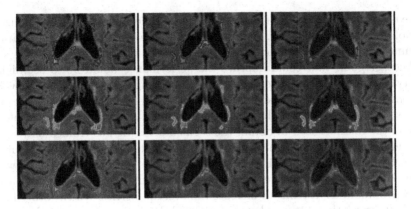

Fig. 6. Lesion segmentation results for one patient in three 3T scanners. Top: EM-model; Middle: base model; Bottom: adversarial model. Adversarial model results appear more consistent, while maintaining physiological meaning.

is similar for the EM- and adversarial models. Figure 6 shows an example of the different lesion segmentations on the different scanners with the three models.

5   Discussion and Future Work

We presented a novel approach to improve the consistency of inter-scanner MS lesion segmentations by using adversarial training on a longitudinal dataset. The proposed solution shows improvements in terms of reproducibility when compared to a base CNN model and to an FDA-approved segmentation method based on an EM approach. The key ingredient in the model is the discriminator, which predicts with 78% accuracy on unseen data whether two lesion segmentations are based on MRI scans acquired using the same scanner. This is a very promising result, since this is not a standard problem.

When evaluated on an unseen dataset of cross-sectional data, the model's performance approximates the EM-model, but decreases slightly after the adversarial training. This indicates a trade-off between performance and reproducibility. One concern was that this would be connected to an under-segmentation due to the consistency constraint learned during the adversarial training. However, evaluating the average predicted lesion volume on a separate test-retest dataset shows no indication of under-segmentation when compared to the EM-model.

Both the adversarial network and the discriminator were trained on longitudinal inter-scanner data. This is not ideal, since MS can have an unpredictable evolution over time, and as such it becomes difficult to distinguish between differences caused by hardware and the natural progression of the disease. We attempt to mitigate this effect by selecting studies within no more than two years interval, but better and more reliable performance could be achieved if the model would be trained on a large dataset with the same characteristics as the test-retest dataset described in Sect. 3. However, large datasets of that type do not exist and would require a very big effort to collect, both from the point of view of patients and logistics. As such, using longitudinal inter-scanner data is a compromise that is cost-efficient and shows interesting results.

Another point that could improve the performance would be to use higher quality images and unbiased segmentations at training time. This would allow for a stronger comparison to other methods in literature and manual delineations. At this moment it is expectable that our model achieves results comparable to those of the method used to obtain the segmentations it was trained on.

Aside from these compromises, some improvements can still be made in future work. Namely, during the training and testing stages of the adversarial network images can be affinely registered to each other instead of using one common atlas space. We would expect this to increase the overlap metrics. On the other hand it was observed that the overlap metrics slightly decrease for the adversarial network with longer training, and as such the weight of the term in the loss function associated with the discriminator can be optimized/lowered to achieve more efficient training and better overlap of the images.

Finally, instead of only freezing the weights of the discriminator to improve the base model, the weights of the base model can also be frozen in a next step to improve the discriminator, so that the base model and discriminator are trained in an iterative process until there are no more performance gains.

Apart from the various optimizations to the model, it would be interesting to apply the same adversarial training to other lesion types, such as the ones resulting from vascular dementia or traumatic brain injuries.

References

1. Biberacher, V., et al.: Intra- and interscanner variability of magnetic resonance imaging based volumetry in multiple sclerosis. NeuroImage **142**, 188–197 (2016). https://doi.org/10.1016/j.neuroimage.2016.07.035
2. Brosch, T., et al.: Deep 3D convolutional encoder networks with shortcuts for multiscale feature integration applied to multiple sclerosis lesion segmentation. IEEE Trans. Med. Imaging **35**(5), 1229–1239 (2016). https://doi.org/10.1109/TMI.2016.2528821
3. Carass, A., et al.: Longitudinal multiple sclerosis lesion segmentation: resource and challenge. NeuroImage **148**(C), 77–102 (2017). https://doi.org/10.1016/j.neuroimage.2016.12.064
4. Ghafoorian, M., et al.: Transfer learning for domain adaptation in MRI: application in brain lesion segmentation. In: Descoteaux, M., Maier-Hein, L., Franz, A., Jannin, P., Collins, D.L., Duchesne, S. (eds.) MICCAI 2017. LNCS, vol. 10435, pp. 516–524. Springer, Cham (2017). https://doi.org/10.1007/978-3-319-66179-7_59
5. Goodfellow, I.J., et al.: Generative adversarial networks (2014)
6. Jain, S., Ribbens, A., Sima, D.M., Van Huffel, S., Maes, F., Smeets, D.: Unsupervised framework for consistent longitudinal MS lesion segmentation. In: Müller, H., et al. (eds.) MCV/BAMBI -2016. LNCS, vol. 10081, pp. 208–219. Springer, Cham (2017). https://doi.org/10.1007/978-3-319-61188-4_19
7. Jain, S., et al.: Automatic segmentation and volumetry of multiple sclerosis brain lesions from MR images. NeuroImage: Clin. **8**, 367–375 (2015). https://doi.org/10.1016/j.nicl.2015.05.003

8. Kamnitsas, K., et al.: Unsupervised domain adaptation in brain lesion segmentation with adversarial networks (December 2016)
9. Kamnitsas, K., et al.: Efficient multi-scale 3D CNN with fully connected CRF for accurate brain lesion segmentation. Med. Image Anal. **36**(C), 61–78 (2017). https://doi.org/10.1016/j.media.2016.10.004
10. Robben, D., Bertels, J., Willems, S., Vandermeulen, D., Maes, F., Suetens, P.: DeepVoxNet: voxel-wise prediction for 3D images (2018). https://lirias.kuleuven.be/retrieve/516811/paper.pdf
11. Shiee, N., et al.: A topology-preserving approach to the segmentation of brain images with multiple sclerosis lesions. NeuroImage **49**(2), 1524–1535 (2010). https://doi.org/10.1016/J.NEUROIMAGE.2009.09.005
12. Shinohara, R.T., et al.: Volumetric analysis from a harmonized multisite brain MRI study of a single subject with multiple sclerosis. AJNR. Am. J. Neuroradiol. **38**(8), 1501–1509 (2017). https://doi.org/10.3174/ajnr.A5254
13. Valverde, S., et al.: Improving automated multiple sclerosis lesion segmentation with a cascaded 3D convolutional neural network approach. NeuroImage **155**, 159–168 (2017). https://doi.org/10.1016/j.neuroimage.2017.04.034
14. Valverde, S., et al.: One-shot domain adaptation in multiple sclerosis lesion segmentation using convolutional neural networks. NeuroImage: Clin. **21**, 101638 (2019). https://doi.org/10.1016/J.NICL.2018.101638
15. Van Leemput, K., et al.: Automated segmentation of multiple sclerosis lesions by model outlier detection. IEEE Trans. Med. Imaging **20**(8), 677–688 (2001). https://doi.org/10.1109/42.938237

Saliency Based Deep Neural Network for Automatic Detection of Gadolinium-Enhancing Multiple Sclerosis Lesions in Brain MRI

Joshua Durso-Finley[1]([envelope]), Douglas L. Arnold[2,3], and Tal Arbel[1]

[1] Centre for Intelligent Machines, McGill University, Montreal, Canada
`joshua.durso-finley@mail.mcgill.ca`
[2] Montreal Neurological Institute, McGill University, Montreal, Canada
[3] NeuroRx Research, Montreal, Canada

Abstract. The appearance of contrast-enhanced pathologies (e.g. lesion, cancer) is an important marker of disease activity, stage and treatment efficacy in clinical trials. The automatic detection and segmentation of these enhanced pathologies remains a difficult challenge, as they can be very small and visibly similar to other non-pathological enhancements (e.g. blood vessels). In this paper, we propose a deep neural network classifier for the detection and segmentation of Gadolinium enhancing lesions in brain MRI of patients with Multiple Sclerosis (MS). To avoid false positive and false negative assertions, the proposed end-to-end network uses an enhancement-based attention mechanism which assigns saliency based on the differences between the T1-weighted images before and after injection of Gadolinium, and works to first identify candidate lesions and then to remove the false positives. The effect of the saliency map is evaluated on 2293 patient multi-channel MRI scans acquired during two proprietary, multi-center clinical trials for MS treatments. Inclusion of the attention mechanism results in a decrease in false positive lesion voxels over a basic U-Net [2] and DeepMedic [6]. In terms of lesion-level detection, the framework achieves a sensitivity of 82% at a false discovery rate of 0.2, significantly outperforming the other two methods when detecting small lesions. Experiments aimed at predicting the presence of Gad lesion activity in patient scans (i.e. the presence of more than 1 lesion) result in high accuracy showing: (a) significantly improved accuracy over DeepMedic, and (b) a reduction in the errors in predicting the degree of lesion activity (in terms of per scan lesion counts) over a standard U-Net and DeepMedic.

Keywords: Segmentation · Gadolinium lesions · Multiple Sclerosis · Attention · Deep learning

1 Introduction

There are many clinical contexts where contrast-enhancing agents, such as Gadolinium, are injected into patients, in order to produce images that better

A. Crimi and S. Bakas (Eds.): BrainLes 2019, LNCS 11992, pp. 108–118, 2020.
https://doi.org/10.1007/978-3-030-46640-4_11

illustrate new pathological activity (e.g. lesions, cancers). In the context of Multiple Sclerosis (MS), Gadolinium enhancing lesions (referred to as "Gad lesions") appearing on T1-weighted MRI indicate new disease activity, disease stage and are important for monitoring treatment efficacy and therefore used extensively in the context of clinical trial analysis for development of new drugs [7,12]. The primary objective of an automatic technique would therefore be to locate and detect **all** Gad lesions in patient MRI, a task that is particularly challenging due to the large variability in their appearances, locations, and sizes which can range from only a few to over 100 voxels in size. In fact, many Gad lesions are very small (e.g. 3–4 voxels in size at a resolution of $1\,mm \times 1\,mm \times 3\,mm$). Furthermore, although the presence of contrast enhancement assists in identifying candidate lesion locations, blood vessels and other normal structures enhance in T1-weighted MRI as well and result in many other similarly appearing false candidates. The problem is further complicated as some lesions do not enhance sufficiently under contrast, and enhancement intensity levels can be inconsistent, making them even more difficult to detect. Figure 1 depicts a case where only 2 Gad lesions are present, but many other MRI enhancements can be seen throughout the brain.

The challenges in detecting and segmenting Gad lesions must be addressed by an automatic method in order for it to be deployed in real clinical practice and in clinical trial analysis, where the stakes for making errors are high. Patients with no Gad lesions are considered *inactive* and those with larger or more Gad lesions, *active*. Even a single false positive or false negative assertion can therefore have significant impact on patient disease assessment and determination of treatment efficacy.

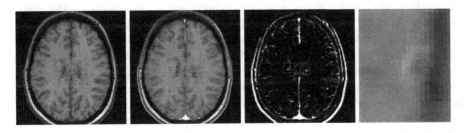

Fig. 1. Example of patient images and Gad lesions. Left to right: T1-weighted precontrast MRI (T1-p); T1-weighted post-contrast MRI (T1-c); Difference image: T1c-T1p with Gad lesions highlighted in yellow boxes. Brighter voxels have more contrast; Zoomed in image of cropped ROI for left Gad lesion. (Color figure online)

Although several methods have been presented for the automatic segmentation of T2 lesions in patients with Multiple Sclerosis [1,4,18], much less work has focused on the context of Gad lesion segmentation and detection. Some early work in Gad lesion segmentation was developed [3,11] but these relied on prior segmentation of other structures. Probabilistic graphical models [8] were developed to address this problem using an adapted Conditional Random Field with promising results. However, this method relied on a series of hand-crafted features and carefully designed CRF models. This work explores how deep neural networks [10] can be adapted to this problem. Although many deep networks have shown promising results for the task of segmenting T2 lesions, they have been shown to have some difficulty segmenting small lesions [13]. Inspired by [8], this work explores how a U-Net [2] can be adapted to this domain through the addition of a saliency based attention mechanism which focuses the network's attention on the candidate regions and then use the bulk of the network to distinguish false positives from true Gad lesions. The method works by generating features at different scales and using those features to segment the remainder of the MRI, rather than jointly extracting features and searching for candidates (which expands the capacity of classification of the network). This allows it to achieve better searching capabilities, particularly for small lesions.

The network was trained on a large proprietary, multi-scanner, multi-center dataset consisting of 5830 multi-channel MRI acquired from 2239 patients with Relapsing-Remitting MS (RRMS) during 2 clinical trials. Experiments were performed on subset of 448 MRI patient scans set aside for testing. Segmentation and detection results were examined through ROC-like curves (TPR vs. FDR). The proposed approach shows improved performance when compared against two common models, the U-Net and DeepMedic, at the tasks of: voxel-level lesion segmentation, lesion detection and estimating the degree of MRI lesion activity per patient scan as determined by Gad lesion counts. Specifically, voxel-level segmentation experiments indicate that the proposed contrast enhancement attention-based mechanism results in an AUC of 0.68, with a decrease in false positive lesion voxels over a basic U-Net [2] (AUC 0.61) and DeepMedic [6] (AUC 0.32). The lesion level detection results show a true positive rate of 0.82 at a false detection rate of 0.2 over all lesions, significantly outperforming the other two methods when detecting small lesions. Experiments aimed at predicting any Gad lesion activity in patient scans result in significantly higher accuracy of the U-Net's (92%) over DeepMedic (77%). Finally, the proposed method shows reduced error in predicting the degree of lesion activity in terms of the number of lesions per patient image over a standard U-Net. On average, the error between the predicted number of lesions present and the actual number of lesions is 0.261 for the proposed attention based method, 0.348 for the U-Net and 0.732 for DeepMedic.

2   Methodology

The proposed framework (Saliency U-Net) consists of a single 3D CNN, which takes as inputs 5 MRI sequences and a labeled T2 weighted lesion mask and produces a binary outcome for each voxel which classifies the voxel as being either a Gad lesion or a non-lesion. The baseline architecture consists of an encoder-decoder (See Fig. 2). The encoding side consists of five blocks where each block executes a convolution, ReLU [14], convolution, ReLU, batch normalization [5] and finally a dropout [17] operation. The upward path executes a transpose convolution, concatenation with the same scale features from across the network followed by a dropout, convolution, ReLU, convolution, ReLU, and finally a batch normalization operation. The upward path contains three blocks following this pattern and finally two more convolutions. The additional attention mechanism [19,21] computes features using a set of $3\times3\times3$, $5\times5\times5$, and $7\times7\times7$ convolutions with an LRelu [20] from both the original set of MRIs and the difference between T1-c and T1-p (see Sect. 3.1) and concatenates them together. A final $3\times3\times3$ convolution followed by a softmax produces a volume of values between 0–1 which multiplies the original set of MRIs. This tensor becomes the input for the remainder of the network. After a pass through the network, the weighted softmax cross entropy loss is computed and backpropagated throughout the network.

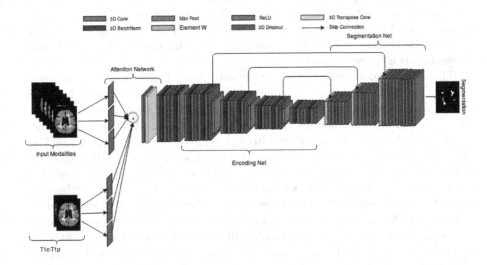

Fig. 2. Network architecture

To derive lesion level detection results from the initial volume of binary predictions we first threshold the volume produced by the network and get the connected components of this binarized image. Next, we remove connected components smaller than 3 voxels. Finally, to compute lesion level statistics we check the overlap of our connected components and the ground truth. Connected components which have a dice score greater than .5 or an overlap of 3 voxels with the ground truth are considered to be correctly detected lesions [13].

3   Experiments

The network is trained to segment Gad lesions on a per voxel basis. The proposed network is compared against a baseline U-Net and the standard DeepMedic [6].

Each of the three networks was trained on the same training sample set of 5830 multi-channel MRI scans from two trials, producing binary lesion/non-lesion outputs at each voxel.

3.1 Data

The data set used in the experiment consists of two large, proprietary, multi-scanner, multi-center, clinical trial image datasets. Patients had the Relapsing Remitting form of Multiple Sclerosis (RRMS). Each MRI sample includes: T1-weighted pre-contrast (T1p), T1-weighted post-contrast (T1c), T2-weighted (T2w), Proton density weighted (PDW), Fluid-attenuated inversion (FLAIR) MRIs each at a resolution of 1 mm × 1 mm × 3 mm. In addition, expert-annotated gadolinium-enhancing (Gad) lesion masks and T2 lesion labels were provided for training and testing. The T2 lesion labels were obtained through a semi-manual procedure, in which labels generated by an automated algorithm were corrected by an expert. Gad lesion masks were obtained manually through consensus between trained experts.

The MRIs underwent brain extraction, Nyul image intensity normalization [15], registration to a common space (ICBM space), and N3 bias field inhomogeneity correction [16]. After cropping, the final dimensions were 192 × 192 × 56 with a resolution of 1 mm × 1 mm × 3 mm. Any patient missing at least one of the scans was removed from the experiment. One clinical trial had MRIs at 1 year intervals and the other at 24 week intervals. All MRI sequences and the T2-weighted lesion map were used as inputs to determine the final Gad lesion map. The 2293 patients for both of the trials were split into fifths with four fifths making up the training set and the final fifth making up the validation and test set. As each patient has approximately 3 sets of MRIs, the total data set contains 6830 sets of scans containing all the required MRIs with 5830 scans set aside for training and validation and the remaining 1000 held out for testing. Division by patient when sampling ensured that the testing set contained no images from the same patients found in the training set.

3.2 Implementation Details

To account for the large imbalance between the background class and the foreground class (about 50000:1), the foreground class was heavily weighted and we slowly decayed the weight over time. The weighting began at 1000x for the foreground class and was reduced by a factor of 1.02 every 2000 samples. Similarly, the learning rate began at 10^{-4} and was reduced by a factor of 1.02 every 2000 samples. The ADAM [9] optimizer was chosen and trained for 240000 samples.

4 Results

Comparisons of the proposed method against the U-Net and DeepMedic are made in three domains. First, the generated labels from the Saliency U-Net are evaluated qualitatively and shown along side generated labels from DeepMedic and a U-Net (both provided with MRI inputs without any saliency maps). Next, quantitative results for instance level detection of different sized lesions are presented as TPR and FPR curves over different lesion sizes and at the voxel level. Finally, results for the count of Gad lesions on a scan-by-scan level are evaluated (regardless of overlap). To demonstrate the network's accuracy we show the confusion matrix for derived lesion counts against ground truth lesion counts for each of the three models and a histogram showing the error from the true label.

4.1 Gad Lesion Detection

Qualitative results in Fig. 3 demonstrate the performance of the Saliency U-Net against the other methods for a series of patient cases. The first, second and fifth case show examples where the proposed method was able to remove false positives over a U-Net. The second and fifth example show that it can accurately detect and segment lesions both when the lesion is isolated and when the lesions are clustered. Overall, the proposed method is shown to overcome several challenges of this domain.

To evaluate the model quantitatively, the ROC-like curve is presented in Fig. 4, depicting the True Positive Rate $(TPR = TP/(TP + FN))$ against the False Discovery Rate $(FDR = FP/(FP + TP))$ for the voxel segmentation task and for Gad lesion level detection. Voxel-level segmentation results show that the proposed method has higher accuracy than the other methods with an AUC 0.68, as compared to a basic U-Net [2] (AUC 0.61) and DeepMedic [6] (AUC 0.32). For the task of lesion level detection, the results are shown for three different lesion sizes: Large (50+ voxels), medium (10 to 50 voxels), and small (3 to 10 voxels). By plotting across lesion sizes, we can more robustly compare results between different trials. The Saliency U-Net shows high overall lesion detection accuracy with a TPR of 0.82 at an operating point of 0.2 FDR (an operating point of clinical relevance) for all lesions, outperforming the other methods, particularly for small lesions. DeepMedic does not perform well overall, and fails to detect small lesions (the worst possible curve for this plot is a vertical line at $x = 1$).

4.2 Quantifying Lesion Activity

The ROC-like Gad lesion segmentation and detection curves are based on the entire collection of Gad lesions in the dataset and not on a per scan basis. Experiments were subsequently performed to test the network's ability to (a) predict the overall patient activity per scan, a binary outcome depicting the presence of any Gad lesions, and (b) to predict the severity of Gad lesion activity per scan, defined by the number of Gad lesions present.

Figure 5 shows the confusion matrices for Gad lesion count prediction per scan for the three methods: Saliency U-Net, U-Net, and DeepMedic. For the task of predicting the binary outcome of active/inactive, error rates per scan were computed by summing the instances when the network predicted the presence of lesions (active) when there were none (inactive), and the instances where the

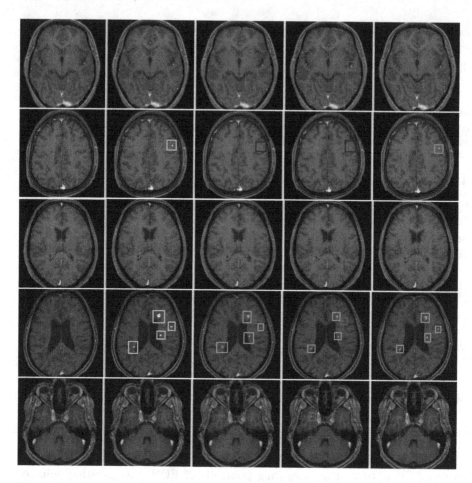

Fig. 3. Qualitative detection results for different patients (rows). Columns from left to right: T1-c, Expert labels (Yellow), DeepMedic Labels, U-Net labels, Saliency U-Net labels. True positive predictions are shown in green, false negatives are shown in red, and false positives are shown in blue. (Color figure online)

network predicted no lesions (inactive) and lesions were present (active). Here, the U-Net and Saliency U-Net perform similarly with 85 and 82 errors respectively. Both, however, perform better than DeepMedic which had 238 errors. Despite similar performance in detecting the presence of lesions, the confusion matrix indicates that the Saliency U-Net is notably more diagonal with a total count of 829 along the diagonal, whereas the U-Net and DeepMedic had counts of 719 and 712, respectively. This indicates an improved performance by the proposed method in predicting the correct degree of lesion activity. To further demonstrate the improvement, the histogram in Fig. 6 depicts the differences between the predicted lesion counts and the lesion counts provided by manual segmentations. Quantitatively, the average error in lesion count on a per scan basis was 0.261 for Saliency U-Net, 0.348 for U-Net, and 0.732 for DeepMedic.

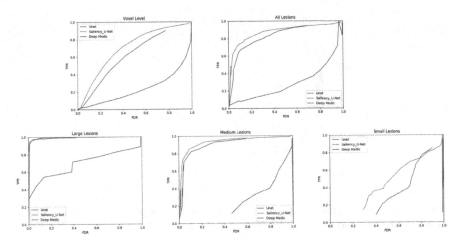

Fig. 4. Gad-enhanced segmentation and detection results: ROC-like TPR vs. FDR curves. Top left: voxel based segmentation results, top right: detection results for all lesions, bottom left: detection results for large (50+ voxels) lesions, bottom center: detection results for medium (10 to 50 voxels), Bottom right: detection results for small (3 to 10 voxels) lesions.

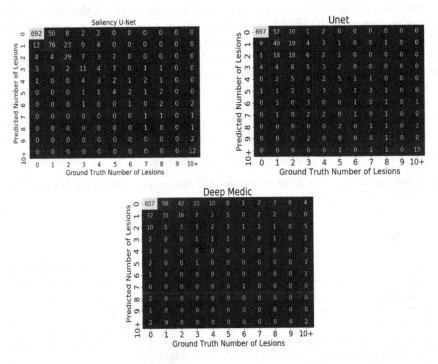

Fig. 5. Per scan lesion count predictions for the saliency U-net, U-net, and DeepMedic.

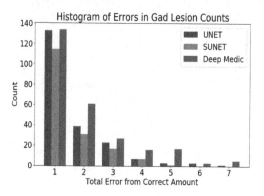

Fig. 6. Histogram of errors in predicted lesion count on a per scan basis relative to manual labels. The x-axis starts at 1 to depict the degree of error when incorrect.

5 Conclusions

This paper presents an end-to-end 3D CNN network for the segmentation and detection of Gadolinium enhanced lesions in MS patient MRI. The model embeds an enhancement-based saliency map which permits the network to quickly focus on candidate regions. Results of experiments on two large multi-center, multi-scanner clinical trial datasets indicate that our proposed method improves the voxel based segmentation and lesion based detection results over a simpler U-Net and DeepMedic, particularly in the detection of small lesions. Furthermore, the proposed method shows improved accuracy in estimating the binary outcome of active/inactive patient scans over DeepMedic and in estimating the correct degree of Gad lesion activity (over both methods). Since Gad lesion detection, and subsequent counts and activity labels, are important markers of treatment efficacy, this method has the potential to improve the speed and accuracy required in clinical trial analysis.

Acknowledgement. This work was supported by an award from the International Progressive MS Alliance (PA-1603-08175).

References

1. Brosch, T., Tang, L.Y.W., Yoo, Y., Li, D.K.B., Traboulsee, A., Tam, R.: Deep 3d convolutional encoder networks with shortcuts for multiscale feature integration applied to multiple sclerosis lesion segmentation. IEEE Trans. Med. Imaging **35**(5), 1229–1239 (2016). https://doi.org/10.1109/TMI.2016.2528821
2. Çiçek, Ö., Abdulkadir, A., Lienkamp, S.S., Brox, T., Ronneberger, O.: 3D U-Net: learning dense volumetric segmentation from sparse annotation. CoRR abs/1606.06650 (2016).http://arxiv.org/abs/1606.06650
3. Datta, S., Sajja, B.R., He, R., Gupta, R.K., Wolinsky, J.S., Narayana, P.A.: Segmentation of gadolinium-enhanced lesions on MRI in multiple sclerosis. J. Magn. Reson. Imaging Off. J. Int. Soc. Magn. Reson. Med. **25**(5), 932–937 (2007)

4. Fleishman, G.M., et al.: Joint intensity fusion image synthesis applied to multiple sclerosis lesion segmentation. In: Crimi, A., Bakas, S., Kuijf, H., Menze, B., Reyes, M. (eds.) BrainLes 2017. LNCS, vol. 10670, pp. 43–54. Springer, Cham (2018). https://doi.org/10.1007/978-3-319-75238-9_4

5. Ioffe, S., Szegedy, C.: Batch normalization: accelerating deep network training by reducing internal covariate shift. CoRR abs/1502.03167 (2015). http://arxiv.org/abs/1502.03167

6. Kamnitsas, K., et al.: DeepMedic for brain tumor segmentation. In: Crimi, A., Menze, B., Maier, O., Reyes, M., Winzeck, S., Handels, H. (eds.) BrainLes 2016. LNCS, vol. 10154, pp. 138–149. Springer, Cham (2016). https://doi.org/10.1007/978-3-319-55524-9_14. https://www.microsoft.com/en-us/research/publication/deepmedic-brain-tumor-segmentation/

7. Kappos, L., et al.: Ocrelizumab in relapsing-remitting multiple sclerosis: a phase 2, randomised, placebo-controlled, multicentre trial. Lancet **378**(9805), 1779–1787 (2011)

8. Karimaghaloo, Z., Rivaz, H., Arnold, D.L., Collins, D.L., Arbel, T.: Temporal hierarchical adaptive texture crf for automatic detection of gadolinium-enhancing multiple sclerosis lesions in brain mri. IEEE Trans. Med. Imaging **34**(6), 1227–1241 (2015). https://doi.org/10.1109/TMI.2014.2382561

9. Kingma, D.P., Ba, J.: Adam: a method for stochastic optimization (2014). arXiv:1412.6980. Published as a conference paper at the 3rd International Conference for Learning Representations, San Diego (2015)

10. LeCun, Y., Bengio, Y., Hinton, G.: Deep learning. Nature **521**, 436–444 (2015). https://doi.org/10.1038/nature14539

11. Linguraru, M.G., Pura, J.A., Chowdhury, A.S., Summers, R.M.: Multi-organ segmentation from multi-phase abdominal CT via 4D graphs using enhancement, shape and location optimization. In: Jiang, T., Navab, N., Pluim, J.P.W., Viergever, M.A. (eds.) MICCAI 2010. LNCS, vol. 6363, pp. 89–96. Springer, Heidelberg (2010). https://doi.org/10.1007/978-3-642-15711-0_12

12. Miller, D., Barkhof, F., Nauta, J.: Gadolinium enhancement increases the sensitivity of MRI in detecting disease activity in multiple sclerosis. Brain **116**(5), 1077–1094 (1993)

13. Nair, T., Precup, D., Arnold, D.L., Arbel, T.: Exploring uncertainty measures in deep networks for multiple sclerosis lesion detection and segmentation. In: Frangi, A.F., Schnabel, J.A., Davatzikos, C., Alberola-López, C., Fichtinger, G. (eds.) MICCAI 2018. LNCS, vol. 11070, pp. 655–663. Springer, Cham (2018). https://doi.org/10.1007/978-3-030-00928-1_74

14. Nair, V., Hinton, G.E.: Rectified linear units improve restricted Boltzmann machines. In: Proceedings of the 27th International Conference on International Conference on Machine Learning, ICML 2010, pp. 807–814. Omnipress, USA (2010). http://dl.acm.org/citation.cfm?id=3104322.3104425

15. Nyul, L.G., Udupa, J.K., Zhang, X.: New variants of a method of MRI scale standardization. IEEE Trans. Med. Imaging **19**(2), 143–150 (2000). https://doi.org/10.1109/42.836373

16. Sled, J.G., Zijdenbos, A.P., Evans, A.C.: A nonparametric method for automatic correction of intensity nonuniformity in MRI data. IEEE Trans. Med. Imaging **17**(1), 87–97 (1998). https://doi.org/10.1109/42.668698

17. Srivastava, N., Hinton, G., Krizhevsky, A., Sutskever, I., Salakhutdinov, R.: Dropout: a simple way to prevent neural networks from overfitting. J. Mach. Learn. Res. **15**, 1929–1958 (2014). http://jmlr.org/papers/v15/srivastava14a.html

18. Valverde, S., et al.: Improving automated multiple sclerosis lesion segmentation with a cascaded 3d convolutional neural network approach. CoRR abs/1702.04869 (2017). http://arxiv.org/abs/1702.04869

19. Vaswani, A., et al.: Attention is all you need. CoRR abs/1706.03762 (2017). http://arxiv.org/abs/1706.03762

20. Xu, B., Wang, N., Chen, T., Li, M.: Empirical evaluation of rectified activations in convolutional network. CoRR abs/1505.00853 (2015). http://arxiv.org/abs/1505.00853

21. Xu, K., et al.: Show, attend and tell: neural image caption generation with visual attention. CoRR abs/1502.03044 (2015). http://arxiv.org/abs/1502.03044

Deep Learning for Brain Tumor Segmentation in Radiosurgery: Prospective Clinical Evaluation

Boris Shirokikh[1,2,3], Alexandra Dalechina[4], Alexey Shevtsov[2,3], Egor Krivov[2,3], Valery Kostjuchenko[4], Amayak Durgaryan[5], Mikhail Galkin[5], Ivan Osinov[4], Andrey Golanov[5], and Mikhail Belyaev[1,2(✉)]

[1] Skolkovo Institute of Science and Technology, Moscow, Russia
m.belyaev@skoltech.ru
[2] Kharkevich Institute for Information Transmission Problems, Moscow, Russia
[3] Moscow Institute of Physics and Technology, Moscow, Russia
[4] Moscow Gamma-Knife Center, Moscow, Russia
[5] Burdenko Neurosurgery Institute, Moscow, Russia

Abstract. Stereotactic radiosurgery is a minimally-invasive treatment option for a large number of patients with intracranial tumors. As part of the therapy treatment, accurate delineation of brain tumors is of great importance. However, slice-by-slice manual segmentation on T1c MRI could be time-consuming (especially for multiple metastases) and subjective. In our work, we compared several deep convolutional networks architectures and training procedures and evaluated the best model in a radiation therapy department for three types of brain tumors: meningiomas, schwannomas and multiple brain metastases. The developed semiautomatic segmentation system accelerates the contouring process by 2.2 times on average and increases inter-rater agreement from 92.0% to 96.5%.

Keywords: Stereotactic radiosurgery · Segmentation · CNN · MRI

1 Introduction

Brain stereotactic radiosurgery involves an accurate delivery of radiation to the delineated tumor. The basis of the corresponding planning process is to achieve the maximum conformity of the treatment plan. Hence, the outcome of the treatment is highly dependent on the clinician's delineation of the target on the MRI. Several papers have been shown that experts defined different tumour volumes for the same clinical case [10]. As there are no margins applied to a contoured target, the differences in contouring could increase normal tissue toxicity or the risk of recurrence.

The process of contouring is the largest source of potential errors and interobserver variations in target delineation [12]. Such variability could create challenges for evaluating treatment outcomes and assessment of the dosimetric

© Springer Nature Switzerland AG 2020
A. Crimi and S. Bakas (Eds.): BrainLes 2019, LNCS 11992, pp. 119–128, 2020.
https://doi.org/10.1007/978-3-030-46640-4_12

impact on the target. Routinely the targets are delineated through slice-by-slice manual segmentation on MRI, and an expert could spend up to one hour delineating an image. However, stereotactic radiosurgery is one-day treatment and it is critical to provide fast segmentation in order to avoid treatment delays.

Automatic segmentation is a promising tool in time savings and reducing inter-observer variability of target contouring [11]. Recently deep learning methods have become popular for a wide range of medical image segmentation tasks. In particular, gliomas auto-segmentation methods are well-developed [1] thanks to BRATS datasets and contests [8]. At the same time, the most common types of brain tumors treated by radiosurgery, namely meningiomas, schwannomas and multiple brain metastases, are less studied. Recently published studies [2,5,6] developed deep learning methods for automatic segmentation of these types of tumors. However, these studies do not investigate the above-mentioned clinical performance metrics: inter-rater variability and time savings.

Our work aimed to fill this gap and evaluate the performance of semi-automatic segmentation of brain tumors in clinical practice. We developed an algorithm based on deep convolutional neural network (CNN) with suggested adjustment to cross-entropy loss, which allowed us to significantly boost quality of small tumors segmentation. The model achieving the state-of-the-art level of segmentation was integrated into radiosurgery planning workflow. Finally, we evaluated the quality of the automatically generated contours and reported the time reduction using these contours within the treatment planning.

2 Related Work

During recent years, various deep learning architectures were developed. For medical imaging, the best results were achieved by 3D convolutional networks: 3D U-Net [3] and V-Net [9]. However, a large size of brain MRI for some tasks places additional restrictions on CNN. A network called DeepMedic [4] demonstrated solid performance in such problems, including glioma segmentation [1].

Some image processing methods were proposed for the other brain tumors as well. For example, authors of [7] developed a multistep approach utilizing classical computer vision tools such as thresholding or super-pixel clustering. In common with other medical image processing tasks, such methods have two key drawbacks: processing speed and quality of small lesions segmentation [6]. Deep learning-based approaches may potentially resolve these issues thanks to its high inference speed and great flexibility. Indeed, several recently published studies validated CNN in the task of nonglial brain tumors segmentation and demonstrated promising results. In [6] authors modified the DeepMedic to improve segmentation quality. Authors of [2] compared various combinations of T1c, T2 and Flair modalities. New patch generation methods were proposed and evaluated on three types of brain tumors in [5]. In [9] authors introduced a novel loss function based on Dice coefficient to improve segmentation results in highly class imbalance tasks.

3 Data

For computational experiments, we used 548 contrast-enhanced T1-weighted MRI with $0.94 \times 0.94 \times 1$ mm image resolution. These cases were characterized by multiple brain tumors (4.5 per patient) of different sizes: from 1.3 mm up to 4.2 cm in diameter. These images were naturally divided into two datasets. The first one, *training* dataset, consisted of 489 unique patients examined before 2017. It was used to train different models and tune their parameters via cross-validation. The second, *hold-out* dataset, was represented by another 59 patients who were treated in 2017. We performed the final comparison of the best methods on the hold-out dataset to avoid overfitting.

Finally, to evaluate the quality of tumor delineation algorithm in clinical practice, we used the third, *clinical*, dataset which consists of four cases of meningioma, two cases of vestibular schwannoma and four cases of multiple brain metastases (ranged from 3 to 19 lesions per case) collected in 2018. Four experts (or users) with experience in brain radiosurgery ranged from 3 to 15 years delineated each of these cases in two setups: manually and using the output of our model as the starting point, see the details in Sect. 4.4.

4 Methods

4.1 CNN

We used vanilla 3D U-Net, V-Net and DeepMedic models as network architectures. We trained all models for 100 epochs, starting with learning rate of 0.1, and reducing it to 0.01 at the epoch 90. Each epoch consists of 200 stochastic gradient descent iterations. At every iteration, we generated training patches of size $64 \times 64 \times 64$ with batches of size 12 for 3D U-Net and 16 for V-Net. For DeepMedic we generated 16 patches of effective size $39 \times 39 \times 39$ in one batch. We used 5-fold cross-validation to split our training data patient-wise. After the train-predict process, we gather test predictions over the 5 splits to form the metric curve and compare experiment results.

For a subset of experiments (see Sect. 5 for the details), we also used a modified loss function, described in the next subsection and Tumor Sampling from [5]. For the Tumor Sampling as well as the original patches sampling procedures we set the probability to choose the central voxel of each patch belonging to the target mask to be 0.5 for all experiments. We reported the results on the hold-out dataset while using the training dataset to fit the models.

4.2 Inversely Weighted Cross-Entropy

We observed that all methods were missing lots of small tumors or inappropriate segmented them. We assumed that such a performance comes from loss function properties: errors on small targets have the same impact on the loss function as small inaccuracies in large lesions. To make all possible errors contribute equally to the BCE (binary cross-entropy) loss function, we construct a tensor of weights, which are equal to inverse relative volumes of regions of interest.

Given the ground truth on the training stage, we generate a tensor of weights for every image in the train set. To form such a tensor for the given image we split the corresponding ground-truth mask into connected components $C_i, i \in \{0..K\}$, where C_0 is the background and K is the number of tumors. Weights of the background component were set to be $w_0 = 1$. The weights for pixels in the connected component C_i $(i \neq 0)$ are equal to:

$$w_i = \beta \cdot \frac{\sum_{k=0}^{K} |C_k|}{|C_i|}, \tag{1}$$

where β is the fraction of positive class in the current training set. The final form of our loss is the same with weighted BCE over n voxels in the propagated sample:

$$\text{iwBCE} = -\frac{1}{n} \sum_{j=1}^{n} \omega_j \cdot (y_j \log p_j + (1 - y_j) \log (1 - p_j)), \tag{2}$$

where ω_j is the weight of the j-th pixel calculated using (1).

We compare proposed loss function with the current state-of-the-art Dice loss [9] as well as with the standard BCE.

4.3   Metric

We highlighted two essential characteristics that could characterize small tumors segmentation: tumor delineation and detection quality. Since delineation could be simply measured by local Dice score and experts could always adjust contours of found tumors, we focus our attention on the detection quality.

We suggested measuring it in terms of tumor-wise precision-recall curves. We adopted the FROC curve from [13] by changing its hit condition between predicted and ground truth tumors. Predicted tumors were defined as connected components above the probability of 0.5, and we treated the maximum probability of a component as a model's certainty level for it. Our hit condition is that the Dice score between real and predicted lesions is greater than zero. We found such lesion-wise PRC (precision-recall curve) to be more interpretable and useful for model comparison than traditional pixel-wise PRC.

4.4   Contouring Quality and Time Reduction

Within a clinical experiment, we implemented the final model as a service which can process Dicom images and generate contours as Dicom RT files. This output was uploaded to a standard planning system and validated and adjusted (if needed) by experts there; we call these contours *CNN-initialized*. In addition, the same cases were annotated manually in the same planning systems by the same four experts.

To perform the quality evaluation of our algorithm we introduced the following three types of comparisons.

- **1 vs 3** – the manual contour of one user comparing to a ground truth estimation which is the averaged contour of the other users. This setting allows us to measure the current inter-rater variability for a specific user.
- **1^+ vs 3** – a CNN-initialized contour of one user comparing to the same ground truth as above. In this setting we estimate the effect of algorithm on the users.
- **1^+ vs 3^+** – the same as previous setting, but the average contour was obtained using CNN-initialized contours for the three corresponding users. The last setting allows us to measure the level of additional standardization provided by CNN.

To investigate the differences in Dice scores we performed the Sign test for pairs of metrics (**1 vs 3, 1^+ vs 3**) and (**1 vs 3, 1^+ vs 3^+**), see Sect. 5.

To evaluate a speed-up provided by our algorithm in routine clinical practice we compared times needed for two contouring techniques: manual delineation of the tumors and user adjustment of the CNN-initialized contours of the same tumors. The time spent on each task was recorded for all users and cases.

We didn't perform comparison types which include pure CNN generated contours, because AI could not be used in a treatment planing solely without user control and verification.

5 Results

5.1 Methods Comparison on the Hold-Out Dataset

Firstly, we compared three network architectures, see Fig. 1. The results suggest the superiority of U-Net-like architectures over the DeepMedic in our task (see Fig. 1). We made the architecture choice in favor of 3D U-Net and changed it in a minor way to fit our inference timings and memory requirements. We used this model for the subsequent experiments and the final model.

We also observed all the models perform poorly on the small tumors (Fig. 1, left). Within the second set of experiments, we aimed to improve recall for small lesions by adding Tumor Sampling and iwBCE to 3D U-Net, the best model from the first experiments. The proposed loss re-weighting strategy (see Sect. 4.2) reduced the number of missed small tumors by a factor of two with the same level of precision (Fig. 2, left) and improve the network performance over all tumors (Fig. 2, right), achieving almost 0.9 recall on the hold-out dataset. It slightly outperformed Dice loss function, so we used iwBCE to train our model for the clinical installation.

The shaded area on the PRC plots shows 95% confidence intervals of bootstrapped curves over 100 iterations choosing 80% of the test patients every time. The median lesion-wise Dice score of 3D U-Net trained with Tumor Sampling and iwBCE is 0.84 for the hold-out dataset.

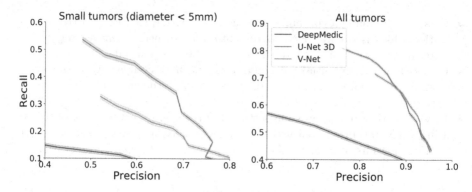

Fig. 1. CNN models comparison. We zoomed all the PRC images from standard $[0; 1]$ scale to better show some model or method had higher recall. We treated *recall as a more important metric than precision* in our task: a radiologist spends few seconds on deleting miss-prediction but much more time on finding and delineating the tumor which CNN didn't predict.

5.2   Clinical Evaluation

We observed better agreement between contours created by the expert and the reference one when the contours were initialized by CNN, even if the reference contour was generated completely manually. Table 1 shows a reduction of inter-rater variability. Improvements for 3 out of 4 experts are statistically significant according to the Sign test p-values. The total median agreement increased from 0.924 to 0.965 in terms of Dice score.

The automatic contours were generated and imported to the treatment planning system in less than one minute. The total median time needed to delineate a case manually was 10.09 min, details for all four experts could be seen in Table 2. On average, the automatic algorithm speeds up the process of the delineation in **2.21** times with the median reduction of time of 5.53 min. We observed speed-up for all users and for all cases they have delineated. We should note that acceleration plays more significant role in the cases of multiple lesions. The total median time needed to delineate a case with multiple metastases manually was 15.7 min (ranged from 15:20 to 44:00 in mm:ss). The automatic tumor segmentation speeded up the delineation of multiple lesions in 2.64 times with median time reduction of 10.23 min.

We also present quality-time plot (see Fig. 3) for both manual and CNN-initialized techniques separately for each user and each case. One can distinguish the global trend of simultaneous improvement of inter-rater agreement and speedup of delineation time. Examples of different contouring techniques for all three types of lesions could be found on the Fig. 4.

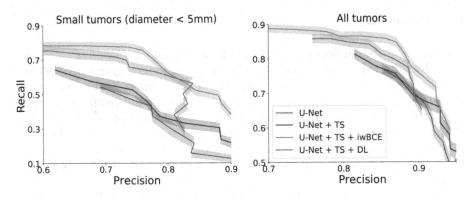

Fig. 2. The best model with TS (Tumor Sampling) and then with iwBCE or DL (Dice Loss).

Table 1. Quality evaluation in tumor contouring. *Case I* evaluated hypothesis that median difference between settings ($\mathbf{1}$ **vs** $\mathbf{3}$) and ($\mathbf{1^+}$ **vs** $\mathbf{3}$) is equal to zero. *Case II* evaluated the same hypothesis for settings ($\mathbf{1}$ **vs** $\mathbf{3}$) and ($\mathbf{1^+}$ **vs** $\mathbf{3^+}$). *All data* contains results for the consolidated set of experiments.

	Median dice scores			p-values	
	1 vs 3	1^+ vs 3	1^+ vs 3^+	I	II
User 1	0.938	0.947	**0.969**	2.85e−1	7.00e−6
User 2	0.930	0.941	**0.968**	7.01e−3	7.00e−6
User 3	0.915	0.920	**0.934**	2.29e−3	2.26e−3
User 4	0.918	0.935	**0.968**	1.40e−2	3.55e−2
All data	0.924	0.941	**0.965**	6.57e−4	3.61e−5

Table 2. Time reduction in tumor delineation. Median time is given per one case.

	Median manual time*	Range	Median time reduction	Range
User 1	13:15	07:00–35:06	06:54	00:40–17:06
User 2	05:30	02:17–15:20	02:16	00:48–08:20
User 3	12:00	03:00–44:00	09:00	01:00–26:00
User 4	06:30	03:00–23:30	05:27	03:00–17:35
All data	10:05	02:17–44:00	05:32	00:40–26:00

*The results are given in *mm:ss*

6   Discussion

For this study, we developed and successfully implemented a deep learning algorithm for automatic brain tumor segmentation into radiosurgery workflow. We demonstrated that our algorithm could achieve near expert-level performance, providing significant time savings in tumor contouring, and reducing the variability in targets delineation at the same time. We should note that within the clinical evaluation, the users initially delineated a case manually, and then they were asked to adjust the CNN-initialized contours of the same case. The adjustment of the CNN-initialized contours typically was performed in one day after manual delineation of the tumor. The fact that the experts had seen tumors previously might have a small impact on the results on the evaluation of time savings.

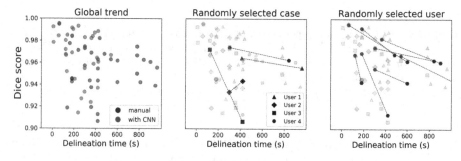

Fig. 3. Plots of inter-rater agreement vs delineation time. *Left*: each point corresponds to a pair lesion-user. Dice scores for blue dots (manual segmentation) were calculated using **1 vs 3** strategy, for red dots - **1 vs 3$^+$**. *Central, right*: dashed lines connect two points for the same pair lesion-user for manual and CNN-initialized delineations. Note that we restricted both time-axis to the maximum of 1000 s and Dice-axis to the minimum of 0.9, therefore few blue points were left outside the plot. (Color figure online)

We proposed a new loss function, called iwBCE, which has not been discussed in all the details. However, it seemed to be a promising approach to improve segmentation quality of modern deep learning tools. We aimed to continue research of the proposed method and compare it with state-of-the-art Dice loss in different setups and on different datasets.

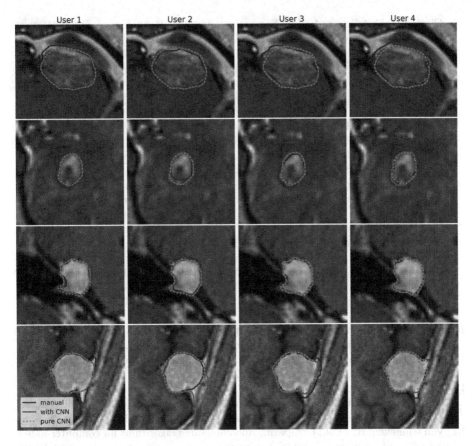

Fig. 4. Segmentation results for two metastatic lesions, one schwannoma and one meningioma in vertical order. **Blue** corresponds to the manual contour, **red** – CNN-initialized contour with user's adjustment, **dashed yellow**—pure CNN contour without user's adjustment. (Color figure online)

Acknowledgements. The Russian Science Foundation grant 17-11-01390 supported the development of the new loss function, computational experiments and article writing.

References

1. Bakas, S., et al.: Identifying the best machine learning algorithms for brain tumor segmentation, progression assessment, and overall survival prediction in the brats challenge. arXiv preprint arXiv:1811.02629 (2018)
2. Charron, O., Lallement, A., Jarnet, D., Noblet, V., Clavier, J.B., Meyer, P.: Automatic detection and segmentation of brain metastases on multimodal MR images with a deep convolutional neural network. Comput. Biol. Med. **95**, 43–54 (2018)

3. Çiçek, Ö., Abdulkadir, A., Lienkamp, S.S., Brox, T., Ronneberger, O.: 3D U-Net: learning dense volumetric segmentation from sparse annotation. In: Ourselin, S., Joskowicz, L., Sabuncu, M.R., Unal, G., Wells, W. (eds.) MICCAI 2016. LNCS, vol. 9901, pp. 424–432. Springer, Cham (2016). https://doi.org/10.1007/978-3-319-46723-8_49

4. Kamnitsas, K., et al.: Efficient multi-scale 3D CNN with fully connected CRF for accurate brain lesion segmentation. Med. Image Anal. **36**, 61–78 (2017)

5. Krivov, E., et al.: Tumor delineation for brain radiosurgery by a ConvNet and non-uniform patch generation. In: Bai, W., Sanroma, G., Wu, G., Munsell, B.C., Zhan, Y., Coupé, P. (eds.) Patch-MI 2018. LNCS, vol. 11075, pp. 122–129. Springer, Cham (2018). https://doi.org/10.1007/978-3-030-00500-9_14

6. Liu, Y., et al.: A deep convolutional neural network-based automatic delineation strategy for multiple brain metastases stereotactic radiosurgery. PLoS One **12**(10), e0185844 (2017)

7. Liu, Y., et al.: Automatic metastatic brain tumor segmentation for stereotactic radiosurgery applications. Phys. Med. Biol. **61**(24), 8440 (2016)

8. Menze, B.H., et al.: The multimodal brain tumor image segmentation benchmark (BRATS). IEEE Trans. Med. Imaging **34**(10), 1993–2024 (2015)

9. Milletari, F., Navab, N., Ahmadi, S.A.: V-Net: fully convolutional neural networks for volumetric medical image segmentation. In: 2016 Fourth International Conference on 3D Vision (3DV), pp. 565–571. IEEE (2016)

10. Roques, T.: Patient selection and radiotherapy volume definition–can we improve the weakest links in the treatment chain? Clin. Oncol. **26**(6), 353–355 (2014)

11. Sharp, G., et al.: Vision 20/20: perspectives on automated image segmentation for radiotherapy. Med. Phys. **41**(5), 1–13 (2014)

12. Torrens, M., et al.: Standardization of terminology in stereotactic radiosurgery: report from the Standardization Committee of the International Leksell Gamma Knife Society: special topic. J. Neurosurg. **121**(Suppl. 2), 2–15 (2014)

13. Van Ginneken, B., et al.: Comparing and combining algorithms for computer-aided detection of pulmonary nodules in computed tomography scans: the ANODE09 study. Med. Image Anal. **14**(6), 707–722 (2010)

Brain Tumor Image Segmentation

3D U-Net Based Brain Tumor Segmentation and Survival Days Prediction

Feifan Wang[1(✉)], Runzhou Jiang[1], Liqin Zheng[1], Chun Meng[1],
and Bharat Biswal[1,2(✉)]

[1] Center for Information in Medicine, School of Life Science and Technology,
University of Electronic Science and Technology of China, Chengdu 611731, China
woodywff@aliyun.com, woodywff@uestc.edu.cn, bbiswal@gmail.com
[2] Department of Biomedical Engineering, New Jersey Institute of Technology,
Newark, NJ, USA

Abstract. Past few years have witnessed the prevalence of deep learning in many application scenarios, among which is medical image processing. Diagnosis and treatment of brain tumors requires an accurate and reliable segmentation of brain tumors as a prerequisite. However, such work conventionally requires brain surgeons significant amount of time. Computer vision techniques could provide surgeons a relief from the tedious marking procedure. In this paper, a 3D U-net based deep learning model has been trained with the help of brain-wise normalization and patching strategies for the brain tumor segmentation task in the BraTS 2019 competition. Dice coefficients for enhancing tumor, tumor core, and the whole tumor are 0.737, 0.807 and 0.894 respectively on the validation dataset. These three values on the test dataset are 0.778, 0.798 and 0.852. Furthermore, numerical features including ratio of tumor size to brain size and the area of tumor surface as well as age of subjects are extracted from predicted tumor labels and have been used for the overall survival days prediction task. The accuracy could be 0.448 on the validation dataset, and 0.551 on the final test dataset.

Keywords: Brain tumor segmentation · 3D U-Net · Survival days prediction

1 Introduction

Human brain stays in a delicate balance under the enclosure of the skull. A brain tumor is a bunch of abnormal brain cells that may harass the balance [1]. Primary brain tumors originate in the brain, while others belong to the secondary or metastatic brain tumors that come from other organs. Brain tumors can also

This work is supported by National Natural Science Foundation (NNSF) of China under Grant 61871420.
The source code of this work is opened on https://github.com/woodywff/brats_2019.

A. Crimi and S. Bakas (Eds.): BrainLes 2019, LNCS 11992, pp. 131–141, 2020.
https://doi.org/10.1007/978-3-030-46640-4_13

be categorized as malignant or benign, the former are cancerous and easy to be spread to the other part of the brain while the later not. Nevertheless, in both cases, the growth of brain tumor in rigid brain space could result in a dysfunction or even life-threatening symptom for human body. Depending on the size and location of the tumor, people may have different symptoms caused by the growing of tumor cells. Some tumors would invade brain tissue directly and some cause pressure on the surrounding brain. As a result, people may suffer from vomiting, blurred vision, confusion, seizures, et al. Magnetic Resonance Imaging (MRI) and resection surgery are the most common diagnosis and treatment means respectively currently used for brain tumors [2]. A priority for a neurosurgeon is to mark the tumor region precisely. Too much or too less surgery may give rise to more loss and suffering. Unfortunately, manually labeling is a laborious and time consuming work for a doctor. Moreover, because of inevitable practical operation factors, it is difficult to replicate a segmentation result exactly the same.

Determining the best computer assistant solutions to the segmentation task, Multimodal Brain Tumor Segmentation Challenge 2019 provides ample MRI scans of patients with gliomas, the most common primary brain tumor, before any kind of resection surgery [3–5]. For training datasets, 259 subjects with high-grade gliomas (HGG) and 76 subjects with low-grade gliomas (LGG) were used [6,7]. Each subject had four $240 \times 240 \times 155$ structural MRI images, including native (T1), post-contrast T1-weighted (T1Gd), T2-weighted (T2), and T2 Fluid Attenuated Inversion Recovery (FLAIR) volumes. Meanwhile, pathologically confirmed segmentation labels for each subject also come as $240 \times 240 \times 155$ images with values of 1 for the necrotic (NCR) and the non-enhancing tumor core(NET), 2 for edema (ED), 4 for enhancing tumor (ET), and 0 for everything else. Further, the segmentation task defines three sub-regions for evaluation, they are 1) the tumor core (TC) including NCR, NET and ET, 2) the ET area, 3) the whole tumor (WT) which is the combination of TC and ED. All these provided MRI images were collected from 19 institutions and had undergone alignment, $1 \times 1 \times 1$ mm resolution resampling and skull stripping. Another task in BraTS 2019 is to predict the overall survival (OS) days of patients after the gross total resection (GTR) surgery. All the OS values are provided together with the age of patients with resection status of GTR.

Deep learning has reentered prosperous ever since AlexNet won the ImageNet competition in 2012 [8,9], which to a great extent attributes to the massively ascending dataset scale and computing power. The advancement of convolutional neural networks came up with a lot of crafted deep learning designs, like VGG-Net [10], Inception networks [11,12] and ResNet [13]. These crafted architectures together with advanced open source frameworks like tensorflow and pytorch energize the development in many research and industrial fields. Semantic segmentation in image processing is to separate the target object from other areas. Fully convalutional networks (FCN) empowers CNN to be able to label each pixel by means of a plain upsampling idea [14]. For medical images, usually they don't share the same features with ordinary pictures from dataset like ImageNet or

CIFAR-10/100 [9,15], which makes it difficult for pre-trained networks on those datasets to be directly used and leaves spaces for specific inventions. U-Net stood out from the IEEE International Symposium on Biomedical Imaging (ISBI) challenge that segments electron microscopy images of the drosophila cell [16]. When it comes to 3D volumetric medical images, the inventors of U-Net also proposed feasible solutions [17].

BraTS initiated by Center for Biomedical Image Computing and Analytics (CBICA) encourages participants to identify competitive solutions to brain tumor segmentation tasks. Most former teams took use of U-Net or put it in ensemble with other modules. For example, the 1st ranked model of BraTS 2018 added a variational auto-encoder (VAE) branch on the U-Net structure to give a new regularization item in the loss function [18]. Isensee et al. argued that a well trained U-Net could be powerful enough and brought out a fine-tuned U-Net model that won the 2nd place in the contest [19]. Moreover, the second task of BraTS is to predict how many days could a patient survive after the GTR operation. Previously, the best record was obtained by performing a linear regression model developed by Feng et al. in BraTS 2018 [20], the features they used include the size and length of tumors and the age information.

In this work, we illustrate our solutions to the two tasks in BraTS 2019. Four modalities (T1, T1Gd, T2, and T2-FLAIR) of structural MRI images collected from patients with gliomas are processed and fed into the network. In particular,

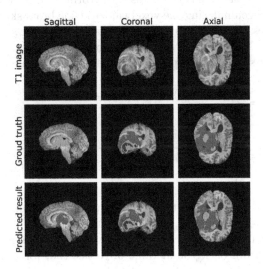

Fig. 1. An example of one modality of input data and the corresponding ground truth and predicted labels, each picture illustrates a slice of 3D MRI images. For ground truth and predicted result, four colors represent four label values. Red for value 1 (NCR and NET), blue for value 2 (ED), yellow for value 4 (ET), black for value 0 (everything else). (Notice that brain areas other than tumor part are also supposed to be black in the six label figures, we draw the brain image just for illustration purpose.) (Color figure online)

rather than normalizing voxel values on the whole image with black background, only the brain wise area has been taken into account for the normalization and scaling. Two phases of training with different patching strategies were undertaken, with one keeping an eye on the black background and the other not, both with a patch covering the center field. An extra parameter was designed for each image to remember the minimum cube that could encapsulate it. All the patching and recovery procedures were maneuvered based on such cubes. Different kinds of tumor tissues would be labeled with different values, as can be seen in Fig. 1 which gives an example of three slices of one T1 brain image in directions of segittal, coronal and axial. These segmentation results have been analyzed and further used for the overall survival task. The proportion of tumor compared to the whole brain and the ratio of length of one sub-region to another have been extracted, together with the phenotypic age information, as the input features.

2   Method

2.1   Preprocessing

All the structural MRI images have been bias field corrected through N4ITK Insight Toolkit hopefully to minimize the bias caused by the differences in multiple scanners, institutions and protocols [21]. Normalization was performed for each modality, by accumulating the voxel values inside brain skull throughout all the training images and calculating the mean value μ and the standard deviation σ. Given $A \in \mathbb{R}^{240 \times 240 \times 155}$ represents an original image, the z-score normalization and min-max scaling are deployed in the following manner.

$$\hat{A}_{ijk} = \begin{cases} (A_{ijk} - \mu)/\sigma & \text{if } A_{ijk} \neq 0 \\ 0 & \text{else,} \end{cases} \tag{1}$$

$$\tilde{A}_{ijk} = \begin{cases} 100 \times \left(\dfrac{\hat{A}_{ijk} - \hat{A}_{\min}}{\hat{A}_{\max} - \hat{A}_{\min}} + 0.1 \right) & \text{if } A_{ijk} \neq 0 \\ 0 & \text{else,} \end{cases} \tag{2}$$

in which $\hat{A}_{\min}$ and $\hat{A}_{\max}$ indicate the minimum and maximum values for all the $\hat{A}_{ijk}$ with corresponding $A_{ijk} \neq 0$, $i, j \in [0, 239]$, $k \in [0, 154]$. Brain area voxel values in the finally preprocessed image would range from 10 to 110, which are discriminated from background value 0. Notice that for validation and test datasets, we still use the μ and σ calculated from the training dataset for the z-score normalization.

2.2   Patching Strategies

Patching strategy would make it possible for less powerful GPU to deal with a large image. In this work we leverage two kinds of patching strategies, both

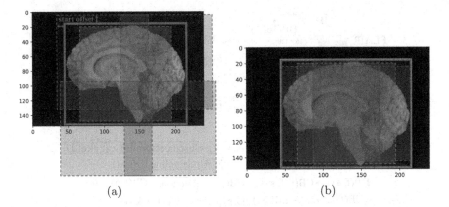

(a)   (b)

Fig. 2. Patching strategies. The green rectangle represents the boundary of the brain, red and blue dashed cubes indicate the patches, in particular, blue for the fixed center one. Start offset marks the largest distance to the boundary a patch could start with. (a) Patching strategy in the first training phase. (b) Patching strategy in the second training phase. (Color figure online)

were maneuvered based on the cuboid boundary of the brain, as shown in Fig. 2, patch size is $128 \times 128 \times 128$. For the convenient of drawing, we use a 2D picture to illustrate the idea behind operations on 3D images.

For the first strategy, as seen in Fig. 2(a), a cubic patch starts with a random distance between 0 to 4 voxels away from the border. The overlap of each two neighbor patches is 32, which stands for each time when the patching window moves 96 voxels. This strategy would generate patches with a bunch of background values. For the second image displayed in Fig. 2(b), all the patches were arranged to the largest extent inside the brain area, each corner of the quadrilateral boundary has a corresponding patch with one corner completely matched. For the training process, two of the patching strategies are employed in sequence. The second strategy is also the one used for the prediction procedure. One extra blue dashed cube that exists in both figures in Fig. 2 refers to a patch we fix in center of the brain area for each image, considering the large amount of information in this part.

2.3   Models

Segmentation Task: Following the framework of U-Net by Isensee et al. [19], the architecture of the segmentation network has been depicted in Fig. 3.

It takes $4 \times 128 \times 128 \times 128$ matrices as input, each of which is stacked by 4 $128 \times 128 \times 128$ patches of different modalities. The Downward Block (DB) squeezes the image size and stretches the channel length. Particularly, as the basic feature embedding element, the Convolutional Block (CB) works with the instance normalization and leaky ReLU modules [22]. As in ResNet, DB also bypasses the front message to the end to fight with the weight decay and

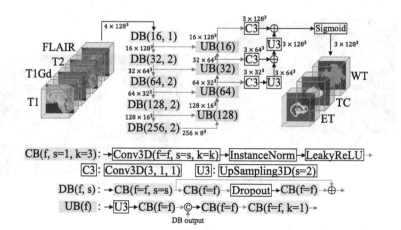

Fig. 3. Schematic of the network for segmentation task. Self-defined modules are listed downstairs. CB refers to Convalutional Block, which takes 3 parameters from Conv3D block inside: f is the number of filters, s is the step length, and k is the kernel size. C3 is short for Conv3D module and U3 for UpSampling3D with upsampling factor equals 2. DB and UB mean Downward Block and Upward Block respectively. In general, black rimmed blocks are original modules in frameworks, while colored blocks and arrows indicate the developed ones. $\oplus$ is element-wise addition, $\copyright$ means concatenation. (Color figure online)

overfitting problems. Upward Blocks (UB) are in charge of reconstructing the location information by means of concatenating the corresponding DB output. They recover the image size to $128 \times 128 \times 128$ and shrink the depth of channel to 3, each of which is the probability matrix that demonstrates the confidence of each voxel belonging to one certain sub-region of the tumor.

The weighted multi-class Dice loss function has been proved to be efficient in former BraTS competitions [23]. As exhibited in Eq. (3),

$$L = -\sum_{c=0}^{2} \frac{\sum_{i,j,k=0}^{127} Y_{cijk} \hat{Y}_{cijk}}{\sum_{i,j,k=0}^{127} Y_{cijk} + \sum_{i,j,k=0}^{127} \hat{Y}_{cijk}}, \tag{3}$$

in which Y indicates the $3 \times 128 \times 128 \times 128$ matrix generated from the ground truth image, $\hat{Y}$ represents the output from the constructed network.

The 3 channels in the output would be mixed to one $128 \times 128 \times 128$ image that each voxel chooses to be the value of one tumor sub-region label or to be zero as the background (by setting up one threshold). Meanwhile, the priority of ET is higher than TC's which in turn is higher than that of WT, which means we compare priorities rather than probabilities once the probability value is over threshold. At the end of the prediction, all the labeled patches with numbers indicating the tumor sub-regions would be concatenated into a brain boundary (the green rectangle in Fig. 2(a)) sized image, in which the overlapped part with more than one values would take the average of them. This image would be further recovered into the original size of 240 × 240 × 155 according to the saved boundary information.

Overall Survival Days Prediction: The predicted tumor images would be further utilized for the estimation of the OS days. In this task, we select seven features from numerous candidates. The first three are the ratios of the volume of each tumor sub-region to the size of the whole brain. Then we calculate the gradient of each kind of tumor matrix and sum up non-zero gradient values to approximate the area of the tumor surface. Last but not the least, age information of subjects have been taken into account. The combination of these seven features have been demonstrated with better performance than other mixtures. The model we choose to solve this problem is straight forward—a fully connected neural network with two hidden layers, each with 64 filters.

3 Results

All the programs developed for this work were written in python and Keras backend by tensorflow, on a single GTX 1080ti GPU installed desktop.

In segmentation task, there are 335 labeled subjects for training, 125 unlabeled subjects for validation and 166 unlabeled subjects for the final test. We trained the network with two patching strategies each for 100 epochs, the second phase use the first phase saved model as pre-trained. Start offset is 4, patch overlap is 32, initial learning rate of Adam optimizer is 5e−4, the learning rate would drop by 50% after 10 steps without loss value decrease. Data augmentation has been undertaken on the fly for each patch, including random rotation, flip and distortion. Table 1 exhibits the mean values of all the required criteria in BraTS 2019, Fig. 4 and Fig. 5 illustrate more details of that. Because we are only afforded the summary stats of results on the final test dataset, here we just illustrate the mean values of the dice coefficient and Hausdorff distance in Table 1.

Table 1. Mean values of different criteria for segmentation task.

Dataset	Dice			Sensitivity			Specificity			Hausdorff95		
	ET	TC	WT	ET	TC	WT	ET	TC	WT	ET	TC	WT
Training	0.830	0.888	0.916	0.856	0.895	0.909	0.998	0.997	0.996	3.073	3.667	4.009
Validation	0.737	0.807	0.894	0.766	0.826	0.897	0.998	0.996	0.995	5.994	7.357	5.677
Final test	0.778	0.798	0.852	–	–	–	–	–	–	3.543	6.219	6.547

For the OS days prediction task, only patients whose 'ResectionStatus' is 'GTR' are taken into consideration, which results in a 101 subjects training set, a 29 subjects validation set and a 107 subjects final test set. In this model, Adam optimizer with initial learning rate as 1e−4 has been deployed, it updates the

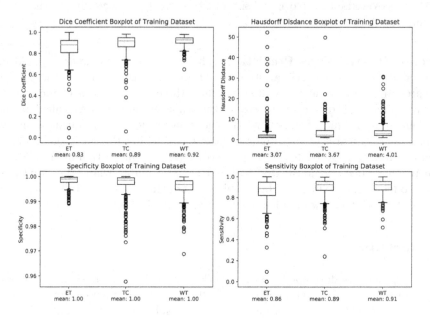

Fig. 4. Dice coefficient, specificity, sensitivity and hausdorff for training dataset.

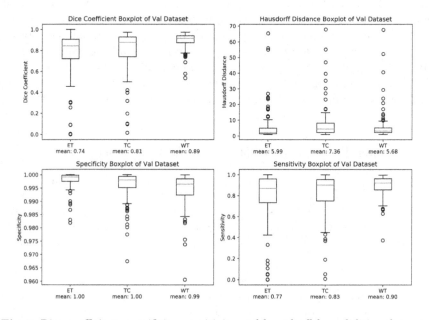

Fig. 5. Dice coefficient, specificity, sensitivity and hausdorff for validition dataset.

learning rate the same way as in task one. For the training process, five-fold cross validation has been employed, which boils down to a configuration with batch size to be 5 and epochs to be more than 500.

Table 2 presents the scores of our predicted survival days. The accuracy is calculated based on a three categories classification, and they define the survival days less than 300 as short-survival, from 300 to 450 as mid-survivor, and more than 450 as long-survival.

Table 2. Mean values of different criteria for OS days prediction task.

Dataset	Accuracy	MSE	MedianSE	stdSE	SpearmanR
Training	0.515	8.73e4	1.96e4	1.83e5	0.472
Validation	0.448	1.0e5	4.93e4	1.35e5	0.25
Final test	0.551	4.10e5	4.93e4	1.23e6	0.323

4 Conclusion

In this work, we introduced a brain-wise normalization and two patching strategies for the training of 3D U-Net for tumor segmentation task. At the same time, we brought about a network taking use of features extracted from predicted tumor labels to anticipate the overall survival days of patients who have undergone gross total resection surgery. Currently on single GPU platform, only one $4 \times 128 \times 128 \times 128$ image could be fed as input each time during training, which probably restricts the capacity of the model. In future works, with more powerful hardwares we would go on with the training and upgrading of this network.

References

1. Zülch, K.J.: Brain Tumors: Their Biology and Pathology. Springer, New York (2013). https://doi.org/10.1007/978-3-642-68178-3
2. Beets-Tan, R.G.H., Beets, G.L., Vliegen, R.F.A., Kessels, A.G.H., Van Boven, H., De Bruine, A., et al.: Accuracy of magnetic resonance imaging in prediction of tumour-free resection margin in rectal cancer surgery. Lancet **357**(9255), 497–504 (2001). https://doi.org/10.1016/S0140-6736(00)04040-X
3. Menze, B.H., Jakab, A., Bauer, S., Kalpathy-Cramer, J., Farahani, K., Kirby, J., et al.: The multimodal brain tumor image segmentation benchmark (BRATS). IEEE Trans. Med. Imaging **34**(10), 1993–2024 (2015). https://doi.org/10.1109/TMI.2014.2377694
4. Bakas, S., Akbari, H., Sotiras, A., Bilello, M., Rozycki, M., Kirby, J.S., et al.: Advancing the cancer genome atlas glioma MRI collections with expert segmentation labels and radiomic features. Nat. Sci. Data **4**, 170117 (2017). https://doi.org/10.1038/sdata.2017.117

5. Bakas, S., Reyes, M., Jakab, A., Bauer, S., Rempfler, M., Crimi, A., et al.: Identifying the best machine Learning algorithms for brain tumor segmentation, progression assessment, and overall survival prediction in the BRATS challenge. arXiv preprint arXiv:1811.02629 (2018)
6. Bakas, S., Akbari, H., Sotiras, A., Bilello, M., Rozycki, M., Kirby, J., et al.: Segmentation labels and radiomic features for the pre-operative scans of the TCGA-GBM collection. Cancer Imaging Arch. (2017). https://doi.org/10.7937/K9/TCIA.2017.KLXWJJ1Q
7. Bakas, S., Akbari, H., Sotiras, A., Bilello, M., Rozycki, M., Kirby, J., et al.: Segmentation labels and radiomic features for the pre-operative scans of the TCGA-LGG collection. Cancer Imaging Arch. (2017). https://doi.org/10.7937/K9/TCIA.2017.GJQ7R0EF
8. Krizhevsky, A., Sutskever, I., Hinton, G.E.: ImageNet classification with deep convolutional neural networks. In: Proceedings of the 25th International Conference on Neural Information Processing Systems, Lake Tahoe, NV, USA, pp. 1097–1105 (2012). https://doi.org/10.1145/3065386
9. Deng, J., Dong, W., Socher, R., Li, L.J., Li, K., Fei-Fei, L.: ImageNet: a large-scale hierarchical image database. In: Proceedings of the IEEE Conference on Computer Vision and Pattern Recognition, Miami, FL, USA, pp. 248–255 (2009). https://doi.org/10.1109/CVPR.2009.5206848
10. Simonyan, K., Zisserman, A.: Very deep convolutional networks for large-scale image recognition. arXiv preprint arXiv:1409.1556 (2014)
11. Szegedy, C., et al.: Going deeper with convolutions. In: Proceedings of the IEEE Conference on Computer Vision and Pattern Recognition, Boston, MA, USA, pp. 1–9 (2015). https://doi.org/10.1109/CVPR.2015.7298594
12. Szegedy, C., Ioffe, S., Vanhoucke, V., Alemi, A.A.: Inception-v4, inception-ResNet and the impact of residual connections on learning. In: Thirty-First AAAI Conference on Artificial Intelligence, San Francisco, CA, USA, pp. 4278–4284 (2017)
13. He, K., Zhang, X., Ren, S., Sun, J.: Deep residual learning for image recognition. In: Proceedings of the IEEE Conference on Computer Vision and Pattern Recognition, Las Vegas, NV, USA, pp. 770–778 (2016). https://doi.org/10.1109/CVPR.2016.90
14. Long, J., Shelhamer, E., Darrell, T.: Fully convolutional networks for semantic segmentation. In: Proceedings of the IEEE Conference on Computer Vision and Pattern Recognition, Boston, MA, USA, pp. 3431–3440 (2015). https://doi.org/10.1109/TPAMI.2016.2572683
15. Krizhevsky, A., Hinton, G.: Learning multiple layers of features from tiny images, vol. 1, no. 4. Technical report, University of Toronto (2009). https://doi.org/10.1.1.222.9220
16. Ronneberger, O., Fischer, P., Brox, T.: U-Net: convolutional networks for biomedical image segmentation. In: Proceedings of the 18th International Conference on Medical Image Computing and Computer Assisted Intervention, Munich, Germany, pp. 234–241 (2015). https://doi.org/10.1007/978-3-319-24574-4_28
17. Çiçek, Ö., Abdulkadir, A., Lienkamp, S.S., Brox, T., Ronneberger, O.: 3D U-Net: learning dense volumetric segmentation from sparse annotation. In: Proceedings of the 19th International Conference on Medical Image Computing and Computer Assisted Intervention, Athens, Greece, pp. 424–432 (2016). https://doi.org/10.1007/978-3-319-46723-8_49
18. Myronenko, A.: 3D MRI brain tumor segmentation using autoencoder regularization. In: Crimi, A., Bakas, S., Kuijf, H., Keyvan, F., Reyes, M., van Walsum, T. (eds.) BrainLes 2018. LNCS, vol. 11384, pp. 311–320. Springer, Cham (2019). https://doi.org/10.1007/978-3-030-11726-9_28

19. Isensee, F., Kickingereder, P., Wick, W., Bendszus, M., Maier-Hein, K.H.: No new-net. In: Crimi, A., Bakas, S., Kuijf, H., Keyvan, F., Reyes, M., van Walsum, T. (eds.) BrainLes 2018. LNCS, vol. 11384, pp. 234–244. Springer, Cham (2019). https://doi.org/10.1007/978-3-030-11726-9_21

20. Feng, X., Tustison, N., Meyer, C.: Brain tumor segmentation using an ensemble of 3D U-Nets and overall survival prediction using radiomic features. In: Crimi, A., Bakas, S., Kuijf, H., Keyvan, F., Reyes, M., van Walsum, T. (eds.) BrainLes 2018. LNCS, vol. 11384, pp. 279–288. Springer, Cham (2019). https://doi.org/10.1007/978-3-030-11726-9_25

21. Tustison, N.J., et al.: N4ITK: improved N3 bias correction. IEEE Trans. Med. Imaging **29**(6), 1310–1320 (2010). https://doi.org/10.1109/TMI.2010.2046908

22. Ulyanov, D., Vedaldi, A., Lempitsky, V.: Instance normalization: the missing ingredient for fast stylization. arXiv preprint arXiv:1607.08022 (2016)

23. Isensee, F., Kickingereder, P., Wick, W., Bendszus, M., Maier-Hein, K.H.: Brain tumor segmentation and radiomics survival prediction: contribution to the BRATS 2017 challenge. In: Crimi, A., Bakas, S., Kuijf, H., Menze, B., Reyes, M. (eds.) BrainLes 2017. LNCS, vol. 10670, pp. 287–297. Springer, Cham (2018). https://doi.org/10.1007/978-3-319-75238-9_25

Aggregating Multi-scale Prediction Based on 3D U-Net in Brain Tumor Segmentation

Minglin Chen[1,2], Yaozu Wu[1,3], and Jianhuang Wu[1(✉)]

[1] Shenzhen Institutes of Advanced Technology, Chinese Academy of Sciences,
Shenzhen, China
jh.wu@siat.ac.cn
[2] University of Chinese Academy of Sciences, Beijing, China
[3] Huazhong University of Science and Technology, Wuhan, China

Abstract. Magnetic resonance imaging (MRI) is the dominant modality used in the initial evaluation of patients with primary brain tumors due to its superior image resolution and high safety profile. Automated segmentation of brain tumors from MRI is critical in the determination of response to therapy. In this paper, we propose a novel method which aggregates multi-scale prediction from 3D U-Net to segment enhancing tumor (ET), whole tumor (WT) and tumor core (TC) from multimodal MRI. Multi-scale prediction is derived from the decoder part of 3D U-Net at different resolutions. The final prediction takes the minimum value of the corresponding pixel from the upsampling multi-scale prediction. Aggregating multi-scale prediction can add constraints to the network which is beneficial for limited data. Additionally, we employ model ensembling strategy to further improve the performance of the proposed network. Finally, we achieve dice scores of 0.7745, 0.8640 and 0.7914, and Hausdorff distances (95th percentile) of 4.2365, 6.9381 and 6.6026 for ET, WT and TC respectively on the test set in BraTS 2019.

Keywords: Brain tumor segmentation · Multi-scale prediction · Deep learning · Multimodal MRI

1 Introduction

A brain tumor is an abnormal mass of tissue in which cells grow and multiply uncontrollably, seemingly unchecked by the mechanisms that control normal cells. They can be categorized as the primary and the metastatic brain tumors. The primary brain tumors originate from the tissues of the brain or the brain's immediate surroundings. The metastatic brain tumors occur when tumors spread to brain from another organ, such as lung or breast. As primary brain tumor, gliomas are the most prevalent type of adult brain tumor, accounting for 78% of malignant brain tumors [1]. They develop from different types of glial cells. Sophisticated imaging techniques, such as magnetic resonance imaging (MRI),

© Springer Nature Switzerland AG 2020
A. Crimi and S. Bakas (Eds.): BrainLes 2019, LNCS 11992, pp. 142–152, 2020.
https://doi.org/10.1007/978-3-030-46640-4_14

can pinpoint brain tumors and assist neurosurgeon in diagnosis and planning treatment for patients.

Automatic segmentation of brain tumor can provide physicians reproducible delineation of different types of tumors and make them more efficient. The outcome of segmentation can be further used in other research (e.g. tumor size calculation, tumor shape analysis, tumor growth analysis). Recently, many approaches based on deep learning have been proposed in various medical imaging tasks and produce the state-of-the-art result [5,8,12].

Multimodal Brain Tumor Segmentation Challenge (BraTS) focuses on the evaluation of state-of-the-art methods for segmentation of brain tumors in multimodal MRI scans [2,3,14,18,19]. In BraTS 2019, the training data comprises 259 High Grade Glioma (HGG) cases and 76 Low Grade Glioma (LGG) cases with corresponding annotations for three glioma tumor subregions (i.e., the enhancing tumor, the peritumoral edema, the necrotic and non-enhancing tumor core). The official validation data contains 125 cases without any published annotations and tumor grade information. All multimodal MRI data describe native (T1), post-contrast T1-weighted (T1Gd), T2-weighted (T2) and T2 Fluid Attenuated Inversion Recovery (T2-FLAIR) volumes with size $155 \times 240 \times 240$. The provided MRI data from BraTS have been pre-processed, i.e. co-registered to the same anatomical template, interpolated to the same resolution ($1\,\text{mm} \times 1\,\text{mm} \times 1\,\text{mm}$) and skull-stripped [2,3,14,18,19]. Three nested tumor subregions including whole tumor (WT), tumor core (TC) and enhancing tumor (ET) are used to evaluate the performance of automatic segmentation method under four types of metrics including dice score, sensitivity, specificity and Hausdorff distance (95%).

In this work, we propose a novel method used for tumor segmentation from 3D multimodal MRI. We aggregate multi-scale prediction from decoder part of 3D U-Net, as features from decoder part of 3D U-Net could produce segmentation masks on different scales. Moreover, some strategies such as test time augmentation (TTA) [23] and models ensembling are employed to make the proposed approach produce more stable and robust result. We evaluate our approaches on the official validation set of BraTS 2019 Challenge and achieve competitive result.

2   Related Work

BraTS 2017. In BraTS 2017, the remarkable competitors included Kamnitsas et al. [9] and Wang et al. [22]. Kamnitsas et el. [9] ensembled three types of models for robust segmentation. The explored Ensembles of Multiple Models and Architectures (EMMA) can reduce the influence of the hyper-parameters of individual approaches and the risk of overfitting. Wang et al. [22] decomposed the multi-class segmentation task into a sequence of three binary segmentation problems. A cascade of fully convolutional neural network was proposed to segment three types of tumor step by step. To boost the segmentation performance, multi-view fusion, dilated convolution technology and multi-scale prediction had been employed.

BraTS 2018. In BraTS 2018, Myronenko et al. [15] won the 1st place, Isensee et al. [8] won the 2nd place, McKinley et al. [13] and Zhou et al. [24] shared the 3rd place. Due to limited training data, Myronenko et al. [15] equipped a variational auto-encoder (VAE) to encoder-decoder network and used the cropped $160 \times 192 \times 128$ MRI volume as input. Instead of modifying the network architecture, Isensee et al. [8] devoted to explore the potential of 3D U-Net by optimizing the training procedure. Furthermore, additional data were used to regularize the network through cotraining. They used a large patch with shape $128 \times 128 \times 128$ as input. McKinley et al. [13] proposed a novel approach embedding densely connected blocks with dilated convolutions into a shallow symmetrical U-Shape network. They also introduced a new loss function called label-uncertainty loss to model label noise and uncertainty. Zhou et al. [24] ensembled multiple deep networks of varied architectures to reduce the risk of overfitting.

Multi-scale Aggregation. As a fully convolutional network (FCN) extracts the hierarchical features, the low-level and the high-level features can be learned in the shallow and the deep layers respectively. Multi-scale prediction from different layers can be combined to derive robust segmentation result [17,20,22]. Instead of taking advantage of multi-scale features in FCN, multi-scale features from decoder part of encoder-decoder architecture can also be exploited [6,7,10]. Predictions from multi-scale features are generally aggregated by element-wise addition after upsampling to the resolution of input. Multi-scale aggregation takes advantage of deep supervision which injects gradient signals deeply into the network.

Comparing to the previous methods, we explore multi-scale prediction aggregation from 3D U-Net [4] by element-wise minimization (see Fig. 1). And we show this aggregation method making multi-scale prediction complementary to each other by visualization. Similar to [15], we use a large crop with shape $160 \times 224 \times 160$ as input. We also employ some optimization strategies such as Test Time Augmentation (TTA) and model ensembling to achieve more robust segmentation.

3   Methods

In this section, we first describe the proposed network architecture with mulit-scale aggregation. Then, we describe element-wise minimization of multi-scale aggregation. Finally, strategies for better performance such as Test Time Augmentation (TTA) and model ensembling are discussed.

3.1   Network Architecture

To perform accurate and robust brain tumor segmentation, the outstanding competitors in BraTS 2018 tend to propose novel method based on 3D U-Shape network [8,15]. Their success in last year challenge demonstrates the power and the potential of U-Shape network for brain tumor segmentation.

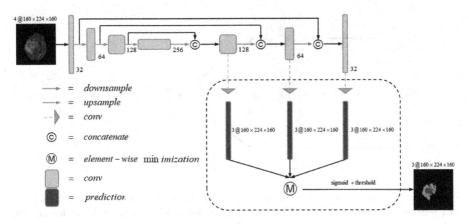

Fig. 1. Schematic visualization of the proposed network architecture. Input is a cropped 3D MRI volume with 4 modalities. Output has the same resolution as the input with 3 channels of the probabilities for WT, TC and ET respectively. Blocks in wathet represent the convolution operations in 3D U-Net [4]. Multi-scale prediction aggregation is shown within the dotted line frame.

The proposed network architecture we used in BraTS 2019 Segmentation Challenge is shown in Fig. 1. We employ 3D U-Net to produce the multi-scale prediction. Due to the limitation of GPU memory, we crop the original MRI into size 224 × 160 in x- and y-axis. And we expand the z-axis size from 155 to 160 by padding zeros in extra area. Thus, the input has the size 160 × 224 × 160. The input MRI volume first goes through encoder part of 3D U-Net which extracts features in each resolution by convolution with ReLU and downsampling. Then, the features with lowest resolution start to restore the resolution as input by processing the concatenation of the features with corresponding resolution from encoder part and themselves after upsampling progressively in decoder part. In each resolution, features are processed under two convolution layers with ReLU. We use max-pooling as downsample and transposed convolution as upsampling in the whole network. Finally, the features in multi-resolution from decoder part are then used in producing multi-scale prediction which is aggregating to derive the segmentation result.

3.2   Multi-scale Prediction Minimization Aggregation

In the decoder part of 3D U-Net, segmentation prediction map can be derived on various scales. However, most of works [8,15] based on 3D U-Net only make prediction on the largest scale features for the minimum resolution loss. These methods put the responsibility of prediction to the last layer of largest scale. Intuitively, prediction can be made on every scales. Unlike other works in [6, 7, 10], we aggregate multi-scale prediction by minimization. Multi-scale prediction minimization aggregation can reduce the responsibility of prediction on the last

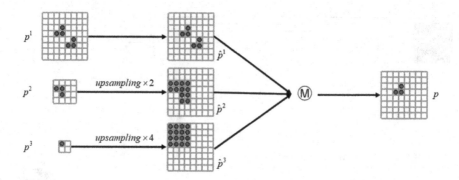

Fig. 2. Demonstration of multi-scale prediction minimization aggregation. Lattice pattern represents prediction map. Red in grid represents the high probability for tumor and white in grid represents the low probability for one. (Color figure online)

layer of largest scale. It also can make predictions on different scales complement each other.

The multi-scale features from decoder part of 3D U-Net are processed by two successive convolution layers with ReLU to produce multi-scale prediction. Let p_j^i and $\hat{p}_j^i$ denote the predictive probability before and after upsampling on scale-i in jth location, respectively. We use nearest neighbor upsampling here. Then, the final predictive probability p_j can be formulated as follows:

$$p_j = \min_i\{\hat{p}_j^i\}$$
$$\hat{p}^i = upsampling \times 2^{(i-1)}(p^i)$$

where $i \in \{1, 2, 3\}$. By minimization aggregation, prediction can be made on any scale. Furthermore, prediction on various scales can complement each other. In other words, prediction on a small scale can refine the one on a large scale by producing fine-grained segmentation result and prediction on a large scale can correct the one on small scale which produces over-segmentation result. (see Fig. 2)

3.3 Optimization Strategies

Test Time Augmentation. Test Time Augmentation (TTA) shows its effectiveness in many applications [23], and it has been used in aleatoric uncertainty estimation under a consistent mathematical framework [21]. TTA combines predictions of multiple transformed versions of a test MRI volume. Let I_i denotes the ith transformed MRI volume from test one I and $p_i = f(I_i)$ denotes the probability from trained network f in case of input I_i, while $i \in \{1, \ldots, N\}$. In this work, we obtain the final prediction o_i by averaging:

$$o_i = \frac{1}{N} \sum_{i=1}^{N} p_i$$

The transformation used in this work is mirror flipping in three axes of MRI volume. In our experiments, TTA not only improves performance but also reduces deviation.

Model Ensembling. Independently trained models are diverse and biased due to the different initial weights and training set. One model may be better than the other models in discriminating some features. Therefore, a more accurate and robust result can be obtained by ensembling multiple diverse models.

After training multiple models with different data from training set, we average the output of each model as the final result. Formally, let p_i^m denotes the predictive probability of target tumor in pixel i from jth model. The ensembling output from M models can be derived as follows:

$$p_i = \frac{1}{M} \sum_{m=1}^{M} p_i^m$$

The final binary result of the target tumor and background in pixel i can be obtained by using a threshold (i.e. 0.5) on the ensembling output.

4   Results and Discussion

4.1   Data and Implementation Details

In BraTS 2019 Segmentation Challenge, the training data comprises 335 multimodal MRI cases and corresponding annotations for glioma tumor. Each multimodal MRI data includes T1, T1 contrast-enhanced, T2 and FLAIR. We randomly split the training data into 268 cases as local training set and 67 cases as local validation set. Our models are trained in local training set with mirror flipping as data augmentation. Local validation set is used to adjust hyperparameters in our models. The validation set in BraTS 2019 contains 125 cases without any public annotations. Online validation tool[1] is applied to evaluate the performances of our methods.

We implement the proposed methods in PyTorch [16] platform and train them on NVIDIA Tesla V100 32 GB GPU. All models are trained in 200 epoches with batch size 1. Adam optimizer [11] with the initial learning rate $1e-4$ and weight decay $1e-5$. The initial learning rate is decayed by 0.1 every 60 epoches.

---

[1] https://ipp.cbica.upenn.edu/.

Table 1. Performance comparison of dice and Hausdorff distance on the BraTS 2019 validation set using a online validation tool.

	basic_channels	Dice			Hausdorff (95%)		
		ET	WT	TC	ET	WT	TC
3D U-Net [4]	10	0.6981	0.8533	0.7719	6.4053	12.2777	10.4436
AMPNet	10	0.7224	0.8651	0.7838	4.8419	7.3529	8.4079
AMPNet+TTA	10	0.7285	0.8718	0.7885	4.5898	7.1399	8.2186
AMPNet+Ensemble	10	0.7530	0.8755	**0.8092**	33.4766	8.8384	12.4542
AMPNet	32	0.7413	0.8934	0.7948	5.2697	6.0294	**7.9257**
AMPNet+TTA	32	0.7416	0.9026	0.7925	**4.5748**	**4.3783**	7.9539
AMPNet+Ensemble	32	**0.7557**	**0.9029**	0.7932	4.7696	4.4932	8.1903

4.2   Performances on BraTS 2019 Validation Set

In BraTS 2019 Segmentation Challenge, four types of metrics, namely dice, sensitivity, specificity and Hausdorff distance (95%) are used to evaluate the segmentation methods in three types of tumor subregions (i.e., ET, WT and TC). Tables 1 and 2 show the performances of our model on the BraTS 2019 validation set. The results are derived from the online validation tool.

Table 2. Performance comparision of sensitivity and specificity on the BraTS 2019 validation set using the online validation tool.

	basic_channels	Sensitivity			Specificity		
		ET	WT	TC	ET	WT	TC
3D U-Net [4]	10	0.7765	**0.9047**	0.7927	0.9976	0.9854	0.9938
AMPNet	10	0.7564	0.8655	0.8027	0.9981	0.9914	0.9944
AMPNet+TTA	10	0.7624	0.8723	0.8062	0.9981	0.9914	0.9944
AMPNet+Ensemble	10	0.7567	0.8692	**0.8103**	0.9904	0.9854	0.9886
AMPNet	32	0.6813	0.8533	0.7144	**0.9993**	0.9974	**0.9988**
AMPNet+TTA	32	0.7385	0.8974	0.7800	0.9987	0.9955	0.9976
AMPNet+Ensemble	32	**0.7769**	0.7769	0.7871	0.9983	**0.9983**	0.9983

We start from basic 3D U-Net architecture. Due to the large input size, we reduce the number of filters in the whole network in a proportional manner. We denote the number of filters in the first convolutional layer as basic_channels. For clarity, our method is denoted as AMPNet.

In Table 1, it can be seen that our AMPNet improves the performance with regards to dice and Hausdorff distance when compared to the 3D U-Net which has less fewer filters in convolutional layer. The TTA improves performance consistently and effectively. The output of our AMPNet is highly uncertain at the blurred edge of brain tumor in MRI. TTA reduces edge uncertainty by averaging the results of multiple transformations of input MRI. In our experiment, TTA can improve dice performance of WT by about 1%. The edge of WT is more obscure than that of ET and TC because of edema. And the model ensembling has only slightly improved performance after using TTA. The uncertainties of different models for predicting different locations are different. We should ensemble the models by weighting the output of different models instead of averaging them. In our experiment, increasing the number of filters in the network can further improve the performance, but it needs more GPU memory. Our AMP-Net increases the dice score by 1–2% when we increase the number of filter in network by about three times.

Figure 3 shows the visualization of predicted results achieved by our AMP-Net.

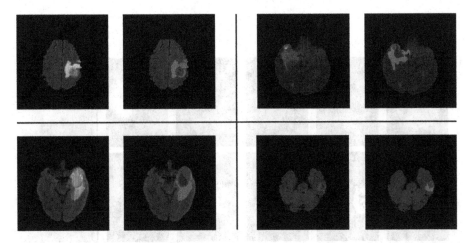

Fig. 3. Visualization of prediction from the proposed AMPNet on BraTS 2019 Validation Set. The left image of each group is FLAIR MRI. The right image of each group is predicted result rendered on FLAIR. (Blue: necrotic and non-enhancing tumor; Green: edema; Red: enhancing tumor). (Color figure online)

4.3   Visualization of Multi-scale Prediction

Figure 4 shows some examples of multi-scale prediction achieved by our AMPNet on BraTS 2019 local validation set. In our experiments, hierarchical features have been learned in each scale. Small-scale prediction focuses on detailed features, while large-scale prediction learns more general relevance information. Multiscale prediction complements and corrects each other to produce the final results.

4.4   Performances on BraTS 2019 Test Set

On testing phase, we used an ensemble of three AMPNet models which were trained with BraTS 2019 training set for the final submission. The test results are shown in Table 3. We compare the results on the test set and the validation set. On the test set, segmentation performances of our method in terms of Dice and 95th Hausdorff Distance in ET and TC are similar to or even better than that on the validation set. However, the performance of WT segmentation gets worse on the test set.

Table 3. The performance of AMPNet on BraTS 2019 test set.

	Dice			Hausdorff (95%)		
	ET	WT	TC	ET	WT	TC
AMPNet+Ensemble	0.7745	0.8640	0.7914	4.2365	6.9381	6.6026

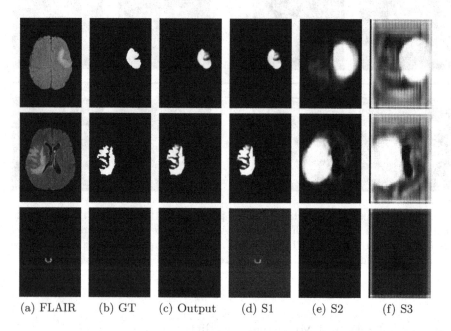

(a) FLAIR   (b) GT   (c) Output   (d) S1   (e) S2   (f) S3

Fig. 4. Visualization of AMPNet output on BraTS 2019 local validation set. Each column from left to right is FLAIR, GT label for WT, the final output of WT from AMPNet, prediction in highest resolution from AMPNet, prediction in middle resolution from AMPNet after upsampling and prediction in lowest resolution from AMPNet after upsampling.

5   Conclusion

In this work, we propose a network structure which aggregates multi-scale predictions from the decoder part of 3D U-Net in brain tumor segmentation. To achieve better performance, two optimization strategies are integrated into our network. The proposed method is applied in BraTS 2019 Segmentation Challenge. Experimental results demonstrate that our method can achieve dice scores of 0.7745, 0.8640 and 0.7914, and Hausdorff distances (95%) of 4.2365, 6.9381 and 6.6026 for ET, WT and TC respectively on the test data in BraTS 2019.

Acknowledgements. This work was supported by the National Natural Science Foundation of China (No. 61672510), and Shenzhen Basic Research Program (No. JCYJ20180507182441903). We appreciate for Xiao Ma and Tong Xia.

References

1. Anonymous: Brain tumors. https://www.aans.org/en/Patients/Neurosurgical-Conditions-and-Treatments/Brain-Tumors. Accessed 21 Dec 2019
2. Bakas, S., et al.: Advancing the cancer genome atlas glioma MRI collections with expert segmentation labels and radiomic features. Sci. Data **4**, 170117 (2017)
3. Bakas, S., et al.: Identifying the best machine learning algorithms for brain tumor segmentation, progression assessment, and overall survival prediction in the brats challenge. arXiv preprint arXiv:1811.02629 (2018)
4. Çiçek, Ö., Abdulkadir, A., Lienkamp, S.S., Brox, T., Ronneberger, O.: 3D U-Net: learning dense volumetric segmentation from sparse annotation. In: Ourselin, S., Joskowicz, L., Sabuncu, M.R., Unal, G., Wells, W. (eds.) MICCAI 2016. LNCS, vol. 9901, pp. 424–432. Springer, Cham (2016). https://doi.org/10.1007/978-3-319-46723-8_49
5. Han, M., et al.: Segmentation of CT thoracic organs by multi-resolution VB-nets. In: SegTHOR@ ISBI (2019)
6. Hu, X., Li, H., Zhao, Y., Dong, C., Menze, B.H., Piraud, M.: Hierarchical multi-class segmentation of glioma images using networks with multi-level activation function. In: Crimi, A., Bakas, S., Kuijf, H., Keyvan, F., Reyes, M., van Walsum, T. (eds.) BrainLes 2018. LNCS, vol. 11384, pp. 116–127. Springer, Cham (2019). https://doi.org/10.1007/978-3-030-11726-9_11
7. Isensee, F., Kickingereder, P., Wick, W., Bendszus, M., Maier-Hein, K.H.: Brain tumor segmentation and radiomics survival prediction: contribution to the BRATS 2017 challenge. In: Crimi, A., Bakas, S., Kuijf, H., Menze, B., Reyes, M. (eds.) BrainLes 2017. LNCS, vol. 10670, pp. 287–297. Springer, Cham (2018). https://doi.org/10.1007/978-3-319-75238-9_25
8. Isensee, F., Kickingereder, P., Wick, W., Bendszus, M., Maier-Hein, K.H.: No new-net. In: Crimi, A., Bakas, S., Kuijf, H., Keyvan, F., Reyes, M., van Walsum, T. (eds.) BrainLes 2018. LNCS, vol. 11384, pp. 234–244. Springer, Cham (2019). https://doi.org/10.1007/978-3-030-11726-9_21
9. Kamnitsas, K., et al.: Ensembles of multiple models and architectures for robust brain tumour segmentation. In: Crimi, A., Bakas, S., Kuijf, H., Menze, B., Reyes, M. (eds.) BrainLes 2017. LNCS, vol. 10670, pp. 450–462. Springer, Cham (2018). https://doi.org/10.1007/978-3-319-75238-9_38
10. Kayalibay, B., Jensen, G., van der Smagt, P.: CNN-based segmentation of medical imaging data. arXiv preprint arXiv:1701.03056 (2017)
11. Kingma, D.P., Ba, J.: Adam: a method for stochastic optimization. arXiv preprint arXiv:1412.6980 (2014)

12. Li, X., Chen, H., Qi, X., Dou, Q., Fu, C.W., Heng, P.A.: H-DenseUNet: hybrid densely connected unet for liver and tumor segmentation from CT volumes. IEEE Trans. Med. Imaging **37**(12), 2663–2674 (2018)
13. McKinley, R., Meier, R., Wiest, R.: Ensembles of densely-connected CNNs with label-uncertainty for brain tumor segmentation. In: Crimi, A., Bakas, S., Kuijf, H., Keyvan, F., Reyes, M., van Walsum, T. (eds.) BrainLes 2018. LNCS, vol. 11384, pp. 456–465. Springer, Cham (2019). https://doi.org/10.1007/978-3-030-11726-9_40
14. Menze, B.H., et al.: The multimodal brain tumor image segmentation benchmark (BRATS). IEEE Trans. Med. Imaging **34**(10), 1993–2024 (2014)
15. Myronenko, A.: 3D MRI brain tumor segmentation using autoencoder regularization. In: Crimi, A., Bakas, S., Kuijf, H., Keyvan, F., Reyes, M., van Walsum, T. (eds.) BrainLes 2018. LNCS, vol. 11384, pp. 311–320. Springer, Cham (2019). https://doi.org/10.1007/978-3-030-11726-9_28
16. Paszke, A., et al.: Automatic differentiation in PyTorch (2017)
17. Soltaninejad, M., Zhang, L., Lambrou, T., Yang, G., Allinson, N., Ye, X.: MRI brain tumor segmentation and patient survival prediction using random forests and fully convolutional networks. In: Crimi, A., Bakas, S., Kuijf, H., Menze, B., Reyes, M. (eds.) BrainLes 2017. LNCS, vol. 10670, pp. 204–215. Springer, Cham (2018). https://doi.org/10.1007/978-3-319-75238-9_18
18. Bakas, S., et al.: Segmentation labels and radiomic features for the pre-operative scans of the TCGA-GBM collection. The Cancer Imaging Archive (2017). https://doi.org/10.7937/K9/TCIA.2017.KLXWJJ1Q
19. Bakas, S., et al.: Segmentation labels and radiomic features for the pre-operative scans of the TCGA-LGG collection. The Cancer Imaging Archive (2017). https://doi.org/10.7937/K9/TCIA.2017.GJQ7R0EF
20. Sun, L., Zhang, S., Luo, L.: Tumor segmentation and survival prediction in glioma with deep learning. In: Crimi, A., Bakas, S., Kuijf, H., Keyvan, F., Reyes, M., van Walsum, T. (eds.) BrainLes 2018. LNCS, vol. 11384, pp. 83–93. Springer, Cham (2019). https://doi.org/10.1007/978-3-030-11726-9_8
21. Wang, G., Li, W., Aertsen, M., Deprest, J., Ourselin, S., Vercauteren, T.: Aleatoric uncertainty estimation with test-time augmentation for medical image segmentation with convolutional neural networks. Neurocomputing **338**, 34–45 (2019)
22. Wang, G., Li, W., Ourselin, S., Vercauteren, T.: Automatic brain tumor segmentation using cascaded anisotropic convolutional neural networks. In: Crimi, A., Bakas, S., Kuijf, H., Menze, B., Reyes, M. (eds.) BrainLes 2017. LNCS, vol. 10670, pp. 178–190. Springer, Cham (2018). https://doi.org/10.1007/978-3-319-75238-9_16
23. Wang, G., Li, W., Ourselin, S., Vercauteren, T.: Automatic brain tumor segmentation using convolutional neural networks with test-time augmentation. In: Crimi, A., Bakas, S., Kuijf, H., Keyvan, F., Reyes, M., van Walsum, T. (eds.) BrainLes 2018. LNCS, vol. 11384, pp. 61–72. Springer, Cham (2019). https://doi.org/10.1007/978-3-030-11726-9_6
24. Zhou, C., Chen, S., Ding, C., Tao, D.: Learning contextual and attentive information for brain tumor segmentation. In: Crimi, A., Bakas, S., Kuijf, H., Keyvan, F., Reyes, M., van Walsum, T. (eds.) BrainLes 2018. LNCS, vol. 11384, pp. 497–507. Springer, Cham (2019). https://doi.org/10.1007/978-3-030-11726-9_44

Brain Tumor Synthetic Segmentation in 3D Multimodal MRI Scans

Mohammad Hamghalam[1,2]([✉]) [ID], Baiying Lei[1], and Tianfu Wang[1]

[1] National-Regional Key Technology Engineering Laboratory for Medical Ultrasound, Guangdong Key Laboratory for Biomedical Measurements and Ultrasound Imaging, School of Biomedical Engineering, Health Science Center, Shenzhen University, Shenzhen 518060, China
m.hamghalam@gmail.com, {leiby,tfwang}@szu.edu.cn
[2] Faculty of Electrical, Biomedical and Mechatronics Engineering, Qazvin Branch, Islamic Azad University, Qazvin, Iran

Abstract. The magnetic resonance (MR) analysis of brain tumors is widely used for diagnosis and examination of tumor subregions. The overlapping area among the intensity distribution of healthy, enhancing, non-enhancing, and edema regions makes the automatic segmentation a challenging task. Here, we show that a convolutional neural network trained on high-contrast images can transform the intensity distribution of brain lesions in its internal subregions. Specifically, a generative adversarial network (GAN) is extended to synthesize high-contrast images. A comparison of these synthetic images and real images of brain tumor tissue in MR scans showed significant segmentation improvement and decreased the number of real channels for segmentation. The synthetic images are used as a substitute for real channels and can bypass real modalities in the multimodal brain tumor segmentation framework. Segmentation results on BraTS 2019 dataset demonstrate that our proposed approach can efficiently segment the tumor areas. In the end, we predict patient survival time based on volumetric features of the tumor subregions as well as the age of each case through several regression models.

Keywords: Tumor segmentation · Synthetic image · GAN · Regression model · Overall survival

1 Introduction

Glioma is the most aggressive and widespread tumor is grouped into low-grade gliomas (LGGs) and high-grade gliomas (HGGs). Multimodal MR channels in BraTS 2019 datasets [1–4,13], included of FLAIR, T1, T1c, and T2, are routinely used to segment internal parts of the tumor, i.e., whole tumor (WT), tumor core (TC), and enhancing tumor (ET). Several segmentation approaches have been proposed to segment regions of interest through classic [7,8,17,18] and modern machine learning methods, especially brain tumor segmentation techniques [10,14].

© Springer Nature Switzerland AG 2020
A. Crimi and S. Bakas (Eds.): BrainLes 2019, LNCS 11992, pp. 153–162, 2020.
https://doi.org/10.1007/978-3-030-46640-4_15

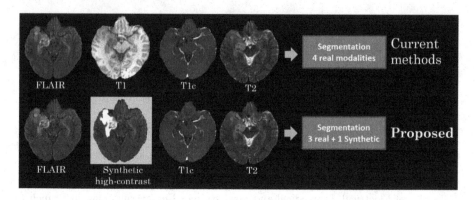

Fig. 1. The pipeline outlines the steps in the current (top) and proposed synthetic (bottom) segmentation techniques. We displace the real T1 channels with the synthetic image.

The focus of current research is to form a generator that increases the contrast within subregions of the brain tissue. The generator, which is a deep neural network model, employes a real channel as input to produce the synthetic one. Our framework comprises two stages: (1) we generate high tissue contrast images based on FLAIR sequence in our convolutional neural network (CNN) model, (2) we train a 3D fully convolutional network (FCN) [5,9,12,16] based on the synthetic images to segment region of interests.

2   Method

Our goal is to segment tumor subregions based on multimodal 3D magnetic resonance (MR) volumes. Figure 1 demonstrates an overview of the proposed method based on synthetic high-contrast images. In contrast to the current methods, we use both real and synthetic volumes for the segmentation task. Following, we first introduce the synthetic image generator module, based on the generative adversarial networks (GANs) model [6], and then 3D FCN architecture for segmentation is discussed.

2.1   Synthetic Image Generator

We extend the image-to-image translation method [11] to deal with the synthesis of high-contrast 2D images. Our model trains on high-contrast images, building based on manual labels, in an adversarial framework. The synthesis model contains a Generator, based on the 2D-U-Net [15], and a Discriminator, build on 2D FCN network. Figure 2 illustrates the image translation framework, where both the generator and the discriminator blocks are trained on FLAIR with a patch size of 128×128 pixels. In implementation details, we follow [11], including the

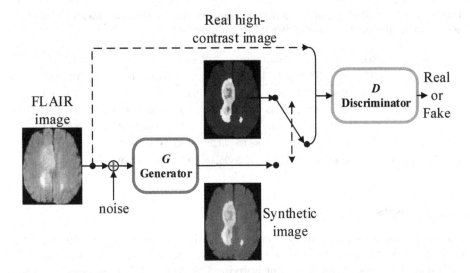

Fig. 2. Deep-learning-based high-contrast synthesis using FLAIR images. After training by GAN, the model outputs the synthetic high tissue contrast images with an inference time of around 20 ms.

number of epochs, the number of layers, and the kernel sizes. For each subject in the BraTS'19 dataset, we provide a 3D synthetic volume for the next stage, segmentation.

2.2 Synthetic Segmentation

The output volumes from synthetic image generator block are concatenated with real modalities (FLAIR, T1c, and T2) and fed into segmentation block to predict region of interests. The segmentation network allows jointly capturing features from FLAIR, synthetic, T1c, and T2 modality. For the 3D segmentation block, we rely on ensembling the 3D FCN on axial, sagittal, and coronal planes.

3 Experimental Results

3.1 Implementation Details

We implement the proposed design employing the KERAS with 12GB NVIDIA TITAN X GPU. We have scaled image patches to sizes 128×128 pixels for translation. The model is trained through the ADADELTA [19] optimizer (learning rate $= 0.9$, $\rho = 0.90$, epsilon $= 1e{-}5$). Dropout is employed to avoid over-fitting over the training ($p_{drop} = 0.4$).

Table 1. DSCs and HD95 of the synthetic segmentation method on BraTS'19 Validation set (training on 335 cases of BraTS'19 training set).

	Dice			Sensitivity			Specificity			HD95 (mm)		
	ET	WT	TC	ET	WT	TC	ET	WT	TC	ET	WT	TC
Mean	76.65	89.65	79.01	76.88	91.32	77.71	99.85	99.39	99.76	4.6	6.9	8.4
Std.	25.86	9.44	23.31	25.35	8.84	26.13	0.23	0.69	0.33	7.2	13.8	12.4
Median	84.73	92.15	89.47	85.47	94.53	90.08	99.93	99.58	99.88	2.2	3.3	4.1
25 quantile	77.88	87.94	74.29	72.82	88.65	73.26	99.82	99.15	99.70	1.4	2.0	2.0
75 quantile	90.21	94.81	93.98	91.97	97.28	95.16	99.98	99.83	99.97	4.1	5.1	10.3

3.2 Datasets

The performance of the proposed method is evaluated on the BraTS'19 dataset, which has two datasets of pre-operative MRI sequences: Training (335 cases) and Validation (125 cases). Each patient is giving $155 \times 240 \times 240$ with four channels: T1, T2, T1c, and FLAIR. In the manual label of BraTS'19, there are three tumor regions: non-enhancing tumor, enhancing tumor, and edema. The evaluation is figured out by CBICA IPP[1] online platforms. Metrics computed by the online evaluation platforms in BraTS'19 are Dice Similarity Coefficient (DSC) and the 95th percentile of the Hausdorff Distance (HD95). DSC is considered to measure the union of prediction and manual segmentation. It is measured as $DSC = \frac{2TP}{FP+2TP+FN}$ where TP, FP, and FN are the numbers of true positive, false positive, and false negative detections, respectively.

3.3 Segmentation Results on BRATS'19

Figure 3 shows examples of brain tumor prediction in LGG and HGG slides on BraTS19 along with corresponding labels, where the subject IDs are "BraTS19-TCIA10-175-1" and "BraTS19-CBICA-APK-1" for LGG and HGG, respectively. The results in Table 1 show that our method performed competitive performance on validation set (125 cases) of BraTS dataset. Results are reported in the online processing platform by BraTS'19 organizer. Moreover, Table 2 reports the average results on 335 training case of the BraTS'19.

4   Overall Survival Prediction Model

BraTS'19 dataset contains 102 gross total resections (GTR) pre-operative scans out of 335 training cases in which the age of patients is available. These subjects are applied for developing a model to predict the overall survival (OS) of the patient. To this end, we measure the volume of WT, TC, and ET after segmentation to create a feature vector to predict patient OS. We also consider patient's

---

[1] https://ipp.cbica.upenn.edu.

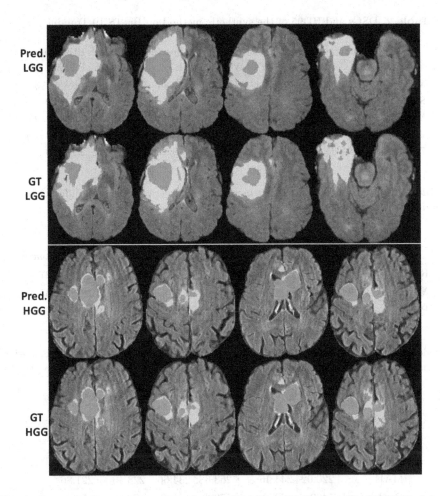

Fig. 3. Segmentation results are overlaid on FLAIR axial slices on BraTS'19 Training Data. The yellow label is edema, blue color means enhancing tumor, and the green one shows the necrotic and non-enhancing tumor core. The first and second rows illustrate LGG brain tumor, prediction (Pred.), and ground truth (GT), respectively. The third and fourth rows are related to HGG tumors. Computed DSCs by the Challenge organizer are reported for the LGG subject as: WT = 96.55% and ET% = 88.85, as well as HGG subject as: TC = 93.80%, WT = 93.97%, and ET = 95.00%. (Color figure online)

age as an input feature to increase survival prediction accuracy. Thus, we have a 4-dimensional normalized feature vector that scaled between 0 and 1. We train different regression models to predict OS through supervised machine learning, including linear models, regression trees, support vector machines (SVMs)

Table 2. DSCs and HD95 of the synthetic method on BraTS'19 Training set.

	Dice			Sensitivity			Specificity			HD95 (mm)		
	ET	WT	TC	ET	WT	TC	ET	WT	TC	ET	WT	TC
Mean	79.26	91.65	90.76	84.49	91.89	90.76	99.86	99.51	99.77	3.5	5.7	3.4
Std.	23.96	05.59	7.13	14.46	08.04	08.17	0.178	0.47	0.34	7.3	11.0	4.6
Median	87.04	93.29	92.88	88.12	94.35	93.22	99.92	99.64	99.88	1.4	2.8	2.0
25 quantile	79.49	89.89	88.34	80.69	88.99	87.96	99.831	99.37	99.74	1.0	1.8	1.4
75 quantile	91.54	95.39	95.28	93.78	97.23	96.43	99.975	99.80	99.95	2.2	4.9	3.6

Table 3. Comparison between linear models and regression trees with different hyper-parameters.

	Linear regression models				Regression trees		
	Linear	Interactions	Robust	Stepwise	Fine	Medium	Coarse
RMSE	316.81	375.23	326.76	314.07	377.46	317.35	327.95
MAE	224.24	250.04	220.04	223.36	277.04	237.8	237.38
Pred. speed	2000	6200	7800	7600	4900	19000	19000

Table 4. Comparison between different SVM kernels. Kernel scales for Gaussian (Gaus.) SVM are considered as 0.5, 2, and 8 for Fine, Medium, and Coarse, respectively.

	SVM					
	Linear	Quadratic	Cubic	Fine Gaus.	Medium Gaus.	Coarse Gaus.
RMSE	323.92	354.44	377.65	349.41	341.52	329.36
MAE	220.02	244.46	263.68	234.66	228.45	221.86
Pred. speed	5400	16000	17000	16000	17000	15000

with different kernel functions, Gaussian process regression (GPR) models, and ensembles of trees. We measure root mean square error (RMSE), maximum absolute error (MAE), and prediction speed during inference (observation/sec) to assess model performance. The 5-fold cross-validation is applied to evaluate these models with four feature vectors.

Table 3 presents linear regression models, including linear, interactions, robust, and stepwise linear models. We also evaluate regression Trees with three minimum leaf sizes, i.e., 4, 12, and 36 in this table.

Table 4 evaluates SVMs models through different Kernel functions and scales. We consider kernel scales 0.5, 2, and 8 for fine, medium, and coarse Gaussian SVM, respectively.

Table 5. Comparison between GPR and ensemble models with several kernel functions. The abbreviation is: (Exp)onential

	Gaussian process regression models				Ensemble trees	
	Squared Exp.	Matern	Exp.	Rational quadratic	Boosted	Bagged
RMSE	332.28	344.9	344.2	332.28	344.16	333.36
MAE	237.93	250.37	249.95	237.93	251.42	240.69
Pred. speed	4900	12000	13000	13000	2600	3400

Table 5 shows GPR and Ensemble Trees models. The former is evaluated with squared exponential, Matern 5/2, exponential, and rational quadratic kernel functions. The boosted Trees and the Bagged Trees are examined for the latter.

Figure 4 displays predicted response versus subject numbers in BraTS'19. The predictions are accomplished with the stepwise linear regression model.

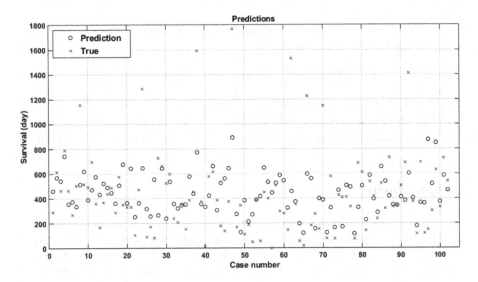

Fig. 4. Survival prediction per day through the stepwise linear regression model. The predicted results versus case number.

Figure 5 also illustrates predicted response based on three features. We removed age feature to evaluate the effect of this feature on OS task. Table 6 compare RMSE with and without age feature for survival task.

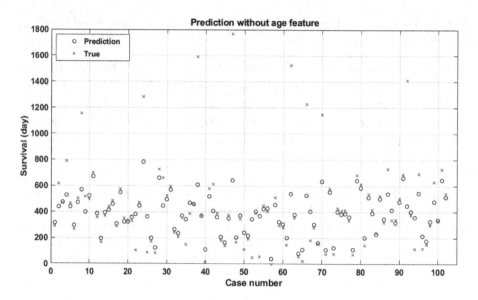

Fig. 5. Survival prediction per day through the stepwise linear regression model. The predicted results versus case number.

Table 6. RMSE with and without age feature.

Feature numbers	Linear	Regression trees	SVM	Ensemble	GPR
RMSE with age feature	314.07	317.35	323.92	333.36	332.26
RMSE without age feature	357.96	361.45	351.99	362.62	352.53

5   Conclusion

This paper provided a framework for the synthetic segmentation that translated FLAIR MR images into high-contrast synthetic MR ones for segmentation. Synthesizing based on the GAN network empowers our model to decrease the number of real channels in multimodal brain tumor segmentation challenge 2019. We also implemented several regression models to predict the OS of each patient. We found that the stepwise linear model overwhelmed other traditional regression models in terms of RMSE. We also observed that patient age as a distinctive feature in the OS prediction tasks.

Acknowledgment. This work was supported partly by National Natural Science Foundation of China (Nos. 61871274, 61801305, and 81571758), National Natural Science Foundation of Guangdong Province (No. 2017A030313377), Guangdong Pearl River Talents Plan (2016ZT06S220), Shenzhen Peacock Plan (Nos. KQTD2016053112 051497 and KQTD2015033016 104926), and Shenzhen Key Basic Research Project (Nos. JCYJ20170413152804728, JCYJ20180507184647636, JCYJ20170818142347 251, and JCYJ20170818094109846).

References

1. Bakas, S., et al.: Segmentation labels and radiomic features for the pre-operative scans of the TCGA-GBM collection. The Cancer Imaging Archive (2017). https://doi.org/10.7937/K9/TCIA.2017.KLXWJJ1Q
2. Bakas, S., et al.: Segmentation labels and radiomic features for the pre-operative scans of the TCGA-LGG collection. The Cancer Imaging Archive (2017). https://doi.org/10.7937/K9/TCIA.2017.GJQ7R0EF
3. Bakas, S., et al.: Advancing the cancer genome atlas glioma MRI collections with expert segmentation labels and radiomic features. Nat. Sci. Data **4**, 170117 (2017). https://doi.org/10.1038/sdata.2017.117
4. Bakas, S., et al.: Identifying the best machine learning algorithms for brain tumor segmentation, progression assessment, and overall survival prediction in the BRATS challenge. arXiv preprint arXiv:1811.02629 (2018)
5. Chen, L., Papandreou, G., Kokkinos, I., Murphy, K., Yuille, A.L.: DeepLab: semantic image segmentation with deep convolutional nets, atrous convolution, and fully connected CRFs. IEEE Trans. Pattern Anal. Mach. Intell. **40**(4), 834–848 (2018)
6. Goodfellow, I., et al.: Generative adversarial nets. In: Advances in Neural Information Processing Systems, pp. 2672–2680 (2014)
7. Hamghalam, M., Ayatollahi, A.: Automatic counting of leukocytes in Giemsa-stained images of peripheral blood smear. In: 2009 International Conference on Digital Image Processing, pp. 13–16 (2009). https://doi.org/10.1109/ICDIP.2009.9
8. Hamghalam, M., Motameni, M., Kelishomi, A.E.: Leukocyte segmentation in Giemsa-stained image of peripheral blood smears based on active contour. In: 2009 International Conference on Signal Processing Systems, pp. 103–106 (2009). https://doi.org/10.1109/ICSPS.2009.36
9. Harley, A.W., Derpanis, K.G., Kokkinos, I.: Segmentation-aware convolutional networks using local attention masks. In: 2017 IEEE International Conference on Computer Vision (ICCV), pp. 5048–5057 (2017)
10. Hatami, T., et al.: A machine learning approach to brain tumors segmentation using adaptive random forest algorithm. In: 2019 5th Conference on Knowledge Based Engineering and Innovation (KBEI), pp. 076–082 (2019). https://doi.org/10.1109/KBEI.2019.8735072
11. Isola, P., Zhu, J., Zhou, T., Efros, A.A.: Image-to-image translation with conditional adversarial networks. In: 2017 IEEE Conference on Computer Vision and Pattern Recognition (CVPR), pp. 5967–5976 (2017)
12. Long, J., Shelhamer, E., Darrell, T.: Fully convolutional networks for semantic segmentation. In: 2015 IEEE Conference on Computer Vision and Pattern Recognition (CVPR), pp. 3431–3440 (2015)
13. Menze, B.H., et al.: The multimodal brain tumor image segmentation benchmark (BRATS). IEEE Trans. Med. Imaging **34**(10), 1993–2024 (2015). https://doi.org/10.1109/TMI.2014.2377694
14. Najrabi, D., et al.: Diagnosis of astrocytoma and globalastom using machine vision. In: 2018 6th Iranian Joint Congress on Fuzzy and Intelligent Systems (CFIS), pp. 152–155 (2018). https://doi.org/10.1109/CFIS.2018.8336661
15. Ronneberger, O., Fischer, P., Brox, T.: U-Net: convolutional networks for biomedical image segmentation. In: Navab, N., Hornegger, J., Wells, W.M., Frangi, A.F. (eds.) MICCAI 2015. LNCS, vol. 9351, pp. 234–241. Springer, Cham (2015). https://doi.org/10.1007/978-3-319-24574-4_28

16. Roy, A.G., Conjeti, S., Sheet, D., Katouzian, A., Navab, N., Wachinger, C.: Error corrective boosting for learning fully convolutional networks with limited data. In: Descoteaux, M., Maier-Hein, L., Franz, A., Jannin, P., Collins, D.L., Duchesne, S. (eds.) MICCAI 2017. LNCS, vol. 10435, pp. 231–239. Springer, Cham (2017). https://doi.org/10.1007/978-3-319-66179-7_27
17. Soleimany, S., et al.: A novel random-valued impulse noise detector based on MLP neural network classifier. In: 2017 Artificial Intelligence and Robotics (IRA-NOPEN), pp. 165–169 (2017). https://doi.org/10.1109/RIOS.2017.7956461
18. Soleymanifard, M., Hamghalam, M.: Segmentation of whole tumor using localized active contour and trained neural network in boundaries. In: 2019 5th Conference on Knowledge Based Engineering and Innovation (KBEI), pp. 739–744 (2019). https://doi.org/10.1109/KBEI.2019.8735050
19. Zeiler, M.D.: ADADELTA: an adaptive learning rate method. CoRR abs/1212.5701 (2012)

Multi-step Cascaded Networks for Brain Tumor Segmentation

Xiangyu Li, Gongning Luo, and Kuanquan Wang[✉]

Harbin Institute of Technology, Harbin, China
wangkq@hit.edu.cn

Abstract. Automatic brain tumor segmentation method plays an extremely important role in the whole process of brain tumor diagnosis and treatment. In this paper, we propose a multi-step cascaded network which takes the hierarchical topology of the brain tumor substructures into consideration and segments the substructures from coarse to fine. During segmentation, the result of the former step is utilized as the prior information for the next step to guide the finer segmentation process. The whole network is trained in an end-to-end fashion. Besides, to alleviate the gradient vanishing issue and reduce overfitting, we added several auxiliary outputs as a kind of deep supervision for each step and introduced several data augmentation strategies, respectively, which proved to be quite efficient for brain tumor segmentation. Lastly, focal loss is utilized to solve the problem of remarkably imbalance of the tumor regions and background. Our model is tested on the BraTS 2019 validation dataset, the preliminary results of mean dice coefficients are 0.886, 0.813, 0.771 for the whole tumor, tumor core and enhancing tumor respectively. Code is available at https://github.com/JohnleeHIT/Brats2019.

Keywords: Brain tumor · Cascaded network · 3D-UNet · Segmentation

1 Introduction

Brain tumor is one of the most serious brain diseases, among which the malignant gliomas are the most frequent occurred type. The gliomas can be simply divided into two categories according to the severity: the aggressive one (i.e. HGG) with the average life expectancy of nearly 2 years and the moderate one (i.e. LGG) with the life expectancy of several years. Due to the considerably high mortality rate, it is of great importance for the early diagnosis of the gliomas, which largely improves the treatment probabilities especially for the LGG. At present, the most possible ways to treat gliomas are surgery, chemotherapy and radiotherapy. For any of the treatment strategies, accurate imaging and segmentation of the lesion areas are indispensable before and after treatment so as to evaluate the effectiveness of the specific strategy.

Among all the existing imaging instruments, MRI has been the first choice for brain tumor analysis for its high resolution, high contrast and present no

A. Crimi and S. Bakas (Eds.): BrainLes 2019, LNCS 11992, pp. 163–173, 2020.
https://doi.org/10.1007/978-3-030-46640-4_16

known health threats. In the current clinical routine, manual segmentation of large amount of MRI images is a common practice which turns out to be remarkably time-consuming and prone to make mistakes for the raters. So, it would be of tremendous potential value to propose an automatic segmentation method. Many researchers have proposed several effective methods based on deep learning or machine learning methods to solve the problem. Among those proposed methods, Zikic et al. [1] used a shallow CNN network to classify 2D image patches which captured from the MRI data volumes in a sliding window fashion. Zhao et al. [2] converted the 3D tumor segmentation task to 2D segmentation in triplanes and introduced multi-scales by cropping different patch sizes. Havaei et al. [3] proposed a cascaded convolutional network, which can capture local and global information simultaneously. Çiçek et al. [4] extended the traditional 2D U-net segmentation network to a 3D implementation which makes the volume segmentation to a voxel-wise fashion. Kamnitsas et al. [5] proposed a dual pathway 3D convolution network named DeepMedic to incorporate multi-scale contextual information, and used the 3D fully connected CRF as the postprocess method to refine the segmentation result. Chen et al. [6] improved DeepMedic by first cropping 3D patches from multiple layers selected from the original DeepMedic and then merging those patches to learn more information in the network, besides, deep supervision was introduced in the network to better propagate the gradient. Ma et al. [7] employed a feature representations learning strategy to effectively explore both local and contextual information from multimodal images for tissue segmentation by using modality specific random forests as the feature learning kernels.

Inspired by Havaei and Çiçek, we proposed a multi-step cascaded network to segment brain tumor substructures. The proposed network uses 3D U-net as the basic segmentation architecture and the whole network works in a coarse-to-fine fashion which can be seen as a kind of spatial attention mechanism.

2   Methodology

Based on the thorough analysis of the substructures of brain tumor, which turns out to be a hierarchical topology (see Fig. 1), We propose a multi-step cascaded network which is tailored for the brain tumor segmentation task. Our proposed method mainly contains three aspects, detailed information are as follows:

2.1   Multi-step Cascaded Network

The proposed multi-step cascaded networks are illustrated in Fig. 2. This method segments the hierarchical structure of the tumor substructures in a coarse-to-fine fashion. In the first step, in order to be consistent with the manual annotations protocol which are detailed descripted in [8], two modalities (Flair&T1ce) of the MRI tumor volumes are utilized. The two-channel data volumes are then fed into the first segmentation network to coarsely segment the whole tumor (WT) which contains all the substructures of the brain tumor; In the second step, similarly, we choose T1ce modality as the data source to segment the tumor

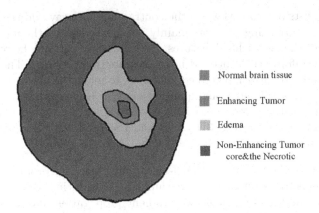

Fig. 1. Schematic diagram of the tumor structures

core (TC) structure. Besides, the result of the first coarse step can be utilized as the prior information for the second step. By multiplying the mask generated in the first step with the T1ce data volume, the second segmentation network will concentrate more on the corresponding masked areas and make it easier to segment the TC structure. Then the masked volumes are processed by the second network, as a result, TC structure (foreground) are introduced. In the last and finest step, by following the same strategies, we can also get the enhancing tumor (ET) substructures from the data volume, and finally by combining the results of the three steps, the final segmentation maps of the brain tumor will be received.

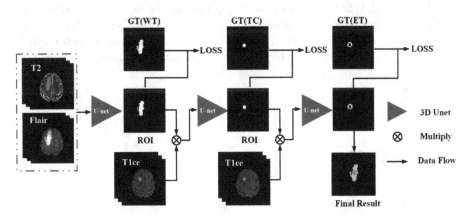

Fig. 2. Overview of the proposed multi-step cascaded network

2.2 3D U-Net Architecture with Deep Supervisions

We take a variant of 3D U-net as the basic segmentation architecture in our multi-step cascaded networks, which is illustrated in Fig. 3. The typical

3D U-net consists of two pathways: the contracting pathway and the expanding pathway. The contracting pathway mainly intends to encode the input volumes and introduces the hierarchical features, the expanding pathway however is used to decode the information encoded in the contracting pathway. The two pathways are connected with skip connections so as to make the network be capable of capturing both local and global information. Our basic segmentation network takes 3D U-net as the prototype, whilst makes some improvements on top of it. The main differences between 3D U-net and the proposed basic segmentation networks are as follows:

(1) Compared to the traditional 3D U-net architecture, our proposed basic segmentation network introduces three auxiliary outputs in the expanding pathway with the intention of better gradient propagation and decreasing the probabilities of vanishing gradient for the relatively deep segmentation networks. As a result, we need to minimize the overall loss functions which comprise both the main branch and the auxiliary loss functions for the basic segmentation process.

(2) We introduce the focal loss [9] as the loss function for the whole training process with the intention of alleviating the considerably imbalance of the positive and negative samples in the training data. The focal loss can be expressed as follows:

$$\text{FL}\,(p_t) = -\alpha_t\,(1 - p_t)^\gamma \log\,(p_t) \tag{1}$$

$$p_t = \begin{cases} p & \text{if } y = 1 \\ 1 - p & \text{otherwise} \end{cases} \tag{2}$$

where $p \in [0, 1]$ is the model's estimated probability for the class with label $y = 1$. $\gamma \geqslant 0$ refers to focusing parameter, it smoothly adjusts the rate at which easy examples are down weighted. α_t refers to balancing factor which balance the importance of the positive and negative samples.

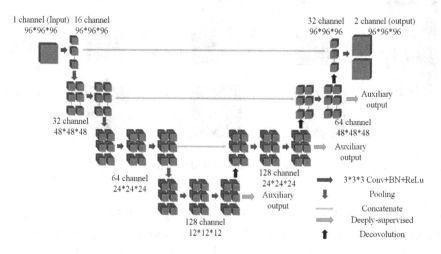

Fig. 3. Schematic of the 3D U-net architecture with deep supervisions

3 Experiments and Results

3.1 Preprocessing

In this paper, we take BraTS 2019 dataset [10–13] as the training data, which comprises 259 HGG and 76 LGG MRI volumes with four modalities (T1, T2, T1ce and Flair) available. According to the official statement of the dataset, all the datasets have been segmented manually following the same annotation protocol. Besides, some preprocessing operations have also been conducted on those datasets, for example, all the MRI volumes have been co-registered to the same anatomical template, interpolated to the same resolution and skull-stripped. Nevertheless, extra preprocessing steps should be done to the raw dataset due to the existence of the intensity nonuniformity in the image data, also called bias field which comes from the imperfect of the MRI machine and the specificity of the patients. This kind of intensity nonuniformity or bias field considerably affects the training process. To eliminate the bias field effect, a great deal of correction methods have been proposed. Among the proposed bias field correction method, the most effective one is the N4 bias field correction [14]. In this paper, N4 bias field correction method is utilized as an important preprocessing step before the segmentation process. At last, we also use the normalization method to normalize all the data to zero mean with unit variance.

3.2 Implementation Details

We mixed all the data in the BraTS 2019 training dataset including HGG and LGG, and then trained our model with the mixed dataset. During training, we first extract the brain region from the volume by getting the largest rectangle which contains the brain. Then we randomly crop the raw data volume to sub-volumes due to memory limitation and we choose the size of the patches as 96*96*96 empirically. We take one patch from a patient's data volume every iteration in the training process. While in the testing phase, for a single data volume we get the sub-volumes in order so as to rebuild the whole volume with those predictions and the patch size is the same as is in the training process. We get different number of patches for each patient data because the brain regions we extracted from the volume are distinct. To reduce overfitting, we introduced some data augmentation methods, for instance, rotating a random angle, flipping horizontally and vertically, and adding guassion blur to the sub-volumes with a certain probability. It turned out that the data augmentation was significant important for the brain tumor segmentation task because the network is prone to be overfitting with relatively less training data. We used Adam optimizer to update the weights of the network. The initial learning rate was set to 0.001 at the very beginning and decayed to 0.0005 when the loss curve plateaued. The batch size was set to 1 in the whole training process.

Our model was trained on a Nvidia RTX 2080 Ti GPU for 50 epochs, which takes around 13 h.

3.3   Segmentation Results

To evaluate our proposed mothed, we tested our algorithm on both training
and validation set by uploading the inference results to the online evaluation
platform (CBICB's IPP), we finally got the evaluation results including Dice
sore, Hausdorff distance, sensitivity and specificity for the whole tumor (WT),
the tumor core (TC) and the enhancing tumor (ET), respectively. The metrics
aforementioned are defined as follows:

$$\text{Dice}(P,T) = \frac{|P_1 \wedge T_1|}{(|P_1| + |T_1|)/2} \tag{3}$$

$$\text{Sensitivity}(P,T) = \frac{|P_1 \wedge T_1|}{|T_1|} \tag{4}$$

$$\text{Specificity}(P,T) = \frac{|P_0 \wedge T_0|}{|T_0|} \tag{5}$$

$$\text{Haus}(P,T) = \max \left\{ \sup_{p \in \partial P_1} \inf_{t \in \partial T_1} d(p,t), \sup_{t \in \partial T_1} \inf_{p \in \partial P_1} d(t,p) \right\} \tag{6}$$

where P refers to the prediction map of the algorithm, and T is the groundtruth
label segmented manually by the experts. $\wedge$ is the logical AND operator, $|\cdot|$ means
the number of voxels in the set, and P_1, P_0 represent the postive and negative
voxels in the prediction map, respectively, and T_1, T_0 denote the positive and
negative voxels in the groundtruth map, respectively. $d(p,t)$ denotes the distance
of the two points p, t. ∂P_1 is the surface of the prediction volume set P_1 and ∂T_1
is the surface of the groundtruth label set T_1.

Table 1 presents the quantitative average results on both training and validation dataset. Not surprisingly, the dice coefficient and sensitivity of the whole
tumor, the tumor core and the enhancing tumor are in a descending order for
both datasets due to the ascending difficulties for those tasks. However, there
still exists small gaps for the evalutation metrics between training and validation
dataset which attributed to the overfitting problem.

Table 1. Quantitative average results on the training and validation dataset

Dataset	Label	Dice	Sensitivity	Specificity	Hausdorff distance
Training	WT	0.915	**0.942**	0.993	4.914
	TC	0.832	0.876	0.996	6.469
	ET	0.791	0.870	0.997	6.036
Validation	WT	0.886	**0.921**	0.992	6.232
	TC	0.813	**0.819**	0.997	7.409
	ET	0.771	**0.802**	0.998	6.033

To better analysis the overall performance of the proposed algorithm, we made the boxplot of all the validation and training results, which can be seen from Fig. 4. It is evident that the proposed method can segment well on almost all the volumes in both datasets except for a few outliers. Besides, by comparing the boxplot of the validation and training dataset, we noticed that the variance of all the evaluation metrics including dice coefficient, sensitivity, specificity and hausdorff distance for the validation dataset is larger than those for training dataset, which means that our method still suffers from the overfitting problem to some extent. Finally, we can see from the 4 subgraphs that the variance of dice coefficient for the whole tumor is smaller than both tumor core and enhancing tumor substructures for both training and validation datasets, the same for sensitivity and hausdorff distance metrics and the opposite for the specificity metrics, which are in line with our expectations. However, what surprise us most is that the variance of the tumor core (TC) is larger than that of enhancing tumor core (ET) on most metrics for the two datasets, the most possible explanation of the fact is that the network sometimes predicts the whole tumor as the tumor core mistakenly with the impact of the LGG tumor samples, which increases the variations sharply.

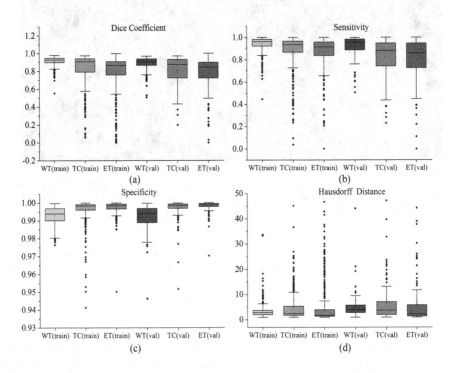

Fig. 4. Boxplot of the overall performance on both training and validation datasets

Qualitative analysis of the segmentation results for the HGG and LGG tumors are also introduced, which can be seen from Fig. 5 and Fig. 6, respectively.

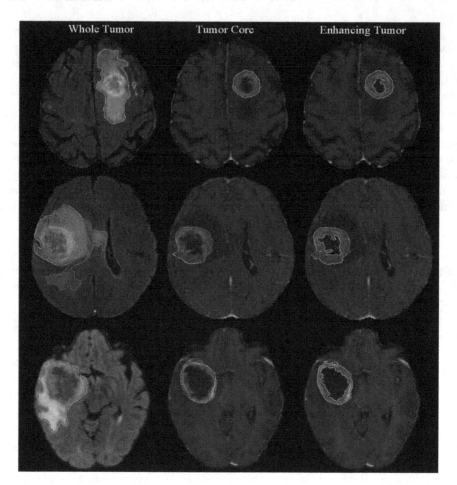

Fig. 5. Segmentation result of the whole tumor (WT), Tumor core (TC) and Enhancing tumor (ET) structures for HGG tumors, each shows the ground truth label (The blue line) and the prediction result (The red line) (Color figure online)

The left row are the flair modality images with the whole tumor ground truth and the prediction result, demonstrated in blue and red curves respectively. The middle row are the T1ce modality images with the tumor core ground truth and the prediction result which are illustrated in the same way as the left row. The right row of course focus on the remaining substructure, i.e. the enhancing tumor.

All of the three regions with great clinical concerns have been well segmented except for some small details. Not surprisingly, our aforementioned guess about the difficulties of the three tasks can be verified again from the visualization result. Specifically, from step one to step three, the task becomes tougher because the contrast between the tumor region and the surrounding background decreases and the segmentation substructures contours become much rougher at the same time.

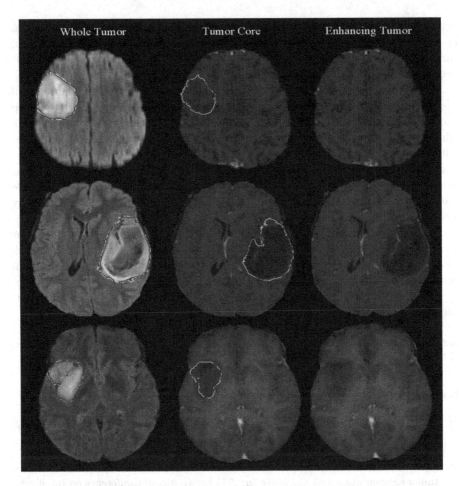

Fig. 6. Segmentation result of the whole tumor (WT), Tumor core (TC) and Enhancing tumor (ET) structures for LGG tumors, each shows the ground truth label (The blue line) and the prediction result (The red line) (Color figure online)

4   Discussing and Conclusion

By visualizing all the validation results, we find it interesting that plenty of bad segmented cases for the tumor core regions are those who mistaken the whole tumor as the tumor core region. The most possible explanation might be the variations between different MRI volumes despite the same modality. So, it is likely that the results would increase if some preprocessing methods which can decrease those variations have been taken before the training process, e.g. histogram equalization.

Besides, we also tried the curriculum learning strategy which trained the network step by step instead of end-to-end training, it turns out that the results are no better than the end-to-end training ones. That is most likely because the network can fit the training data better if all the parameters in the network can be updated. Lastly, we tried to weight the three steps of the cascaded network, surprisingly, we find that the final results present no big difference for increment, decrement or even weights of the training steps.

In conclusion, we present a very efficient multi-step network to segment all the tumor substructures. We first choose specific modalities for each step to keep the automatic segmentation process to be consistent with the manual protocol which improves our result a lot compared to the method to use all the modalities. After that, we preprocess the input volumes with N4 bias field correction and normalization. Due to the memory limitation, we randomly crop volume patches from the original data and introduce data augmentation on those patches, We find the data augmentation is quite important for reducing overfitting especially when the training data is scarce.

At last, the training patches are trained in the multi-step network which has proved to be more effective than the one-step couterpart as it trains the network in a coarse-to-fine fashion and seperates the tough multi-classification problem to three much easier binary-classification issuse.

We evaluated the proposed mothod on the BraTS 2019 validation dataset, the results show that our method performance well on all three substructures.

Acknowledgments. This work was supported by the National Key R&D Program of China under Grant 2017YFC0113000.

References

1. Zikic, D., Ioannou, Y., Brown, M., Criminisi, A.: Segmentation of brain tumor tissues with convolutional neural networks. In: Proceedings MICCAI-BRATS, pp. 36–39 (2014)
2. Zhao, L., Jia, K.: Deep feature learning with discrimination mechanism for brain tumor segmentation and diagnosis. In: 2015 International Conference on Intelligent Information Hiding and Multimedia Signal Processing (IIH-MSP), pp. 306–309. IEEE (2015)
3. Havaei, M., et al.: Brain tumor segmentation with deep neural networks. Med. Image Anal. **35**, 18–31 (2017)
4. Çiçek, Ö., Abdulkadir, A., Lienkamp, S.S., Brox, T., Ronneberger, O.: 3D U-Net: learning dense volumetric segmentation from sparse annotation. In: Ourselin, S., Joskowicz, L., Sabuncu, M.R., Unal, G., Wells, W. (eds.) MICCAI 2016. LNCS, vol. 9901, pp. 424–432. Springer, Cham (2016). https://doi.org/10.1007/978-3-319-46723-8_49
5. Kamnitsas, K., et al.: Efficient multi-scale 3D CNN with fully connected CRF for accurate brain lesion segmentation. Med. Image Anal. **36**, 61–78 (2017)
6. Chen, S., Ding, C., Liu, M.: Dual-force convolutional neural networks for accurate brain tumor segmentation. Pattern Recogn. **88**, 90–100 (2019)
7. Ma, C., Luo, G., Wang, K.: Concatenated and connected random forests with multiscale patch driven active contour model for automated brain tumor segmentation of MR images. IEEE Trans. Med. Imaging **37**(8), 1943–1954 (2018)
8. Menze, B.H., et al.: The multimodal brain tumor image segmentation benchmark (BRATS). IEEE Trans. Med. Imaging **34**(10), 1993–2024 (2014)

9. Lin, T.Y., Goyal, P., Girshick, R., He, K., Dollár, P.: Focal loss for dense object detection. In: Proceedings of the IEEE International Conference on Computer Vision, pp. 2980–2988 (2017)
10. Bakas, S., et al.: Advancing The Cancer Genome Atlas glioma MRI collections with expert segmentation labels and radiomic features. Sci. Data **4**, 170117 (2017)
11. Bakas, S., et al.: Identifying the best machine learning algorithms for brain tumor segmentation, progression assessment, and overall survival prediction in the BRATS challenge. arXiv preprint arXiv:1811.02629 (2018)
12. Bakas, S., et al.: Segmentation labels and radiomic features for the pre-operative scans of the TCGA-LGG collection. Cancer Imaging Arch. **286** (2017)
13. Bakas, S., et al.: Segmentation labels and radiomic features for the pre-operative scans of the TCGA-GBM collection. Cancer Imaging Arch. (2017)
14. Tustison, N.J., et al.: N4ITK: improved N3 bias correction. IEEE Trans. Med. Imaging **29**(6), 1310 (2010)

TuNet: End-to-End Hierarchical Brain Tumor Segmentation Using Cascaded Networks

Minh H. Vu$^{(\boxtimes)}$, Tufve Nyholm, and Tommy Löfstedt

Department of Radiation Sciences, Umeå University, Umeå, Sweden
minh.vu@umu.se

Abstract. Glioma is one of the most common types of brain tumors; it arises in the glial cells in the human brain and in the spinal cord. In addition to having a high mortality rate, glioma treatment is also very expensive. Hence, automatic and accurate segmentation and measurement from the early stages are critical in order to prolong the survival rates of the patients and to reduce the costs of the treatment. In the present work, we propose a novel end-to-end cascaded network for semantic segmentation in the Brain Tumors in Multimodal Magnetic Resonance Imaging Challenge 2019 that utilizes the hierarchical structure of the tumor sub-regions with ResNet-like blocks and Squeeze-and-Excitation modules after each convolution and concatenation block. By utilizing cross-validation, an average ensemble technique, and a simple post-processing technique, we obtained dice scores of 88.06, 80.84, and 80.29, and Hausdorff Distances (95th percentile) of 6.10, 5.17, and 2.21 for the whole tumor, tumor core, and enhancing tumor, respectively, on the online test set. The proposed method was ranked among the top in the task of Quantification of Uncertainty in Segmentation.

1 Introduction

Glioma is among the most aggressive and dangerous types of cancer [11], leading, for instance, to around 80% and 75% of all malignant brain tumors diagnosed in the United States [8] and Sweden [1], respectively. Gliomas with different prognosis and numerous heterogeneous histological sub-regions, such as edema, necrotic core, and enhancing and non-enhancing tumor core, are classified into four world health organisation (WHO) grades according to their aggressiveness: low grade glioma (LGG) (class I and II, considered as slow-growing), and high grade glioma (HGG) (class III and IV, considered as fast-growing).

The aim of the Brain Tumors in Multimodal Magnetic Resonance Imaging Challenge 2019 (BraTS19) [2–5,16] is in evaluating and finding new state-of-the-art methods for tumor segmentation and to set a gold standard for the segmentation of intrinsically heterogeneous brain tumors. BraTS19 provided a large data set comprising multi-institutional pre-operative magnetic resonance imaging (MRI) scans with four sequences: post-contrast T1-weighted (T1c),

© Springer Nature Switzerland AG 2020
A. Crimi and S. Bakas (Eds.): BrainLes 2019, LNCS 11992, pp. 174–186, 2020.
https://doi.org/10.1007/978-3-030-46640-4_17

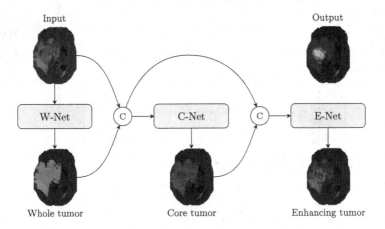

Fig. 1. Schematic visualization of TuNet. The input is the MRI volumes as four different channels, while the output is the predicted masks for the three labels: the necrotic and non-enhancing tumor core (NCR/NET—label 1, green), the peritumoral edema (ED—label 2, yellow) and the GD-enhancing tumor (ET—label 3, red). First, the multi-modal input is fed into the W-Net to generate a probability map of the whole tumor region (orange) including labels 1, 2 and 3. Second, the concatenation of the whole tumor probability map and the input is passed through the C-Net to produce the tumor core probability map (blue) including labels 1 and 3. Third, the two obtained maps from W-Net and C-Net are concatenated with the multi-modal input and then fed into the E-Net to generate an enhancing tumor probability map including label 3. Last, the outputs of W-Net, C-Net, and E-Net are merged to produce the final brain tumor mask. (Color figure online)

T2-weighted (T2w), T1-weighted (T1w), and T2 Fluid Attenuated Inversion Recovery (FLAIR). Masks were annotated manually by one to four raters followed by improvements by expert raters. The segmentation performance of the participants was measured using Sørensen-Dice coefficient (DSC), sensitivity, specificity, and 95th percentile of the Hausdor distance (HD95).

Traditional discriminative approaches, such as Support-vector Machines [6], have been widely used in medical image segmentation. In recent years, Convolutional Neural Networks (CNNs) have achieved state-of-the-art performance in numerous computer vision tasks. In the field of medical image segmentation, a fully convolutional encoder-decoder neural network named U-Net, introduced by Ronneberger et al. [18], has received a lot of attention in recent years. With its success, it is unsurprising that the U-Net has motivated many top-ranking teams in segmentation competitions in previous years [7,13,17,19].

Kitrungrotsakul et al. [15] presented CasDetNet-CLSTM to detect mitotic events in 4D microscopic images by introducing a connection between a region-based convolutional neural network and convolutional long short-term memory. In the BraTS18, instead of proposing a new network, Isensee et al. [13] focused on the training process, and made only minor modifications to the U-Net, used additional training data, and applied a simple post-processing technique. Myro-

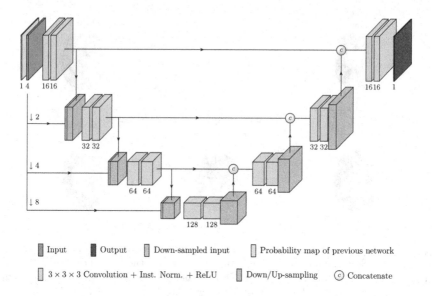

Fig. 2. Our C-Net and E-Net with a U-Net backbone. The W-Net does not include the probability map produced by a previous sub-network. To enrich the feature map at each level in the encoder part, we concatenate the down-sampled original input (light green) with the down-sampled output (grey) of the convolution block (yellow). Each convolution block comprises a convolution operation, an instance normalization followed by a RELU activation function. The W-Net constrains the C-Net, while the C-Net constrains the E-Net. The E-Net functions as a regularizer for the C-net, while the C-Net plays the same role for the W-Net. (Color figure online)

nenko [17] was the winner of BraTS18 by introducing a U-Net-like network with an asymmetrically large encoder and a Variational Auto-Encoder (VAE) branch to reconstruct the input image to add guidance and regularization to the encoder part of the network.

In another work, Wang *et al.* [19] proposed an encoder-decoder cascaded anisotropic CNN, that won the second place in the BraTS17, that hierarchically segments the whole tumor, tumor core, and enhancing tumor core sequentially by separating the complex brain tumor segmentation problem into three binary segmentation problems. We argue that: (1) the training process employed in [19] might be time-consuming as there are three separate binary segmentation problems, and (2) the lack of regularization could lead to overfitting.

Motivated by the successes of the cascaded anisotropic network, introduced in [19], and the VAE branch, presented in [17], we propose a novel architecture, denoted End-to-end Hierarchical Tumor Segmentation using Cascaded Networks (TuNet). TuNet exploits separating a complex problem into less challenging sub-problems, and also attempts to avoid overfitting by using heavy regularization through multi-task learning. In the present work, we constructed three U-Net-like networks, each was used to segment a specific tumor region, *i.e.* whole tumor, core tumor, and enhancing tumor. We connected the three sub-networks to form an end-to-end cascaded network in the hope that three sub-networks could mutually regularize each other to prevent overfitting and reduce the training time.

2   Methods

Motivated by the drawbacks in [19] and the VAE branch in [17], we thus propose a framework that we expect not only utilizes region-based optimization, but we also hypothesis that it will prevent or reduce overfitting.

2.1   Cascaded Architecture

Figure 1 illustrates the proposed end-to-end cascaded architecture. As in [19], we employed three encoder-decoder networks to cope with the three aforementioned tumor regions. We denote them W-Net (whole tumor network), C-Net (core tumor network), and E-Net (enhancing tumor network). In our proposed approach, instead of using three separate networks, we joined them together to form an end-to-end cascaded network.

2.2   Segmentation Network

Figure 2 shows the proposed symmetric encoder-decoder C-Net and E-Net. The patch size was set to $80 \times 96 \times 64$. Note that the probability map (pink) was concatenated with the input image as a first step for the C-Net and the E-Net, but not for the W-Net. We also concatenated the max-pooled feature maps with the down-sampled input (light green) in order to enhance the feature maps at the beginning of each level in the encoder part. The base number of filters was set to 16 and the number of filters were doubled at each level. Skip-connections were used like in the U-Net. The probability maps produced at the end of the decoder part of each sub-network had the same spatial size as the original image and were activated by a logistic sigmoid function.

2.3   ResNet Block

He *et al.* [9] proposed the ResNet as a way to make training deep networks easier, and won the 1st place in the ImageNet Large Scale Visual Recognition Challenge (ILSVRC) 2015 classification task. It has been shown that ResNet-like networks are not only easier to optimize but also boost accuracy from considerably deeper networks [9,17]. Inspired by the winner of BraTS18 [17], we replaced the convolution blocks used in the traditional U-Net by ResNet blocks in two of our models (see Fig. 2 and Table 1). An illustration of the ResNet-like block can be seen in Fig. 3.

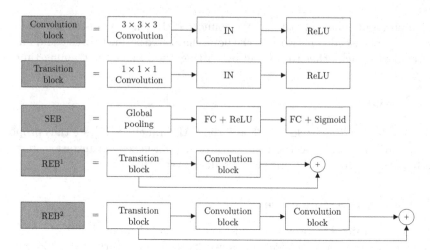

Fig. 3. The convolution blocks used in the different experiments (see Table 1). SEB, REB[1] and REB[2] denote Squeeze-and-Excitation block and two variations of ResNet blocks, respectively. Here, IN and FC stand for instance normalization and fully-connected layer, respectively; while rectified linear unit (ReLU) and Sigmoid are activation functions.

2.4 Squeeze-and-Excitation Block

We added a Squeeze-and-Excitation block (SEB) as developed by Hu *et al.* [12] after each convolution and concatenation block. The SEB is a computationally efficient means to incorporate channel-wise inter-dependencies, and has been widely used to improve network performances by significant margins [12]. SEB is also illustrated in Fig. 3.

2.5 Preprocessing and Augmentation

We normalized all input images to have mean zero and unit variance. To increase the data set size, we employed simple on-the-fly data augmentation by randomly rotating the images within a range of −1 to 1 degrees and random mirror flips (on the x-axis) with a probability of 0.5. We also experimented with median denoising, but it did not demonstrate any additional improvements.

2.6 Post-processing

One of the most difficult tasks of BraTS19 is to detect small vessels in the tumor core and to label them as edema or as necrosis. To cope with the fact that LGG patients may have no enhancing tumor region, Isensee *et al.* [13] proposed to replace all enhancing tumor regions with less than 500 voxels by necrosis. We employed that approach in the present work as well. Additionally, we kept decreasing the threshold segmentation from 0.5 to 0.3, 0.1, 0.05, and 0.01, respectively, if no core tumor region was found. This technique resolved several cases where the networks failed to distinguish between core and whole tumor.

2.7 Ensemble of Multiple Models

We employed five-fold cross-validation when training on the 335 cases (259 HGG + 76 LGG) of BraTS19, and did not use any additional in-house data. We trained 30 models in total and used 15 of them, that correspond to TuNet + SEB, TuNet + REB^1 + SEB and TuNet + REB^2 + SEB, to cast votes when predicting the final label maps. We used an average ensemble approach, such that

$$p_c = \frac{1}{M} \sum_{m=1}^{M} f_{mc}, \tag{1}$$

where $p_c \in \mathbb{R}^{|C|}$ and $f_{mc} \in \mathbb{R}^{|C|}$ denote the final probability of label c and the probability of label c generated by model $m = 1, \ldots, M$ at an arbitrary voxel, respectively. Here, $C = \{1, 2, 3\}$ is the set of tumor labels, and thus $c \in C$.

The ensemble that we used was computed as the average of the prediction outputs of TuNet + SEB, TuNet + REB^1 + SEB and TuNet + REB^2 + SEB (see Table 1 and Table 2).

2.8 Task 3: Quantification of Uncertainty in Segmentation

In BraTS19, the organizers decided to include a new task that focuses on exploring uncertainty measures in the context of tumor segmentation on three glioma regions: whole, core, and enhancing. This task, called "Quantification of Uncertainty in Segmentation", aimed at rewarding participating methods with resulting predictions that are: (a) confident when correct and (b) uncertain when incorrect. Participants were called on to generate uncertainty maps associated with the resulting labels at every voxel with values in $[0, 100]$, where 0 represents the most certain prediction and 100 represents the most uncertain.

Our proposed approach, TuNet, benefits from this task since, fundamentally, it segments the brain tumor MRI images into three aforementioned tumor regions instead of partitioning into three labels. We define an uncertainty score, $u_{i,j,k}^r$, at voxel (i, j, k) as:

$$u_{i,j,k}^r = \begin{cases} 200(1 - p_{i,j,k}^r) & \text{if } p_{i,j,k}^r \geq 0.5 \\ 200 p_{i,j,k}^r & \text{if } p_{i,j,k}^r < 0.5 \end{cases} \tag{2}$$

where $u_{i,j,k}^r \in [0, 100]^{|\mathcal{R}|}$ and $p_{i,j,k}^r \in [0, 1]^{|\mathcal{R}|}$ denote the uncertainty score map and probability map (the network's likelihood outputs) corresponding to tumor region, $r \in \mathcal{R}$, as produced by the TuNet (see Fig. 1), where $\mathcal{R}$ is the set of tumor regions, i.e. whole, core, and enhancing region.

3 Experiments

We implemented our network in Keras 2.2.4[1] using TensorFlow 1.12.0[2] as the backend. This research was conducted using the resources of the High Performance Computing Center North (HPC2N)[3] at Umeå University, Umeå, Sweden, and the experiments were run on NVIDIA Tesla V100 16 GB GPUs.

We report the results on the BraTS19 training set (335 cases) using cross-validation and on the validation set (125 cases) by uploading our predicted masks and corresponding uncertainty maps to the evaluation server. The evaluation metrics of Task 1—Segmentation included: DSC, sensitivity, specificity, and HD95; while the evaluation metrics of Task 3—Quantification of Uncertainty in Segmentation were Dice Area Under Curve (DAUC) and Ratio of Filtered True Positives (RFTPs).

3.1 Implementation Details and Training

For evaluation of the segmentation performance, we employed the DSC, defined as

$$D(X,Y) = \frac{2|X \cap Y|}{|X| + |Y|}, \tag{3}$$

where X and Y denote the output segmentation and its corresponding ground truth, respectively.

The HD95 is defined as the largest value in the set of closest distances between two structures, or

$$H(X,Y) = \max\left\{ \max_{x \in X} \min_{y \in Y} d(x,y), \max_{y \in Y} \min_{x \in X} d(y,x) \right\}, \tag{4}$$

where $d(x,y)$ denotes the Euclidian distance between two points $x \in X$ and $y \in Y$. It is common practice to report the 95th percentile instead of the maximum to compensate for outliers.

The loss function used for training contained three terms,

$$\mathcal{L}(x,y) = \mathcal{L}_{whole}(x,y) + \mathcal{L}_{core}(x,y) + \mathcal{L}_{enh}(x,y), \tag{5}$$

where $\mathcal{L}_{whole}$, $\mathcal{L}_{core}$, and $\mathcal{L}_{enh}$ where the soft dice loss of whole, core, and enhancing tumor regions, respectively, and where the soft dice loss is defined as

$$\mathcal{L}_{dice}(x,y) = \frac{-2\sum_i x_i y_i}{\sum_i x_i + \sum_i y_i + \epsilon}, \tag{6}$$

in which u is the softmax output of the network, v is a one-hot encoding of the ground truth segmentation map, and ϵ is a small constant added to avoid division by zero.

---

[1] https://keras.io.
[2] https://tensorflow.org.
[3] https://www.hpc2n.umu.se/.

Table 1. Mean DSC (higher is better) and HD95 (lower is better) and their standard deviations (SDs) (in parentheses) computed from the five-folds cross-validation on the training set (335 cases) for the different models. Details of the variations of the convolution blocks used for each model is shown in Fig. 3.

Model	DSC			HD95		
	whole	core	enh.	whole	core	enh.
TuNet	89.89 (2.07)	84.08 (4.02)	74.92 (7.24)	6.22 (2.66)	6.82 (2.93)	4.72 (2.21)
TuNet + SEB	91.38 (2.02)	85.93 (3.77)	78.11 (7.29)	4.50 (2.34)	6.70 (2.72)	4.24 (2.07)
TuNet + REB^1	90.68 (1.98)	86.07 (3.84)	75.88 (7.36)	5.14 (2.07)	**5.16 (1.99)**	4.20 (1.99)
TuNet + REB^1 + SEB	90.86 (2.12)	86.30 (3.72)	76.20 (7.32)	**5.10 (2.29)**	5.72 (2.62)	3.89 (1.61)
TuNet + REB^2	90.77 (2.18)	85.84 (3.96)	75.22 (7.30)	5.42 (2.50)	6.00 (2.38)	4.56 (2.09)
TuNet + REB^2 + SEB	91.90 (2.00)	86.09 (3.81)	77.43 (7.28)	5.07 (2.38)	6.20 (2.45)	3.98 (1.87)
Ensemble	91.92 (2.18)	86.35 (3.81)	78.01 (7.30)	5.30 (2.19)	5.80 (2.69)	3.45 (1.93)
Ensemble + post-process	**91.92 (2.18)**	**86.45 (3.79)**	**78.72 (7.10)**	5.30 (2.19)	5.75 (2.65)	**3.08 (1.90)**

We used the Adam optimizer [14] with a learning rate of $1 \cdot 10^{-4}$ and momentum parameters of $\beta_1 = 0.9$ and $\beta_2 = 0.999$. We also used L_2 regularization with a penalty parameter of $1 \cdot 10^{-5}$, that was applied to the kernel weight matrices, for all convolutional layers to cope with overfitting. The activation function of the final layer was the logistic sigmoid function.

All models were trained for 200 epochs, with a mini-batch size of four. To prevent over-fitting, we selected a patience period and dropped the learning rate by a factor of 0.2 if the validation loss did not improve over six epochs. Further, the training process was stopped if the validation loss did not improve after 15 epochs. The training time for a single model was around 35 h on an NVIDIA Tesla V100 GPU.

4   Results and Discussion

The key strengths of the proposed method are: (1) it takes advantage of the hierarchical structure of the tumor sub-regions, since the whole tumor region must contain the core tumor, and the core tumor region must contain the enhancing tumor region, and (2) it consists of three connected sub-networks that mutually regularize each other, *i.e.* the E-Net functions as a regularizer for the C-net, C-Net plays that role for the W-Net, and the W-Net and the E-Net constrain each other through the C-Net.

Table 1 shows the mean DSC and HD95 scores and SDs computed from the five-folds of cross-validation on the training set. As can be seen in Table 1, with DSC of 89.89/84.08/74.92 (whole/core/enh.) using cross-validation on the training set and 89.70/76.96/77.90 (whole/core/enh.) on the validation set, our *baseline* (TuNet) produced acceptable results. Adding SEB to TuNet (TuNet + SEB) improved the DSC on all tumor regions; however, only core and core+enhancing DSC scores were boosted when adding SEB to two variations, that used ResNet-like blocks, *i.e.* TuNet + REB^1 + SEB and TuNet + REB^2 + SEB.

We gained a few DSC points when the average ensemble technique (Ensemble), *i.e.* DSC reached 91.92/86.45/78.72 (whole/core/enh). The post-processing

Table 2. Results of Segmentation Task on BraTS19 validation data (125 cases). The results were obtained by computing the mean of predictions of five models trained over the folds. "UmU" denotes the name of our team and the ensemble of TuNet + SEB, TuNet + REB^1 + SEB and TuNet + REB^2 + SEB. The metrics were computed by the online evaluation platform. All the predictions were post-processed before submitting to the server. Bottom rows correspond to the top-ranking teams from the online system.

Model	DSC			HD95		
	whole	core	enh.	whole	core	enh.
TuNet + SEB	90.41	81.67	78.97	4.35	6.12	3.35
TuNet + REB^1 + SEB	90.06	79.43	78.12	4.66	7.98	3.34
TuNet + REB^2 + SEB	90.34	79.14	77.38	4.29	8.80	3.57
UmU	90.34	81.12	78.42	4.32	6.28	3.70
ANSIR	90.09	84.38	80.06	6.94	6.00	4.52
lfn_	90.91	85.48	80.24	4.35	**5.32**	3.88
SVIG1	**91.16**	85.79	81.33	**4.10**	5.92	4.21
NVDLMED	91.01	**86.22**	**82.28**	4.42	5.46	**3.61**

Table 3. Results of Segmentation Task on BraTS19 test data (166 cases).

Model	DSC			HD95		
	whole	core	enh.	whole	core	enh.
UmU	88.06	80.84	80.29	6.10	5.17	2.20

step only improved slightly the Ensemble model on the core region, but boosted the DSC of the enhancing region by a large margin, from 78.01 to 78.72, making it (Ensemble + post-processing) the best-performing model of all proposed models on the training set.

Table 2 shows the mean DSC and HD95 scores on the validation set by uploading our predicted masks to the evaluation server[4] (team name *UmU*). What is interesting in this table is: (1) DSC of core region on the validation set was lower compared to the cross-validation score on the training set, and (2) TuNet + SEB, perhaps surprisingly, performed slightly better than the ensemble model of TuNet + SEB, TuNet + REB^1 + SEB and TuNet + REB^2 + SEB. Table 3 shows that our BraTS19 testing dataset results are 88.06, 80.84 and 80.29 average dice for whole tumor, tumor core and enhanced tumor core, respectively.

Table 4 provides the mean DAUC and RFTPs scores on the validation set obtained after uploading our predicted masks and corresponding uncertainty maps to the evaluation server[5]. Similar to Table 2, it can be seen in Table 4 that TuNet + SEB is the best-performing model of our models. Though our best-

---

[4] https://www.cbica.upenn.edu/BraTS19/lboardValidation.html.
[5] https://www.cbica.upenn.edu/BraTS19/lboardValidationUncertainty.html.

Table 4. Results of Quantification of Uncertainty Task on BraTS19 validation data (125 cases) including mean DAUC (higher is better) and RFTPs (lower is better). The results were obtained by computing the mean of predictions of five models trained over the folds. "UmU" denotes the name of our team and the ensemble of TuNet + SEB, TuNet + REB^1 + SEB and TuNet + REB^2 + SEB. The metrics were computed by the online evaluation platform. The bottom rows correspond to the top-ranking teams from the online system. The proposed method was ranked among the top that was evaluated on the online test set.

Model	DAUC			RFTPs		
	whole	core	enh.	whole	core	enh.
TuNet + SEB	87.52	79.90	75.97	4.50	12.97	6.59
TuNet + REB^1 + SEB	86.13	80.01	75.13	5.80	16.50	8.02
TuNet + REB^2 + SEB	87.40	79.32	75.90	5.50	15.30	8.53
UmU	87.47	79.86	75.89	5.83	13.72	8.46
ANSIR	89.42	88.72	87.67	1.23	6.16	4.63
NVDLMED	89.95	88.03	84.67	1.68	**2.43**	2.86
TEAM_ALPACA	**90.93**	**88.08**	**87.90**	1.54	3.82	**2.59**

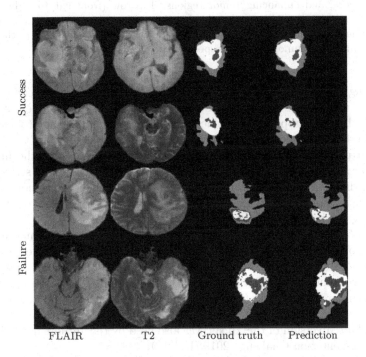

Fig. 4. A comparison of the ground truth masks and the results. The two examples show the input T1c (left), the ground truth masks (middle) and the results of the proposed method (right).

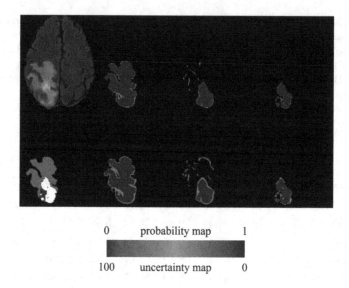

0   probability map   1

100   uncertainty map   0

Fig. 5. The uncertainty score maps generated from probability maps corresponding to whole, core, and enhancing tumor regions. Top row (from left to right): input, probability map of the whole, core and enhancing tumor regions, respectively. Bottom row (from left to right): multi-class label map (output), uncertainty map of the whole, core and enhancing tumor regions, respectively.

performing model performs slightly worse than the top-ranking teams on the validation set, it was one of the top-performing models on the test set This indicates that the proposed ensemble model might generalize the problem well.

Figure 4 illustrates four qualitative examples generated from the Ensemble + post-process model on the training set. As can be seen in the first and second rows, our model detected all three regions well. However, it struggled to correctly differentiate between the necrotic and non-enhancing tumor core and the enhancing tumor (third and last rows). This is most likely due to difficulties in labeling the homogeneous areas.

Figure 5 illustrates the uncertainty score maps generated from the corresponding output probability maps, that are produced by the TuNet, for three tumor regions (whole, core, and enhancing). As can be seen in Fig. 5, the uncertainty scores tends to be: (i) more obvious at the borderline or overlapping areas between tumor regions, and (ii) less apparent on the background or non-tumor regions.

An adapted version of the TuNet was also used by the authors in the Kidney Tumor Segmentation Challenge 2019 (KiTS19) [10].

5   Conclusion

In conclusion, we developed a cascaded architecture by connecting three U-Net-like networks to segment glioma sub-regions from multimodal brain MRI images.

We separated the complex brain tumor segmentation into three simpler binary tasks to segment the whole tumor, tumor core, and enhancing tumor core, respectively. Our network used an encoder-decoder structure with ResNet-like blocks and Squeeze-and-Excitation blocks after each convolution and concatenation block. Dice scores on the training set and validation set were 91.92/86.45/78.72 and 90.34/81.12/78.42 for the whole tumor, tumor core, and enhancing tumor core, respectively.

References

1. Asklund, T., BjörBjör, O., Malmström, A., Blomquist, E., Henriksson, R.: Överlevnanden vid maligna gliom har ökat senaste tio åren. analys av kvalitetsregisterdata. Läkartidningen **109**(17–18), 875–878 (2012)
2. Bakas, S., et al.: Segmentation labels and radiomic features for the pre-operative scans of the TCGA-GBM collection. Cancer Imaging Arch. (2017). https://doi.org/10.7937/K9/TCIA.2017.KLXWJJ1Q
3. Bakas, S., et al.: Segmentation labels and radiomic features for the pre-operative scans of the TCGA-LGG collection. Cancer Imaging Arch. **286** (2017). https://doi.org/10.7937/K9/TCIA.2017.GJQ7R0EF
4. Bakas, S., et al.: Advancing the cancer genome atlas glioma MRI collections with expert segmentation labels and radiomic features. Sci. Data **4**, 170117 (2017)
5. Bakas, S., et al.: Identifying the best machine learning algorithms for brain tumor segmentation, progression assessment, and overall survival prediction in the BRATS challenge. arXiv preprint arXiv:1811.02629 (2018)
6. Boser, B.E., Guyon, I.M., Vapnik, V.N.: A training algorithm for optimal margin classifiers. In: Proceedings of the Fifth Annual Workshop on Computational Learning Theory, pp. 144–152. ACM (1992)
7. Chen, L.-C., Zhu, Y., Papandreou, G., Schroff, F., Adam, H.: Encoder-decoder with atrous separable convolution for semantic image segmentation. In: Ferrari, V., Hebert, M., Sminchisescu, C., Weiss, Y. (eds.) ECCV 2018. LNCS, vol. 11211, pp. 833–851. Springer, Cham (2018). https://doi.org/10.1007/978-3-030-01234-2_49
8. Gholami, A., Mang, A., Biros, G.: An inverse problem formulation for parameter estimation of a reaction-diffusion model of low grade gliomas. J. Math. Biol. **72**(1–2), 409–433 (2016). https://doi.org/10.1007/s00285-015-0888-x
9. He, K., Zhang, X., Ren, S., Sun, J.: Deep residual learning for image recognition. In: Proceedings of the IEEE Conference on Computer Vision and Pattern Recognition, pp. 770–778 (2016)
10. Heller, N., et al.: The KiTS19 challenge data: 300 kidney tumor cases with clinical context, CT semantic segmentations, and surgical outcomes (2019)
11. Holland, E.C.: Progenitor cells and glioma formation. Curr. Opin. Neurol. **14**(6), 683–688 (2001)
12. Hu, J., Shen, L., Sun, G.: Squeeze-and-excitation networks. In: Proceedings of the IEEE Conference on Computer Vision and Pattern Recognition, pp. 7132–7141 (2018)
13. Isensee, F., Kickingereder, P., Wick, W., Bendszus, M., Maier-Hein, K.H.: No New-Net. In: Crimi, A., Bakas, S., Kuijf, H., Keyvan, F., Reyes, M., van Walsum, T. (eds.) BrainLes 2018. LNCS, vol. 11384, pp. 234–244. Springer, Cham (2019). https://doi.org/10.1007/978-3-030-11726-9_21

14. Kingma, D.P., Ba, J.: Adam: a method for stochastic optimization. arXiv preprint arXiv:1412.6980 (2014)
15. Kitrungrotsakul, T., et al.: A cascade of CNN and LSTM network with 3D anchors for mitotic cell detection in 4D microscopic image. In: 2019 IEEE International Conference on Acoustics, Speech and Signal Processing (ICASSP 2019), pp. 1239–1243. IEEE (2019)
16. Menze, B.H., et al.: The multimodal brain tumor image segmentation benchmark (BRATS). IEEE Trans. Med. Imaging **34**(10), 1993–2024 (2014)
17. Myronenko, A.: 3D MRI brain tumor segmentation using autoencoder regularization (2018). http://arxiv.org/abs/1810.11654
18. Ronneberger, O., Fischer, P., Brox, T.: U-Net: convolutional networks for biomedical image segmentation. In: Navab, N., Hornegger, J., Wells, W.M., Frangi, A.F. (eds.) MICCAI 2015. LNCS, vol. 9351, pp. 234–241. Springer, Cham (2015). https://doi.org/10.1007/978-3-319-24574-4_28
19. Wang, G., Li, W., Ourselin, S., Vercauteren, T.: Automatic brain tumor segmentation using cascaded anisotropic convolutional neural networks. In: Crimi, A., Bakas, S., Kuijf, H., Menze, B., Reyes, M. (eds.) BrainLes 2017. LNCS, vol. 10670, pp. 178–190. Springer, Cham (2018). https://doi.org/10.1007/978-3-319-75238-9_16

Using Separated Inputs for Multimodal Brain Tumor Segmentation with 3D U-Net-like Architectures

N. Boutry[1]([✉])[iD], J. Chazalon[1][iD], E. Puybareau[1][iD], G. Tochon[1][iD], H. Talbot[2][iD], and T. Géraud[1][iD]

[1] EPITA Research and Development Laboratory (LRDE),
Le Kremlin-Bicêtre, France
nicolas.boutry@lrde.epita.fr
[2] CentraleSupelec - Université Paris-Saclay, Gif-sur-Yvette, France
https://lrde.epita.fr, https://hugues-talbot.github.io/

Abstract. The work presented in this paper addresses the MICCAI BraTS 2019 challenge devoted to brain tumor segmentation using magnetic resonance images. For each task of the challenge, we proposed and submitted for evaluation an original method. For the tumor segmentation task (Task 1), our convolutional neural network is based on a variant of the U-Net architecture of Ronneberger et al. with two modifications: first, we separate the four convolution parts to decorrelate the weights corresponding to each modality, and second, we provide volumes of size $240 * 240 * 3$ as inputs in these convolution parts. This way, we profit of the 3D aspect of the input signal, and we do not use the same weights for separate inputs. For the overall survival task (Task 2), we compute explainable features and use a kernel PCA embedding followed by a Random Forest classifier to build a predictor with very few training samples. For the uncertainty estimation task (Task 3), we introduce and compare lightweight methods based on simple principles which can be applied to any segmentation approach. The overall performance of each of our contribution is honorable given the low computational requirements they have both for training and testing.

Keywords: Biomedical imaging · Brain tumor segmentation · Glioblastoma · CNN · U-Net

1 Introduction

The work presented in this paper was realized in the context of MICCAI BraTS 2019 Challenge [1–4,12], which aims at stimulating brain tumor detection, segmentation and analysis. This challenge is composed of 3 tasks, and we propose a contribution for each of them which will be described in separate sections.

Task 1 – Tumor segmentation. Given a set of unknown brain scans with four modalities, segment tumoral regions. We propose a deep architecture which fully decorrelates each modality with partial 3D convolutions.

© Springer Nature Switzerland AG 2020
A. Crimi and S. Bakas (Eds.): BrainLes 2019, LNCS 11992, pp. 187–199, 2020.
https://doi.org/10.1007/978-3-030-46640-4_18

Task 2 – Survival prediction. Predict the patient overall survival time. We propose a predictor based on kernel PCA, Random Forests and a custom brain atlas.

Task 3 – Quantification of uncertainty in segmentation. Assess how reliable the results from Task 1 are. We propose a set of lightweight techniques based on intrinsic confusion and geometry properties of the segmentation.

2   Brain Tumor Segmentation—Task 1

Starting from a set of 335 brain images where tumors are segmented by neuroradiologists, the aim of Task 1 is to segment new brain images whose ground truth is not known. The provided modalities are magnetic resonance images (T1/T1CE/T2 and FLAIR). The resolution of the provided images is $240 * 240 * 155$ voxels of 1 mm^3. These images result from captures of different protocols, magnetic fields strengths and MRI scanners.

Previous Work. For BraTS 2018 challenge, the first place was won by Myronenko [13] who used a semantic segmentation network based on a encoder-decoder architecture. Due to limited training dataset size, he connected a variational auto-encoder (able to reconstruct the initial image) to this network during the training procedure. This way, some constraints are added on the layers of the shared encoder which is in some way "regularized" and also less sensible to the random initialization. A crop size of $160 \times 192 \times 128$ has been used, which implied a batch size of 1 due to GPU memory limitations. Isensee et al. [7] won the second place and proved that a U-Net-like architecture with slight modifications (like using the LeakyReLU instead of the usual ReLU activation function and using instance normalization [18]) can be very efficient and hard to beat. They used a batch size of 2, a crop size of $128 \times 128 \times 128$, and a soft Dice loss function [7]. They also used an additional training data from their own institution to optimize the enhancing tumor dice. McKinly et al. [11] shared the third place with Zhou et al. [20]. On one side, McKinly et al. [11] proposed an embedding of a DenseNet [6] structure using dilated convolutions into a U-Net [15] architecture, to obtain their segmentation CNN. On the other side, Zhou et al. [20] ensembled different networks in cascade.

For the BraTS 2017 challenge, the first place was won by Kamnitsas et al. [8] who ensembled several models (trained separately) for robust segmentation (EMMA): they combined DeepMedic [9], FCN [10], and U-Net [15] models. During the training procedure, they used a batch size of 8 and a crop size of $64 \times 64 \times 64$ 3D patch. Wang et al. [19] won the second place. They segmented tumor regions in cascade using anisotropic dilated convolutions with 3 networks for each tumor subregion.

Proposed Architecture. Because the U-Net architecture [15] has demonstrated good performance in matter of biomedical image analysis, we propose

Table 1. Our U-Net-like multimodal 3D architecture, with 4 contractive branches.

Layer name	Operation	Output shape	Input(s)	
mod1_input	Input	240, 240, 3, 1		
mod1_conv1-1	Conv3D	240, 240, 3, 64	mod1_input	
mod1_conv1-2	Conv3D	240, 240, 3, 64	mod1_conv1-1	
mod1_conv2	BLOCK_A	120, 120, 3, 128	mod1_conv1-2	
mod1_conv3	BLOCK_A	60, 60, 3, 256	mod1_conv2	
mod1_conv4	BLOCK_A	30, 30, 3, 512	mod1_conv3	
mod1_conv5	BLOCK_A	15, 15, 3, 1024	mod1_conv4	
The branch for mod1 is repeated for each input (modality).				
concatenate_1	Concatenate	15, 15, 3, 4096	mod$i$_conv5	$\forall i \in [1,4]$
up_samp3d	UpSampling3D	30, 30, 3, 4096	concatenate_1	
conv3d_1	Conv3D	30, 30, 3, 512	up_samp3d	
conv3d_2	BLOCK_B	60, 60, 3, 256	mod$i$_conv4	$\forall i \in [1,4]$
			conv3d_1	
conv3d_3	BLOCK_B	120, 120, 3, 128	mod$i$_conv3	$\forall i \in [1,4]$
			conv3d_2	
conv3d_4	BLOCK_B	240, 240, 3, 64	mod$i$_conv2	$\forall i \in [1,4]$
			conv3d_3	
concatenate_2	Concatenate	240, 240, 3, 320	mod$i$_conv1-2	$\forall i \in [1,4]$
			conv3d_4	
conv3d_5	Conv3D	240, 240, 3, 64	concatenate_2	
conv3d_6	Conv3D	240, 240, 3, 64	conv3d_5	
conv3d_7	Conv3D	240, 240, 3, 4	conv3d_6	
output	Conv3D	240, 240, 3, 4	conv3d_7	$k\colon 1 \times 1 \times 1$

Table 2. Detail of the contractive block BLOCK_A.

Layer name	Operation	Output shape	Input(s)	
b1_input	Input	H, W, 3, C		
b1_mp	MaxPooling3D	H/2, W/2, 3, C	b1_input	*pool:* $2 \times 2 \times 1$
b1_conv	Conv3D	H/2, W/2, 3, 2*C	b1_mp	
b1_output	Conv3D	H/2, W/2, 3, 2*C	b1_conv	

here to re-adapt this architecture for multimodal biomedical image analysis. The complete architecture of our network is detailed in Tables 1, 2 and 3. We associate each modality to one input in our network. Then, each input is followed with a sequence of five layers made of two successive convolutional layers plus a max pooling and a dropout layer (contractive paths). Then, from the bottleneck, we apply five deconvolutional layers, each made of an upscaling layer followed

Table 3. Detail of the expanding block BLOCK_B.

Layer name	Operation	Output shape	Input(s)	
b2_input_mod$i$	Input	H, W, 3, C		$\forall i \in [1,4]$
b2_input_prev	Input	H, W, 3, C		
b2_concatenate	Concatenate	H, W, 3, 5*C	b2_input_mod$i$	$\forall i \in [1,4]$
			b2_input_prev	
b2_conv3d_1	Conv3D	H, W, 3, C	b2_concatenate	
b2_conv3d_2	Conv3D	H, W, 3, C	b2_conv3d_1	
b2_up_samp3d	UpSampling3D	2*H, 2*W, 3, C	b2_conv3d_2	*pool:* $2 \times 2 \times 1$
b2_output	Conv3D	2*H, 2*W, 3, C/2	b2_up_samp3d	

with two convolutional layers (expanding path). Finally, skip connections are used to connect the contractive path to the expanding path at each scale. Note that the number of skip connections is multiplied by a factor of four due to the structure of our network. To ensure continuity in the segmentation results, we propose also to provide partial volumes as inputs in our network (we use the *Conv3D* layers of Keras on volumes of size $W * H * 3$ with kernels of shape $3 \times 3 \times 3$).

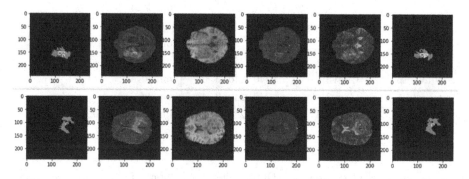

Fig. 1. Segmentation results with our architecture: on the left side, the ground truths, then the four modalities, and then our segmentation results.

Note that we know that the T1 modality and the T1CE one are strongly related, like the T2 one and the FLAIR one, but we are convinced that using separated weights for each inputs allows to improve segmentation results. This way we force the network to optimize different weights for each modality during the learning procedure.

Our motivation for our 3D approach (we provide partial volumes of size $3 * 240 * 240$) is twofold: first, the winners of the BraTS of 2018 used a full-3D approach [13], and second, we obtain smoother results thanks to the 3D convolutional layer (2D approaches generally lead to discontinuities along the z axis when slices are along x and y).

Note that we do not do any particular pre-processing, we just normalize each brain in the following manner like in [7]:

$$X_{norm} := \frac{X - \mu}{\sigma},$$

where μ and σ are respectively the statistical mean and standard deviation of the modality X corresponding to some patient. Also, we consider only the volumes (when we consitute the data set for the training procedure) where the number of voxels of the brain is greater than or equal to $(240 * 240 * 3)/6$. We do not use any post-processing.

Finally, we chose the standard parameters for our model: the number of filters are 64, 128, 256, 512, and 1024 for the 5 bi-convolutional layers, the number of filters are 512, 256, 128 and 64 for the bi-deconvolutional layers. Also, the learning rate is equal to 10^{-4}, we use categorical cross-entropy. We use the *selu* activation for all hidden layers, and use sigmoidal activation for the output layer.

Results. Table 4 summarizes the results obtain by the proposed method on Task 1 and Fig. 1 illustrates them. At test time, segmentations are predicted on a single-pass without any augmentation. Furthermore, not post-processing was applied on the results we report. The proposed approach exhibits a reasonable performance regarding the computational constraints required for training: indeed, a single GPU card with 16 GB of memory was sufficient to conduct our experiments. The DICE measure suggests that for some volumes or for some specific areas, the method fails to detect the correct elements but succeeds most of the time. The Hausdorff measure suggests that the boundary of the detected regions are not very precise and that more regularization at inference time could improve the method.

Table 4. Mean values of the segmentation metrics for each region, for the validation set and the test set. ↑ (resp. ↓) indicates that a higher (resp. lower) value is better.

Dataset	DICE (%) ↑			Hausdorff95 (voxels) ↓		
	WT	TC	ET	WT	TC	ET
Validation	68.4	87.8	74.7	10.2	10.9	14.8
Test	73.7	86.2	75.1	5.6	10.7	15.4

3   Survival Prediction—Task 2

The second task of the MICCAI 2019 BraTS challenge is concerned with the prediction of patient overall survival from pre-operative scans (only for subjects with gross total resection (GTR) status). The classification procedure is conducted by labeling subjects into three classes: short-survivors (less than 10 months), mid-survivors (between 10 and 15 months) and long-survivors (greater than 15 months). For post-challenge analyses, prediction results are also compared in terms of mean and median square error of survival time predictions, expressed in days. For that reason, our proposed patient survival prediction algorithm is organized in two steps:

1 We first predict the overall survival class, *i.e.* short-, mid- or long-survival (hereafter denoted by class/label 1, 2 and 3, respectively).
2 We then adjust our prediction within the predicted class by means of linear regression, in order to express the survival time in days.

Definition and Extraction of Relevant Features. Extracting relevant features is critical for classification purposes. Here, we re-use the features implemented by our team in the framework of the patient survival prediction task of MICCAI 2018 BraTS challenge, which ranked tie second [14]. Those features were chosen after in-depth discussions with a practitioner and are the following:

feature 1: the patient age (expressed in years).
feature 2: the relative size of the necrosis (labeled 1 in the groundtruth) class with respect to the brain size.
feature 3: the relative size of the edema class (labeled 2 in the groundtruth) with respect to the brain size.
feature 4: the relative size of the active tumor class (labeled 4 in the groundtruth) with respect to the brain size.
feature 5: the normalized coordinates of the binarized enhanced tumor (thus only considering necrosis and active tumor classes).
feature 6: the normalized coordinates of the region that is the most affected by necrosis, in a home made brain atlas.

For the training stage, features 2, 3 and 4 are computed thanks to the patient ground truth map for each patient. As this information is unknown during the test stage, the segmented volumes predicted by our Deep FCN architecture are used instead. In any case, these size features are expressed relatively to the total brain size (computed as the number of voxels in the T2 modality whose intensity is greater than 0).

In addition, we also re-use the home-made brain atlas that we also developed for the 2018 BraTS challenge. This atlas is divided into 10 crudely designed regions accounting for the frontal, parietal, temporal and occipital lobes and the cerebellum for each hemisphere (see [14] for more details regarding this atlas and how it is adjusted to each patient brain size). Feature 6 is defined as the coordinates of the centroid of the region within the atlas that is the most affected by the necrosis class (*i.e.*, the region that has the most voxels labeled as necrosis with respect to its own size). Note that this feature, as well as feature 5, is then normalized relatively to the brain bounding box. This leads to a feature vector with 10 components per patient (since both centroids coordinates are 3-dimensionals).

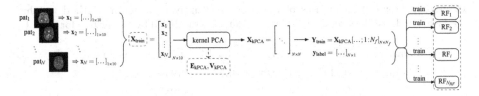

Fig. 2. Workflow of the proposed class-based training procedure. The information stored after the training phase (necessary for the test phase) is written in red or encircled in dashed red. (Color figure online)

Fig. 3. Workflow of the proposed test procedure.

Training Phase. For the training phase, we modified our previous work [14] in the following way: while we maintained the final learning stage through random forest (RF) classifiers [17], we replaced the principal component analysis (PCA) transformation, acting as preprocessing step for the learning stage, by its kernel counterpart (kPCA) [16]. The rationale is that we hope to increase the RFs performances in terms of classification/prediction as the input features are highly non-linear in terms of survival labels.

More specifically, the training stage of our prediction algorithm is as follows:

1. The feature vector $x_i \in \mathbb{R}^{10}$ of each of the N patients in the training set is extracted as described in the previous Sect. 3. All those feature vectors are then stacked in a $N \times 10$ feature matrix $\mathbf{X}_{\text{train}}$.
2. A kPCA is performed on $\mathbf{X}_{\text{train}}$, yielding the $N \times N$ matrix $\mathbf{X}_{\text{kPCA}}$. This matrix is obtained through the computation, normalization and diagonalization of the so-called *kernel matrix* which represents the dot product between the N features vectors when mapped in the feature space through a kernel function (here defined as a polynomial kernel with degree $d = 3$).
4. The $N \times N_f$ matrix $\mathbf{Y}_{\text{train}}$ is defined from $\mathbf{X}_{\text{train}}$ by retaining the first N_f columns (corresponding to the leading N_f features in the feature space, here set to $N_f = 10$). N_{RF} RF classifiers [17] are finally trained on all rows of $\mathbf{Y}_{\text{train}}$ to learn to predict the survival class of each training patient using the true label vector $\mathbf{y}_{\text{label}}$ as target values. The used RF parameters (number of decision trees per RF, splitting criterion, total number of RFs N_{RF}) are defined as in [14].
5. Three linear regressors (one per survival class) are finally trained using the patient age and its whole tumor size (relatively to its brain size) as explanatory variables and its true survival time (expressed in days) as measured variable.

Table 5. Classification metrics of the proposed survival prediction method for the validation and test data sets, given the segmentation produced by our system for Task 1.

Data set	Accuracy	MSE	medianSE	stdSE	SpearmanR
Validation	0.414	158804	80437	194618	0.272
Test	0.505	464492	60237	1408716	0.363

Steps 1 to 4 are depicted by the workflow in Fig. 2. In addition to the three linear regressors, we also store (for the test phase) the training feature matrix $\mathbf{X}_{train}$, the eigenvector matrix V_{kPCA} and eigenvalues E_{kPCA} of the kernel matrix, and the number of retained features N_f after kPCA.

Test Phase. The test phase is conducted in a similar fashion as the training phase. Given some input test patient, its overall survival class is first predicted, before being refined and expressed in terms of number of days. More specifically:

1. The features vector $\mathbf{x}_{test}$ of the test patient is retrieved as described previously.
2. This feature vector is then projected onto the principal axes learnt by the kPCA during the training phase. For that purpose, a new kernel matrix is computed and centered (hence the need for $\mathbf{X}_{train}$) before proper projection (through $\mathbf{V}_{kPCA}$) and scaling (with $\mathbf{E}_{kPCA}$).
3. This results in the projected vector $\mathbf{x}_{kPCA} \in \mathbb{R}^N$ from which the first N_f features are retained, yielding the test vector $\mathbf{y}_{test}$. This vector is then fed to the N_{RF} RF classifiers, leading to N_{RF} independent class label predictions. The final label prediction y_{pred} (1, 2 and 3 for short-, mid- and long-survivors, respectively) is eventually obtained by majority voting.
4. Once the survival class has been established, the final patient survival rate is predicted by means of the appropriate learnt linear regressor.

Steps 1 to 3 are illustrated by the workflow in Fig. 3.

Results. Table 5 presents the various classification performance metrics, namely the class-based accuracy, the mean, median and standard deviation square errors and Spearman R coefficient for survival predictions expressed in days, for the proposed prediction algorithm for the validation data set and the test data set. The validation and test data sets are comprised of $N = 27$ and $N = 107$ patients, respectively.

Results reported in Table 5 exhibit a slight improvement over the class-based classification accuracy between the validation set (0.414) and the test set (0.505).

4   Uncertainty Estimation in Segmentation—Task 3

The last task of the challenge is a new task which consists in estimating the uncertainty of the segmentation predictions produced in Task 1. The sub-regions considered for evaluation are: (i) the "enhancing tumor" (ET); (ii) the "tumor core" (TC); and (iii) the "whole tumor" (WT).

Participants had to produce uncertainty maps for each glioma sub-region. Each map contains integer values ranging from 0 (certain) to 100 (uncertain), and indicates the confidence of a decision to classify a particular voxel as belonging or not belonging to the a particular sub-region.

Results are reported using two metrics. (i) The area under the curve formed by the DICE scores computed for each uncertainty threshold (DICE score computed only on voxels for which the uncertainty is strictly inferior to the current threshold). This metric is the principal metric used for ranking. (ii) The area under the curve formed by the ratio of filtered true positive for each uncertainty threshold (wrongly discarded as being uncertain).

Uncertainty Estimation Methods. We focused on the study of lightweight uncertainty estimation techniques relying on two aspects of the predictions made by our segmentation system: (i) the consistency between independent predictions made for each classes; and (ii) the instability at the spatial boundary between two regions predicted as belonging to different classes. We believe that such approaches can be complementary to approaches based on the stability of the prediction under perturbations like Monte Carlo Dropout [5] which tend to be computationally demanding.

To take into account the consistency between independent predictions made for each classes, we propose a simple indicator called "weighted score difference" (abbreviated "WSDIFF") which estimates the uncertainty by computing the difference of activation between the most likely (maximally activated) class and the others, weighted by the absolute value of the greatest activation (in order to penalize cases where there is no clear activation of any class). This requires that the segmentation network outputs predictions for each class in an independent way (therefore it *cannot use a softmax* which would constrain predictions to be mutually exclusive).

Let c_i be the activation maps for each class i belonging to the sub-region R to consider, then the WSDIFF indicator for this sub-region R is computed as:

$$\text{WSDIFF}_R = (1 - max(s_R, s_{\overline{R}}) |s_R - s_{\overline{R}}|) * 100,$$

where:

$$s_R = max_{\forall i \in R}(c_i) \qquad \text{and} \qquad s_{\overline{R}} = max_{\forall i \notin R}(c_i).$$

SDIFF ("score difference") is the variant of this indicator without weighting:

$$\text{SDIFF}_R = (1 - |s_R - s_{\overline{R}}|) * 100.$$

As shown later in the results, the weighting factor increased the performance of this indicator in our tests. Other attempts using the sum of the activation maps for each set of classes gave poor results and were harder to normalize.

Regarding the instability of the spatial boundary between two regions predicted as belonging to different classes, we designed an indicator (abbreviated "BORDER") which assigns a maximal uncertainty (100) at the boundary between two regions, and linearly decreases this uncertainty to the minimal value (0) at a given distance to the boundary. This distance defines the (half) width of an "uncertainty border" between two regions.

It is calibrated independently for each class and was estimated with respect to the 95th percentile of the Hausdorff distance metric reported for our segmentation method for this particular class. In practice, we used the following parameters: for the whole tumor (WT) we used a half-width of 9 voxels, for the tumor core (TC), 12 voxels, and for the enhancing tumor (ET), 7 voxels.

To compute this indicator, we first compute the Boundary Distance Transform BDT $= max(\mathrm{DT}(R), \mathrm{DT}(\overline{R}))$ using the Distance Transform DT to the given sub-region R and its complement $\overline{R}$. Then, we invert, shift and clip the BDT such that the map is maximal on the boundary and have 0 values at a distance greater or equal to the half-width of the border. We finally scale the resulting map so its values are comprised between 0 (far from the boundary) and 100 (on the boundary). The resulting uncertainty map for a given class exhibits a triangular activation shape on the direction perpendicular to the boundary of the objects detected by the segmentation stage.

Results and Discussion. Experimental results regarding the different uncertainty estimation methods are reported in Table 6. They indicate the results obtained for the validation set computed by the official competition platform.

Table 6. Mean values of the metrics for each region, computed by the official competition platform on the validation set for Task 3 (uncertainty estimation), for each of our uncertainty estimation approaches. ↑ (resp. ↓) indicates that a higher (resp. lower) value is better. Best values are in **bold** face.

Metric	DICE_AUC (%) ↑			FTP_RATIO_AUC (%) ↓		
	WT	TC	ET	WT	TC	ET
(original DICE score)	*85.2*	*62.0*	*59.9*	-	-	-
SDIFF	85.9	76.2	72.0	**20.7**	**17.8**	**16.8**
WSDIFF	85.9	77.5	**74.1**	30.7	24.0	21.4
BORDER	**87.8**	**80.6**	68.3	58.1	68.7	70.9
MEAN BORDER WSDIFF	86.7	79.5	73.3	44.1	45.7	45.9

Regarding the *DICE AUC* metric, the BORDER approach exhibits better results for glioma sub-regions WT and TW, while the WSDIFF approach performs better for the ET sub-region. The integration of a weighting of the uncertainty according to the activation of a given class provided some improvement to the WSDIFF method over the SDIFF one.

Regarding the *FTP Ratio AUC* metric, the BORDER method filters true positives quite aggressively and gives very high measures. The SDIFF method, on the other side of the spectrum, filters much less true positives. The WSDIFF method presents an interesting compromise in terms of true positive filtering. We can also notice that mean of the BORDER and WSDIFF indicators yields some form of compromise (sub-optimal results, but less aggressive filtering). The best balance seems to use the BORDER indicator for WT and TC regions, and the WSDIFF indicator for ET regions: *this the strategy we used.*

Figure 4 illustrates the responses of the uncertainty estimation methods on a case for which the segmentation step performed reasonably well. We can see that while the BORDER method generates a lot of false positives, it successfully captures erroneous regions with a high uncertainty score. A better calibration of this method may improve its performance. For ET regions, the WSDIFF method is more selective and yields a lower amount of false positives.

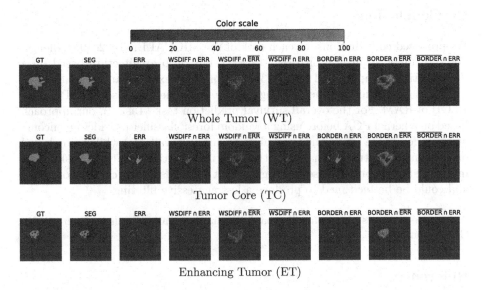

Fig. 4. Comparison of the WSDIFF and BORDER indicators on a reasonably well segmented case. Each row illustrates the response of uncertainty estimation methods for a different glioma region. The GT column is the ground truth, SEG the predicted segmentation, ERR the prediction error, and for each uncertainty estimation METHOD $\in$ {WSDIFF, BORDER}: METHOD $\cap$ ERR shows the uncertainty values for erroneous areas (true positives – higher is better), METHOD $\cap$ $\overline{\text{ERR}}$ shows the uncertainty values for well-classified areas (false positives – lower is better), and $\overline{\text{METHOD}}$ $\cap$ ERR shows the inverted $(100 - x)$ uncertainty values for erroneous areas (false negative – lower is better).

When comparing our results with other approaches from the public leader board for Task 3 (for the validation set), it should be noted that direct comparison is hard because the performance at Task 3 is directly linked to the performance at Task 1, hence a measure of a relative gain or loss might provide some hint, but ultimately each uncertainty estimator should be tested on every segmentation method. Nevertheless, we identified three interesting trends among those results. (1) Methods with high performance in both Task 1 and Task 3: the uncertainty estimation may be great but the score at Task 3 is indubitably boosted by the one at Task 1. (2) Methods with average scores at Task 1 but with a noticeable improvement with respect to the DICE AUC score at Task 3: those methods seem to have an efficient uncertainty estimation strategy. Such methods may have: (2.1) a good score for the FTP Ratio AUC metric of Task 3, indicating an efficient approach; (2.2) an average score for this metric: we believe our approach belongs to this category.

Those results let us believe that our uncertainty estimation methods are better suited for cases were the underlying segmentation method already performs quite well. Because of their simplicity and fast computation, they may be a natural baseline for more complex methods to be compared against.

5   Conclusion

We proposed contributions for each task of the MICCAI BraTS 2019 challenge. For the tumor segmentation task (Task 1), our deep architecture based on a decorrelation of inputs and partial 3D convolutions exhibits an honorable performance given the fact the training can be performed on a single GPU with 16 GB of RAM. For the overall survival prediction task (Task 2), our approach based on a kernel PCA before using a random forest classifier provides an encouraging performance (given the few training examples available) while being based on explainable features. Finally, for the uncertainty estimation task (Task 3), we introduced and compared several lightweight methods which can be combined and could be better tuned to produce a less aggressive filtering.

Acknowledgments. We would like thank NVidia Corporation for their *Quadro P6000* GPU donation.

References

1. Bakas, S., et al.: Segmentation labels and radiomic features for the pre-operative scans of the TCGA-GBM collection. Cancer Imaging Arch. **286** (2017)
2. Bakas, S., et al.: Segmentation labels and radiomic features for the pre-operative scans of the TCGA-LGG collection. Cancer Imaging Arch. **286** (2017)
3. Bakas, S., et al.: Advancing the cancer genome atlas glioma MRI collections with expert segmentation labels and radiomic features. Sci. Data **4**, 170117 (2017)
4. Bakas, S., et al.: Identifying the best machine learning algorithms for brain tumor segmentation, progression assessment, and overall survival prediction in the brats challenge. arXiv preprint arXiv:1811.02629 (2018)
5. Gal, Y., Ghahramani, Z.: Dropout as a Bayesian approximation: representing model uncertainty in deep learning. In: Proceedings of the 33rd International Conference on Machine Learning (ICML 2016), pp. 1050–1059, June 2015
6. Huang, G., Liu, Z., Van Der Maaten, L., Weinberger, K.Q.: Densely connected convolutional networks. In: Proceedings of the IEEE Conference on Computer Vision and Pattern Recognition, pp. 4700–4708 (2017)

7. Isensee, F., Kickingereder, P., Wick, W., Bendszus, M., Maier-Hein, K.H.: No new-net. In: Crimi, A., Bakas, S., Kuijf, H., Keyvan, F., Reyes, M., van Walsum, T. (eds.) BrainLes 2018. LNCS, vol. 11384, pp. 234–244. Springer, Cham (2019). https://doi.org/10.1007/978-3-030-11726-9_21

8. Kamnitsas, K., et al.: Ensembles of multiple models and architectures for robust brain tumour segmentation. In: Crimi, A., Bakas, S., Kuijf, H., Menze, B., Reyes, M. (eds.) BrainLes 2017. LNCS, vol. 10670, pp. 450–462. Springer, Cham (2018). https://doi.org/10.1007/978-3-319-75238-9_38

9. Kamnitsas, K., et al.: Efficient multi-scale 3D CNN with fully connected CRF for accurate brain lesion segmentation. Med. Image Anal. **36**, 61–78 (2017)

10. Long, J., Shelhamer, E., Darrell, T.: Fully convolutional networks for semantic segmentation. In: Proceedings of the IEEE Conference on Computer Vision and Pattern Recognition, pp. 3431–3440 (2015)

11. McKinley, R., Meier, R., Wiest, R.: Ensembles of densely-connected CNNs with label-uncertainty for brain tumor segmentation. In: Crimi, A., Bakas, S., Kuijf, H., Keyvan, F., Reyes, M., van Walsum, T. (eds.) BrainLes 2018. LNCS, vol. 11384, pp. 456–465. Springer, Cham (2019). https://doi.org/10.1007/978-3-030-11726-9_40

12. Menze, B.H., et al.: The multimodal brain tumor image segmentation benchmark (BRATS). IEEE Trans. Med. Imaging **34**(10), 1993–2024 (2014)

13. Myronenko, A.: 3D MRI brain tumor segmentation using autoencoder regularization. In: Crimi, A., Bakas, S., Kuijf, H., Keyvan, F., Reyes, M., van Walsum, T. (eds.) BrainLes 2018. LNCS, vol. 11384, pp. 311–320. Springer, Cham (2019). https://doi.org/10.1007/978-3-030-11726-9_28

14. Puybareau, E., Tochon, G., Chazalon, J., Fabrizio, J.: Segmentation of gliomas and prediction of patient overall survival: a simple and fast procedure. In: Crimi, A., Bakas, S., Kuijf, H., Keyvan, F., Reyes, M., van Walsum, T. (eds.) BrainLes 2018. LNCS, vol. 11384, pp. 199–209. Springer, Cham (2019). https://doi.org/10.1007/978-3-030-11726-9_18

15. Ronneberger, O., Fischer, P., Brox, T.: U-Net: convolutional networks for biomedical image segmentation. In: Navab, N., Hornegger, J., Wells, W.M., Frangi, A.F. (eds.) MICCAI 2015. LNCS, vol. 9351, pp. 234–241. Springer, Cham (2015). https://doi.org/10.1007/978-3-319-24574-4_28

16. Schölkopf, B., Smola, A., Müller, K.-R.: Kernel principal component analysis. In: Gerstner, W., Germond, A., Hasler, M., Nicoud, J.-D. (eds.) ICANN 1997. LNCS, vol. 1327, pp. 583–588. Springer, Heidelberg (1997). https://doi.org/10.1007/BFb0020217

17. Svetnik, V., Liaw, A., Tong, C., Culberson, J.C., Sheridan, R.P., Feuston, B.P.: Random forest: a classification and regression tool for compound classification and QSAR modeling. J. Chem. Inf. Comput. Sci. **43**(6), 1947–1958 (2003)

18. Ulyanov, D., Vedaldi, A., Lempitsky, V.: Instance normalization: the missing ingredient for fast stylization. arXiv preprint arXiv:1607.08022 (2016)

19. Wang, G., Li, W., Ourselin, S., Vercauteren, T.: Automatic brain tumor segmentation using cascaded anisotropic convolutional neural networks. In: Crimi, A., Bakas, S., Kuijf, H., Menze, B., Reyes, M. (eds.) BrainLes 2017. LNCS, vol. 10670, pp. 178–190. Springer, Cham (2018). https://doi.org/10.1007/978-3-319-75238-9_16

20. Zhou, C., Chen, S., Ding, C., Tao, D.: Learning contextual and attentive information for brain tumor segmentation. In: Crimi, A., Bakas, S., Kuijf, H., Keyvan, F., Reyes, M., van Walsum, T. (eds.) BrainLes 2018. LNCS, vol. 11384, pp. 497–507. Springer, Cham (2019). https://doi.org/10.1007/978-3-030-11726-9_44

Two-Step U-Nets for Brain Tumor Segmentation and Random Forest with Radiomics for Survival Time Prediction

Soopil Kim, Miguel Luna, Philip Chikontwe, and Sang Hyun Park[✉]

Department of Robotics Engineering, DGIST, Daegu, South Korea
{soopilkim,shpark13135}@dgist.ac.kr

Abstract. In this paper, a two-step convolutional neural network (CNN) for brain tumor segmentation in brain MR images with a random forest regressor for survival prediction of high-grade glioma subjects are proposed. The two-step CNN consists of three 2D U-nets for utilizing global information on axial, coronal, and sagittal axes, and a 3D U-net that uses local information in 3D patches. In our two-step setup, an initial segmentation probability map is first obtained using the ensemble 2D U-nets; second, a 3D U-net takes as input both the MR image and initial segmentation map to generate the final segmentation. Following segmentation, radiomics features from T1-weighted, T2-weighted, contrast enhanced T1-weighted, and T2-FLAIR images are extracted with the segmentation results as a prior. Lastly, a random forest regressor is used for survival time prediction. Moreover, only a small number of features selected by the random forest regressor are used to avoid overfitting. We evaluated the proposed methods on the BraTS 2019 challenge dataset. For the segmentation task, we obtained average dice scores of 0.74, 0.85 and 0.80 for enhanced tumor core, whole tumor, and tumor core, respectively. In the survival prediction task, an average accuracy of 50.5% was obtained showing the effectiveness of the proposed methods.

Keywords: Brain tumor segmentation · Survival prediction · Convolutional neural network · Radiomics · Random forest

1 Introduction

Glioma is the most frequently occurring primary brain tumor in the human brain [11]. It contains subregions that are heterogeneous, each with a different pattern on the brain MRI scan. Information such as the shape, size, and location of these subregions is vital to both surgery and treatment planning, as well as eventual diagnosis of disease progression. Glioma is further divided into glioblastoma (GBM/HGG) and lower grade glioma (LGG); with HGG being an aggressive and life-threatening tumor. Thus, accurate prediction of survival time of HGG patients is valuable for physicians to determine treatment planning.

Accordingly, the problem of segmenting glioma tumors into three areas such as peritumoral edema, necrotic and non-enhancing tumor core, and enhancing

© Springer Nature Switzerland AG 2020
A. Crimi and S. Bakas (Eds.): BrainLes 2019, LNCS 11992, pp. 200–209, 2020.
https://doi.org/10.1007/978-3-030-46640-4_19

tumor core were considered in the BraTS 2012 challenge [5]; whereas the BraTs 2017 challenge [8] not only addressed the segmentation task but also survival prediction.

Many methods have been proposed to address the survival prediction problem. In most cases, tumor regions are first segmented from multi-modal MR images, and then features extracted from the segmentation results are used for survival time prediction. For the tumor segmentation problem, convolutional neural networks (CNN) are commonly used and show good segmentation performance [10]. Specifically, methods using 2D CNN [16], 3D CNN [12,20], or ensemble of multiple CNNs [6,22] have been proposed. Moreover, the use of different types of networks [22], loss functions [9,13], or regularization [15] techniques have been explored. Notably, for survival prediction; deep learning techniques alone did not achieve high performance as compared to the segmentation task. The majority of state-of-the-art methods extract features from multi-modal MR images with segmentation results, and use such features with machine learning methods to predict survival time. Specifically, methods using features defined by authors [6,21], or using thousands of Radiomics features with feature selection algorithms [1,18] have been proposed.

In this work, following similar procedures of recent state-of-the-art methods, we address brain tumor segmentation and survival prediction sequentially. For tumor segmentation, we propose a two-step CNNs consisting of three 2D U-nets to utilize global information in the axial, coronal, and sagittal axes, and a 3D U-net for utilizing local information in 3D patches. Segmentation is predicted through the 2D U-nets and the 3D U-net sequentially. Lastly, for survival time prediction; a random forest regressor with radiomics features, extracted from the T1-weighted (T1), T2-weighted (T2), contrast enhanced T1-weighted (T1ce), and Fluid Attenuation Inversion Recovery (T2-FLAIR) images with segmentation results, is used for final prediction. To avoid overfitting, a feature selection scheme is employed.

2   Method

The overall procedure consists of preprocessing, tumor segmentation, and survival time prediction (see Fig. 1). First, the multi-modal MR images are normalized to have values ranging between 0 and 1. Second, the normalized images are used to obtain tumor subregion segmentation probability maps predicted by three 2D U-nets on axial, coronal and sagittal axes, respectively. The probability maps are further averaged and aggregated as 3D segmentation maps. Then, 3D patches of the segmentation maps and multi-modal MR images are input to a 3D U-net to predict the segmentation of the 3D patches. The final segmentation of the whole image is obtained by aggregating the local 3D patch predictions. Given the final 3D segmentation maps and the multi-modal MR images, selective radiomics features are extracted. Finally, survival time is predicted by the

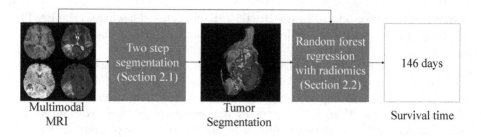

Fig. 1. The overall procedure of the survival time prediction.

random forest regressor. Also, a feature selection model is used to determine which features need to be chosen from the radiomics features. Details are described in following subsections.

2.1 Tumor Segmentation Model

The proposed 2D-3D two-step CNN framework is shown in Fig. 2. In the first step, three identical 2D CNNs are trained on axial, coronal, and sagittal axes, respectively, using half of the training data. The network architectures follow the classic U-Net structure [17] taking as input 4-channel 2D slices from 4 different modality MR images with output labels: enhancing tumor, peritumoral edema, necrotic and non-enhancing tumor, and background, respectively. Each CNN predicts a 2D probability map with 4 channels where each channel indicates the probability of tumor subregions and background, respectively. Finally, 2D probability maps on the three axes are averaged into a 3D probability map with 4 channels.

In the second step, a 3D CNN is trained with the remainder of the training data. First, a 3D probability map is generated using the ensemble 2D CNNs from the first step. Following, 4-channel patches are extracted from both the probability map and 3D multi-modal MR images to obtain a single 8-channel patch as input for the 3D CNN. A U-Net style architecture [17] is employed for the segmentation model and is trained with the segmentation labels of tumor subregions. The patch size was empirically set as $16 \times 16 \times 16$ voxels.

During inference, patch-wise 3D probability maps are sequentially generated using ensemble 2D CNNs and the trained 3D CNN. Finally, the patch predictions are aggregated to reconstruct the whole probability map. The final segmentation is determined by taking the label with the maximum probability.

2.2 Overall Survival Time Prediction

Overall procedure of survival time prediction is described in Fig. 3.

Feature Extraction: Radiomics features are extracted from multi-modal MR images and tumor segmentation obtained by the proposed two-step network. The radiomics features were extracted from T1, T2, T1ce, T2-FLAIR images on

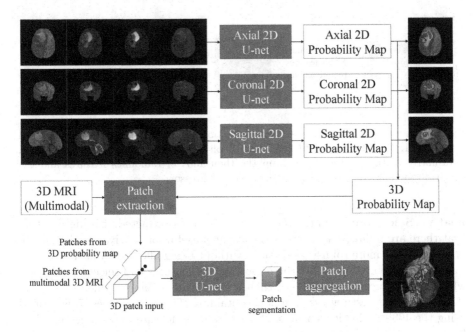

Fig. 2. The proposed 2D-3D two-step segmentation networks.

each of the enhancing tumor (ET), tumor core (TC), and whole tumor (WT) regions using the Pyradiomics toolbox [19]. TC refer to the area including the necrotic and the non-enhancing tumor core, while WT refer to the area where all subregions are included.

Given an image and its segmentation, the Pyradiomics toolbox extracts shape features, intensity features, and texture features, respectively. Specifically, 14 shape features regarding tumor shapes such as tumor volume, surface area, 3D diameter, 2D diameter, and so on, are extracted from the segmentation. Whereas, 18 first order statistic features related to the tumor region intensity such as mean intensity, standard deviation of the intensity values, median intensity, and so on, are extracted from the image. In total, 75 texture features containing 24 gray-level co-occurrence matrix (GLCM) features, 16 gray-level run length matrix (GLRLM) features, 16 gray-level size zone matrix (GLSZM) features, 5 neighboring gray-tone difference matrix (NGTDM) features and 14 gray-level dependence matrix (GLDM) features are also extracted from the image.

Intensity and the texture features are extracted from the input image as well as the images filtered by Laplacian of Gaussian (LoG) filters with different sigma values i.e. 1, 2, and 3. Thus, a total $(18+75) \times 4 + 14 = 386$ features are extracted per a single modality image on a single subregion. The radiomics features are extracted 12 times from four different modality MR images on three subregions, respectively. In total, $386 \times 12 = 4632$ features are extracted.

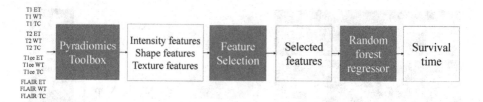

Fig. 3. The procedure of survival time prediction. Radiomics features are extracted from the T1, T2, T1ce, and T2-FLAIR images on three subregions, respectively. Then, informative features are selected from the thousands of radiomics features and the survival time is predicted by the random forest regressor.

Feature Selection: From the thousands of features extracted, it is highly likely that there are redundant or irrelevant features useful for survival prediction. In order to create more robust predictions without being affected by such unnecessary information, we compute the importance of features and select the features whose importance value is larger than a certain importance value. To achieve this, a random forest regression model containing 100 decision trees [7] is trained using the 4632 radiomics features. During training, feature importance is calculated from the variation of the variance based on the selected features. The higher the feature importance, the more important the feature is selected in the regression analysis. We empirically extract 17 features using a threshold and use them for the survival prediction.

Overall Survival Time Prediction: Finally, we train a random forest regression model containing 100 decision trees [7] using the selected 17 features. During inference, the trained model is used to predict survival time following the extraction of 17 features from a given test image and its segmentation.

3   Experimental Results

3.1   Dataset

We evaluated our method on the dataset in BraTS 2019 challenge. The dataset contains 3T multi-modal brain MR images routinely taken for clinical diagnosis [2–4, 14]. To observe the tumor subregions of glioma, four modality MR images such as T1, T2, T1ce, and T2-FLAIR were acquired. These images were skull-stripped, registered on an anatomical template, and resampled with isotropic $1\,mm^3$ resolution. The final dimension of the images was $240 \times 240 \times 155$. The ground truths for these images were provided by board-certified neuroradiologists.

The dataset contains training, validation, and testing datasets. The training dataset includes the images and its ground truths acquired from 259 HGG subjects and 76 LGG subjects. Among them, the survival times of the HGG subjects were given for the survival time prediction task. The validation dataset contained 125 subjects for the segmentation task and 29 subjects for the survival time prediction task. The testing dataset contained 166 subjects for the segmentation task and 107 subjects for the survival time prediction task.

Table 1. Segmentation results on validation data.

Measure	Mean	Std	Median
Dice ET	0.672	0.316	0.819
Dice WT	0.876	0.097	0.912
Dice CT	0.764	0.246	0.883
Sensitive ET	0.763	0.268	0.864
Sensitive WT	0.887	0.106	0.918
Sensitive TC	0.765	0.269	0.894
Specificity ET	0.998	0.004	0.998
Specificity WT	0.991	0.027	0.996
Specificity TC	0.996	0.009	0.999
95% Hausdorff ET	8.843	18.976	2.236
95% Hausdorff WT	14.175	23.528	4.000
95% Hausdorff TC	11.667	18.400	5.099

Table 2. Segmentation results on testing data.

Measure	Mean	Std	Median
Dice ET	0.743	0.258	0.836
Dice WT	0.858	0.154	0.906
Dice CT	0.804	0.266	0.910
95% Hausdorff ET	4.381	13.673	1.732
95% Hausdorff WT	16.454	28.240	4.977
95% Hausdorff TC	7.241	13.767	3.000

3.2   Results of Tumor Segmentation

The segmentation performance on the validation set was evaluated by dice score, sensitivity, specificity, and 95% Hausdorff distance for each of the ET, TC, and WT regions, while the performance on the testing set was evaluated by dice score and 95% Hausdorff distance. The segmentation results on the validation set and the testing set are shown in Tables 1 and 2, respectively. Moreover, Fig. 4 shows the box plots of segmentation accuracy on the validation dataset. Generally, the results on the testing set were consistent with those of the validation data set. According to the dice scores, the highest accuracy was obtained on the

WT among 3 subregions since the appearance of WT was consistent compared to the heterogeneous subregions such as ET and TC. In the case of ET and TC, the standard deviation of DSC and sensitivity was also higher than that of WT. Figure 5 shows representative segmentation results on the validation dataset obtained by the proposed method. In most cases, our method achieved good segmentation performances.

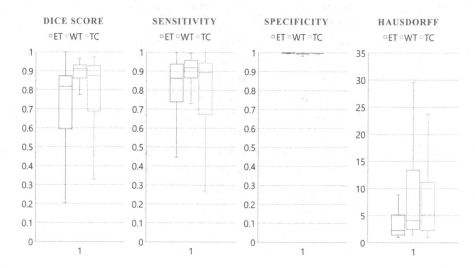

Fig. 4. Box plots of segmentation accuracy on the validation dataset. The top, center and bottom lines of each box represent upper quartile, median, and lower quartile scores, respectively.

Table 3. Survival prediction results on the validation and testing sets.

	Accuracy	Mean square error	Median square error
Validation	0.483	121778.6	20736
Testing	0.505	409680.9	50625

3.3 Results of Survival Time Prediction

The survival time prediction was evaluated with accuracy, mean square error, and median square error. Table 3 shows the performances of our method for the validation and testing sets. Accuracy of regression model was measured by classification of subjects as long-survivors (i.e., >15 months), short-survivors (i.e., <10 months), and mid-survivors (i.e., between 10 and 15 months). The proposed method achieved an average accuracy of 50.5%, mean square error of 409680.9, and median square error of 50625 on the testing set.

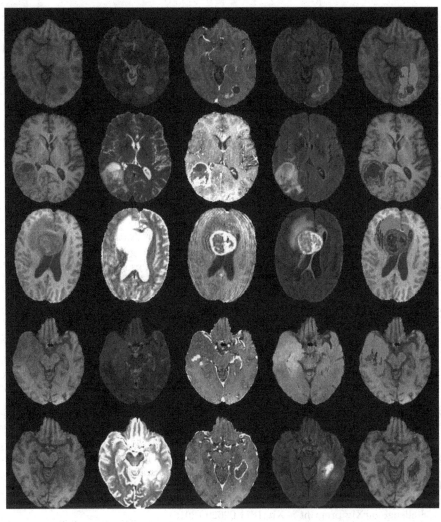

T1   T2   T1ce   T2-FLAIR   Segmentation

Fig. 5. Multi-modal MR images and those segmentations for 5 representative cases on the validation set. In the segmentation, green, red, and yellow indicates peritumoral edema, enhancing and non-enhancing tumor core, necrotic core, respectively. (Color figure online)

4   Conclusion

In this work, we addressed the task of tumor segmentation and survival prediction of HGG subjects in multi-model MR images with a two-step CNN framework and a random forest regression model. In the segmentation task, ensemble 2D CNNs trained with global multi-model information predict segmentation maps that are further aggregated and feed as input to a 3D-CNN that takes advantage of local information to produce a final segmentation map. Following this, in the survival task; a trained random forest regressor with selected features is used to predict the survival time of patients. Experimental results demonstrate the effectiveness of the proposed method with dice scores 0.74, 0.85 and 0.80 for enhancing tumor core, whole tumor, and tumor cores, respectively. Moreover, empirical results for survival prediction also show that by using less features for the regressor we can obtain improved performance i.e. an accuracy of 50.5% and reduce overfitting. The combination of deep learning methods and handcrafted features in this multi-stage framework shows potential viability of the proposed methods. As a future point of research, it would be interesting to further explore the use of a single model for both survival prediction and segmentation that is invariant to modality shifts.

Acknowledgement. This work was supported by the National Research Foundation of Korea(NRF) grant funded by the Korea government(MSIT) (No. 2019R1C1C1008727) and by the international cooperation program managed by NRF of Korea (NRF-2018K2A9A2A06020642).

References

1. Baid, U., et al.: Deep learning radiomics algorithm for gliomas (DRAG) model: a novel approach using 3D UNET based deep convolutional neural network for predicting survival in gliomas. In: Crimi, A., Bakas, S., Kuijf, H., Keyvan, F., Reyes, M., van Walsum, T. (eds.) BrainLes 2018. LNCS, vol. 11384, pp. 369–379. Springer, Cham (2019). https://doi.org/10.1007/978-3-030-11726-9_33
2. Bakas, S., et al.: Segmentation labels and radiomic features for the pre-operative scans of the TCGA-LGG collection. Cancer Imaging Arch. **286** (2017). https://doi.org/10.7937/K9/TCIA.2017.GJQ7R0EF
3. Bakas, S., et al.: Advancing the cancer genome atlas glioma MRI collections with expert segmentation labels and radiomic features. Sci. Data **4**, 170117 (2017)
4. Bakas, S., et al.: Identifying the best machine learning algorithms for brain tumor segmentation, progression assessment, and overall survival prediction in the brats challenge. arXiv preprint arXiv:1811.02629 (2018)
5. Bauer, S., Fejes, T., Slotboom, J., Wiest, R., Nolte, L.P., Reyes, M.: Segmentation of brain tumor images based on integrated hierarchical classification and regularization. In: MICCAI BraTS Workshop. Miccai Society, Nice (2012)
6. Feng, X., Tustison, N., Meyer, C.: Brain tumor segmentation using an ensemble of 3D U-nets and overall survival prediction using radiomic features. In: Crimi, A., Bakas, S., Kuijf, H., Keyvan, F., Reyes, M., van Walsum, T. (eds.) BrainLes 2018. LNCS, vol. 11384, pp. 279–288. Springer, Cham (2019). https://doi.org/10.1007/978-3-030-11726-9_25
7. Ho, T.K.: Random decision forests. In: Proceedings of 3rd International Conference on Document Analysis and Recognition, vol. 1, pp. 278–282. IEEE (1995)
8. Isensee, F., Kickingereder, P., Wick, W., Bendszus, M., Maier-Hein, K.H.: Brain tumor segmentation and radiomics survival prediction: contribution to the BRATS 2017 challenge. In: Crimi, A., Bakas, S., Kuijf, H., Menze, B., Reyes, M. (eds.) BrainLes 2017. LNCS, vol. 10670, pp. 287–297. Springer, Cham (2018). https://doi.org/10.1007/978-3-319-75238-9_25

9. Isensee, F., Kickingereder, P., Wick, W., Bendszus, M., Maier-Hein, K.H.: No new-net. In: Crimi, A., Bakas, S., Kuijf, H., Keyvan, F., Reyes, M., van Walsum, T. (eds.) BrainLes 2018. LNCS, vol. 11384, pp. 234–244. Springer, Cham (2019). https://doi.org/10.1007/978-3-030-11726-9_21

10. Kuijf, H.J., et al.: Standardized assessment of automatic segmentation of white matter hyperintensities; results of the WMH segmentation challenge. IEEE Trans. Med. Imaging **38**(11), 2556-2568 (2019)

11. Lee, J.H.: Meningiomas: Diagnosis, Treatment, and Outcome. Springer, Heidelberg (2008). https://doi.org/10.1007/978-1-84628-784-8

12. Luna, M., Park, S.H.: 3D patchwise U-Net with transition layers for MR brain segmentation. In: Crimi, A., Bakas, S., Kuijf, H., Keyvan, F., Reyes, M., van Walsum, T. (eds.) BrainLes 2018. LNCS, vol. 11383, pp. 394–403. Springer, Cham (2019). https://doi.org/10.1007/978-3-030-11723-8_40

13. McKinley, R., Meier, R., Wiest, R.: Ensembles of densely-connected CNNs with label-uncertainty for brain tumor segmentation. In: Crimi, A., Bakas, S., Kuijf, H., Keyvan, F., Reyes, M., van Walsum, T. (eds.) BrainLes 2018. LNCS, vol. 11384, pp. 456–465. Springer, Cham (2019). https://doi.org/10.1007/978-3-030-11726-9_40

14. Menze, B.H., et al.: The multimodal brain tumor image segmentation benchmark (BRATS). IEEE Trans. Med. Imaging **34**(10), 1993–2024 (2014)

15. Myronenko, A.: 3D MRI brain tumor segmentation using autoencoder regularization. In: Crimi, A., Bakas, S., Kuijf, H., Keyvan, F., Reyes, M., van Walsum, T. (eds.) BrainLes 2018. LNCS, vol. 11384, pp. 311–320. Springer, Cham (2019). https://doi.org/10.1007/978-3-030-11726-9_28

16. Pereira, S., Pinto, A., Alves, V., Silva, C.A.: Brain tumor segmentation using convolutional neural networks in MRI images. IEEE Trans. Med. Imaging **35**(5), 1240–1251 (2016)

17. Ronneberger, O., Fischer, P., Brox, T.: U-Net: convolutional networks for biomedical image segmentation. In: Navab, N., Hornegger, J., Wells, W.M., Frangi, A.F. (eds.) MICCAI 2015. LNCS, vol. 9351, pp. 234–241. Springer, Cham (2015). https://doi.org/10.1007/978-3-319-24574-4_28

18. Sun, L., Zhang, S., Luo, L.: Tumor segmentation and survival prediction in glioma with deep learning. In: Crimi, A., Bakas, S., Kuijf, H., Keyvan, F., Reyes, M., van Walsum, T. (eds.) BrainLes 2018. LNCS, vol. 11384, pp. 83–93. Springer, Cham (2019). https://doi.org/10.1007/978-3-030-11726-9_8

19. Van Griethuysen, J.J., et al.: Computational radiomics system to decode the radiographic phenotype. Cancer Res. **77**(21), e104–e107 (2017)

20. Wang, G., Li, W., Ourselin, S., Vercauteren, T.: Automatic brain tumor segmentation using cascaded anisotropic convolutional neural networks. In: Crimi, A., Bakas, S., Kuijf, H., Menze, B., Reyes, M. (eds.) BrainLes 2017. LNCS, vol. 10670, pp. 178–190. Springer, Cham (2018). https://doi.org/10.1007/978-3-319-75238-9_16

21. Weninger, L., Rippel, O., Koppers, S., Merhof, D.: Segmentation of brain tumors and patient survival prediction: methods for the BraTS 2018 challenge. In: Crimi, A., Bakas, S., Kuijf, H., Keyvan, F., Reyes, M., van Walsum, T. (eds.) BrainLes 2018. LNCS, vol. 11384, pp. 3–12. Springer, Cham (2019). https://doi.org/10.1007/978-3-030-11726-9_1

22. Zhou, C., Chen, S., Ding, C., Tao, D.: Learning contextual and attentive information for brain tumor segmentation. In: Crimi, A., Bakas, S., Kuijf, H., Keyvan, F., Reyes, M., van Walsum, T. (eds.) BrainLes 2018. LNCS, vol. 11384, pp. 497–507. Springer, Cham (2019). https://doi.org/10.1007/978-3-030-11726-9_44

Bag of Tricks for 3D MRI Brain Tumor Segmentation

Yuan-Xing Zhao[1,2(✉)], Yan-Ming Zhang[1], and Cheng-Lin Liu[1,2,3]

[1] NLPR, Institute of Automation, Chinese Academy of Sciences, Beijing, China
{yuanxing.zhao,ymzhang,liucl}@nlpr.ia.ac.cn
[2] University of Chinese Academy of Sciences, Beijing, China
[3] CAS Center for Excellence of Brain Science and Intelligence Technology, Beijing, China

Abstract. 3D brain tumor segmentation is essential for the diagnosis, monitoring, and treatment planning of brain diseases. In recent studies, the Deep Convolution Neural Network (DCNN) is one of the most potent methods for medical image segmentation. In this paper, we review the different kinds of tricks applied to 3D brain tumor segmentation with DNN. We divide such tricks into three main categories: data processing methods including data sampling, random patch-size training, and semi-supervised learning, model devising methods including architecture devising and result fusing, and optimizing processes including warming-up learning and multi-task learning. Most of these approaches are not particular to brain tumor segmentation, but applicable to other medical image segmentation problems as well. Evaluated on the BraTS2019 online testing set, we obtain Dice scores of 0.810, 0.883 and 0.861, and Hausdorff Distances (95th percentile) of 2.447, 4.792, and 5.581 for enhanced tumor core, whole tumor, and tumor core, respectively. Our method won the second place of the BraTS 2019 Challenge for the tumor segmentation.

Keywords: Brain tumor segmentation · Deep neural network

1 Introduction

Gliomas are the most common primary brain malignancies, with different degrees of aggressiveness. 3D brain tumor segmentation plays a vital role in addressing the diagnosis, monitoring, and treatment planning of brain diseases. Although Deep Convolutional Neural Networks (DCNN) have shown great success in solving general computer vision problems, when applied to MRI image segmentation, they face two special challenges.

First, annotated MRI images are very scarce due to privacy concerns and the high cost of human annotation. For example, the training set of BraTS2019 contains only 355 annotated cases. Given that DCNN typically has millions of parameters, the scarcity of annotated data can hardly guarantee the generalization performance of DCNN. Second, the training of DCNN consumes large GPU memory, while the large volume of 3D MRI image data makes the training of DCNN with large patches and large batches impossible. As a result, by training with small MRI patches, DCNN

A. Crimi and S. Bakas (Eds.): BrainLes 2019, LNCS 11992, pp. 210–220, 2020.
https://doi.org/10.1007/978-3-030-46640-4_20

cannot gain enough receptive fields to model the global structure of brains and the spatial relations between different anatomical regions. Besides the specific difficulties of 3D MRI segmentation, the phenomenon of class imbalance is also a common problem that must be faced in semantic segmentation. Most of the tricks described in this paper are aimed at conquering the problems described above.

Since the introduction of U-Net [1] in 2015, DCNN has become the dominating approach for medical image segmentation. Various new approaches have been proposed based on the original U-Net. Myronenko, A. [2] uses auto-encoder to reconstruct the input image itself and regularize the optimizing process. Isensee, F. [3] takes advantage of other labeled data, using a co-training method. McKinley, R. [4] proposes label-uncertainty loss to models to label noise and uncertainty. These methods were shown to be effective in improving the segmentation, yet further improvements can be achieved by considering and combining various strategies in data processing, network architecture, and learning-algorithm design. In this work, we introduce several useful tricks in model learning and combine them to boost the overall accuracy of the model.

The rest of this paper is organized as follows. Section 2 introduces tricks used in 3D MRI brain-tumor segmentation. Section 3 presents the implementation details and experimental results, and Sect. 4 provides concluding remarks.

2   Methods

Many tricks in general DCNN design and training for the image can also be applied to the 3D brain-image segmentation. We divide such tricks into three categories: data processing methods, model designing methods, and optimizing methods.

2.1   Data Processing Methods

Sampling. Data imbalance has always been a hot topic for segmentation. Commonly, that training data contains an overwhelming number of background voxels, most of which are easy for the classifier to predict, and only a few are difficult. Here, we use two methods to cope with this problem.

Heuristic Sampling. To reduce the effect of background voxels, we use a heuristic sampling method to select more informative patches. More concretely, several patches are randomly cropped from the input MRI, and the one that contains the most foreground voxels is selected to feed the model.

Hard Sample Mining. A standard solution was known as hard sample mining. At every iteration, the voxels with the largest loss values are selected as training voxels, and their gradients are back-propagated to update the model's parameters. The gradients of other voxels are discarded directly. The percentage of the selected voxels is set to and decreases as the number of iteration increases.

The method described above can be seen as a method that uses a hard threshold to select the difficult sample. Lin, T.Y. [5] proposed a new loss called focal loss to conquer the class imbalance problem through a soft threshold and outperforms the alternatives of training with the hard sample mining. The loss is defined as (2).

$$p_t = \begin{cases} p & if \quad y = 1 \\ 1 - p & otherwise \end{cases} \tag{1}$$

$$CE(p_t) = -(1 - p_t)^\gamma \log(p_t) \tag{2}$$

In the above, $y \in \{\pm 1\}$ specifies the ground-truth class, and $p \in [0, 1]$ is the model's estimated probability for the class with $y = 1$. $\gamma > 0$ is a tunable focusing parameter. When an example is misclassified and p_t is small, the modulating factor is near 1, and the loss is unaffected. As $p_t \rightarrow 1$, the factor goes to 0, and the loss for well-classified examples is down-weighted. The focusing parameter γ smoothly adjusts the rate at which easy examples are down-weighted. Experiment results demonstrate that focal loss is better than general hard sample mining.

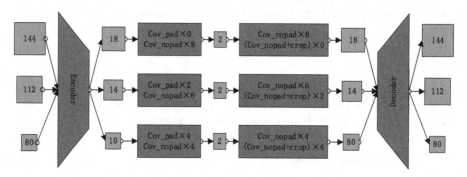

Fig. 1. Random patch-size training strategy. At each iteration, we select a training batch with random patch size and using padding and cropping to adjust the size of the feature map. The number in each square is the size of the patch.

Random Patch-Size Training. Using large patches could include more contextual information but leads to the small batch size, which increases the variance of stochastic gradients and hurts optimization. On the other hand, using a large batch size could facilitate the optimization but leads to small patches, which results in less contextual information. To take benefit of both sizes of patches, we construct a training batch pool with batches of different patch sizes. Note that if the size of the patch is large, the corresponding batch size will be small.

This method is illustrated in Fig. 1. Using the different numbers of padding and cropping layers between the convolution layer, this model can learn global information from the largest patch and informative texture from the small patch with the same parameter. For each iteration during training, we randomly select a batch from the pool to update the model. We take advantage of both the large patches and the large batch size. In practice, we found this simple strategy very efficient.

Semi-supervised Learning. To tackle the lack of annotated data, we use a semi-supervised method called the multi-space semi-supervised method. At the first iteration, the manually labeled dataset is used as a training set, *and* different student-models, s_i^0 are trained on the training set under some different conditions, such as the different subset of training set or the different subspace of features, etc. Then, all student-models are combined as a teacher model, so that the teacher model is defined as

$$T^0 = \frac{1}{n} \sum_{i=1}^n s_i^0 \qquad (3)$$

Finally, the teacher model, T^0, is used to label the unlabeled dataset.

After the first iteration, we combine the manually labeled dataset and model labeled dataset as the new training set and then repeat the training process as the first iteration. We repeated the process until the accuracy of the student model is stable. Such a process is summarized in Fig. 2.

Algorithm 1:   Multi-View Semi-Supervised Segmentation Algorithm

---

1. **Input:**
   a. Overlapped MRI labeled patches $\{(x_l, y_l)\}$ and unlabeled patches $\{x_u\}$
   b. Data transformation functions: $\{(T_v, T_v^{-1})\}$
2. **Training the ensemble model:**
   a. Initialize pseudo-labels $\{y_u\}$ by some supervised, trained model
   b. Repeat
      - For each transformation function T_v, update the model f_v:

$$f_v = \underset{f}{\operatorname{argmin}} \sum_l L(T_v(x_l), T_v(y_l), f) + \alpha \sum_u L(T_v(x_u), T_v(y_u), f)$$

      - Update the ensemble model by Eq. (4)
      - Update pseudo labels $\{y_u\}$ by the method described in Sec.2.2
      Until convergence
3. **Training the student model:**
   a. Learn a student model based on F, $\{(x_l, y_l)\}$, $\{x_u\}$

Fig. 2. Multi-view semi-supervised segmentation algorithm

2.2   Model Devising Methods

Architecture Devising. The whole architecture used in this paper is shown in Fig. 3. For many computer vision tasks, a classic way to boost the accuracy is to combine multiple prediction results made at different scales. Inspired by this observation, we introduce a new architecture, named self-ensemble, that makes predictions at each scale of U-Net and then joins them to obtain the final prediction. The simple way to combine predictions of different sizes is to resize them to the size of the input image, as illustrated on the left side of Fig. 4. However, it is highly memory consuming since it up-samples every prediction tensor to the largest size. Instead, we propose to combine the predictions in a recursive manner which is illustrated at the right side of Fig. 4 and formulated as

$$y_s = \mathrm{Up}(y_{s+1}, 2) + \widetilde{y}_s \qquad s = S-1, S-2, \ldots, 1 \tag{4}$$

$$y = y_1 \tag{5}$$

where $\widetilde{y}_s$ is the prediction tensor of the s-th scale, S is the number of scales in U-Net, and $\mathrm{Up}(y, t)$ is a function that up-samples y by rate t. The prediction of the current scale y_s is based on y_{s+1} and only needs to model the residual of y_{s+1}. The final result of $y = y_1$, combining predictions at each scale, outperforms any single prediction.

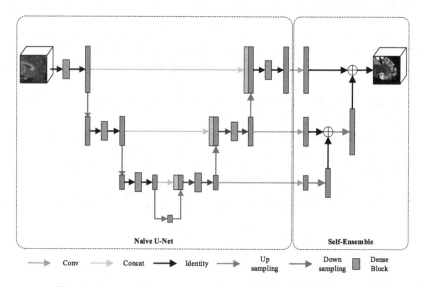

Fig. 3. The overall architecture of the self-ensemble U-Net model.

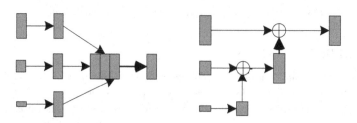

Fig. 4. The left figure is the naïve way to combine predictions of different scales. The right figure is our self-ensemble method.

Fusing Result. We use two methods for combining different results to improve the final prediction. The overall architecture is shown in Fig. 5. The ensemble model fusing the prediction of different models is shown at the top of Fig. 5, and the method of fusing the prediction of overlapped patches is illustrated at the bottom of Fig. 5.

Fusing the Prediction of Different Models. The method combined different models like [3]. We evaluate our model by running five-folds cross-validation on the training cases. Then we use the average of all the five models as the final ensemble model.

Fusing the Prediction of the Overlapped Patch. The model may predict the different results of the same voxel because of the voxel located in a different position related to the different patches. Base on this phenomenon, we crop the input MRI into overlapped patches and then combine these patches as a batch predicted by the model. In this way, the overlapped voxel is predicted more than one time. A more accurate result would be obtained by averaging these predictions.

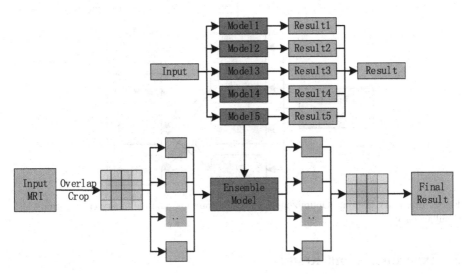

Fig. 5. The overall architecture of the result is fusing. At the top of the figure is the method for fusing the predictions of the different models, and at the bottom of the figure is the method fusing the prediction of overlapped patches.

2.3   Optimizing Methods

Gradual Warming Up Learning Rate. The gradual warming up learning rate, which was first proposed in [6], gradually increases the learning rate from a small value to a large value. In practice, with a mini-batch of size, we start from a learning rate of η and increase it by a constant amount at each iteration until it reaches $\widehat{\eta} = k\eta$ after several epochs. After the warmup phase, we go back to the original learning rate schedule.

Multitask Learning. Multitask learning can be seen as a regulation method. It can affect the process of optimizing and provide additional information related to the learning problem. In this paper, we first calculate the cross-entropy loss between the softmax result and the label provided by the organizer, including the background (BG), the necrotic and the non-enhancing tumor core (NCR/NET), the peritumoral edema (ED), and the enhancing tumor (ET). Then the predicted result and ground truth are reorganized as four independent categories. These include background, enhancing tumor, whole tumor (WT), and tumor core (TC). Finally, the binary cross-entropy loss is calculated between the reorganized prediction and ground truth. The overall process can be seen as Fig. 6.

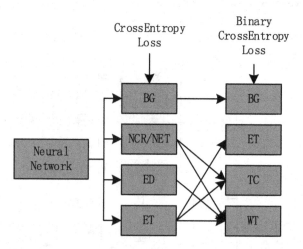

Fig. 6. Multitask learning. Optimizing the cross-entropy loss and the binary cross-entropy loss simultaneously.

3   Experiments and Results

3.1   Datasets and Evaluation Metrics

Datasets. We use two datasets in our experiments.

BraTS2019 [7–11]. It contains 355 cases whose corresponding manual segmentation is provided. Each case has four MRI sequences that are named T1, T1 contrast-enhanced, T2, and FLAIR, respectively.

Decathlon [12]. It comprises 750 cases collected from older BraTS challenges. We use this dataset as the unlabeled dataset.

Evaluation Metrics. The segmentation performance was quantitatively assessed using the mean Dice coefficient (DSC). Let A and B denote the manual label and predicted label, respectively. The mean Dice similarity coefficient is defined as

$$DSC = \frac{1}{n} \sum_{i=1}^{n} \frac{2|A_i B_i|}{|A_i| + |B_i|} \tag{6}$$

where $|A_i|$ denotes the number of positive elements in the binary segmentation A_i, and $|A_i B_i|$ is the number of positive elements shared by A_i and B_i. n = 4 is the number of labels.

3.2 Preprocessing

In our approach, before feeding the data to the deep neural network, each MRI sequence of a case is normalized independently. Specifically, all voxels of an MRI sequence are normalized to range from 0 to 1. We also apply a random axis mirror along the horizontal axis.

3.3 Implementation Details

We use the summation of cross-entropy and average Dice similarity as the loss function. The patch size is randomly selected from 64, 80, 96, 112, 128, 144, and the corresponding batch size is 15, 8, 4, 2, 1, 1. SGD with momentum is used as the optimizer, and the learning rate is set to 0.4. The step of warming up is set to 20 epochs. The model is trained in an end-to-end way, and no additional preprocessing or post-processing is performed. The method is implemented by Pytorch, and all experiments are conducted on two TITAN GPUs with 12G RAM. It took around 21 h to train the model.

3.4 Results

To better understand our method, we conduct ablation experiments to examine how some trick affects the final performance. We evaluate the performance under different experimental settings: (1) BL, the original U-Net: equipped with dense block structure and self-ensemble structure. (2) BL+warmup: model 1 with warming-up learning rate. (3) BL+warmup+fuse: model 2 with the resultant fusing of five different models trained by fivefold cross-validation. (4) BL+warmup+fuse+semi: model 3 with semi-supervised learning. From the results listed in Table 1, we can observe that our method steadily improves accuracy with the addition of each component. Three examples are visualized in Fig. 7. We can see that most voxels are segmented correctly, while some errors occur in the small regions and boundaries between regions. Table 2 shows the results of our model on the BraTS 2019 testing dataset.

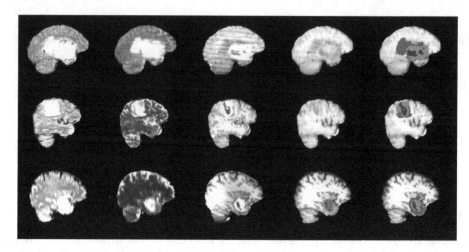

Fig. 7. The visualization result of the validation set of the BraTS2019 dataset. From left to right, the column is the original FLAIR image, the original T2 image, the original T1ce image, the original T1 image, and the segmentation result overlaid over the T1 image.

Table 1. Results of the BraTS2019 validation data (125 cases). Metrics are computed by the online evaluation platform.

Method	Dice			Hausdorff Dist.		
	ET	WT	TC	ET	WT	TC
BL	0.702	0.893	0.800	4.766	5.078	6.472
BL+warmup	0.729	0.904	0.802	3.832	4.141	8.099
BL+warmup+fuse	0.737	0.908	0.823	4.089	4.599	6.433
BL+warmup+fuse+psudo label	0.754	0.910	0.835	3.844	4.569	5.581

Table 2. Results of the BraTS2019 testing data (166 cases). Metrics are computed by the online evaluation platform.

	Dice			Hausdorff Dist.		
	ET	WT	TC	ET	WT	TC
Mean	0.810	0.883	0.861	2.447	4.792	4.217
StdDev	0.193	0.145	0.225	4.030	6.619	7.503
Median	0.850	0.924	0.928	1.732	3.0	2.236
25quantile	0.783	0.875	0.882	1.0	1.494	1.414
75quantile	0.915	0.951	0.960	2.236	4.899	3.606

4   Conclusion

In this paper, we review useful tricks for training DCNN to improve the accuracy of brain tumor segmentation and evaluate their performance. Our empirical results on the BraTS2019 indicate that these tricks improve model accuracy consistently. In particular, stacking all of them together leads to significantly higher accuracy. On the BraTS2019 online validation set, our combined method achieved average Dice scores of 0.754, 0.910, 0.835 for the enhancing tumor, whole tumor, and tumor core, respectively. However, our model tends to make false predictions for small anatomical regions.

In the future, we will investigate methods for accurately segmenting small regions and apply them to other tasks such as prediction of patient overall survival from pre-operative scans.

Acknowledgments. This work is supported by the National Natural Science Foundation of China(NSFC) Grants 61773376, 61721004, 61836014, as well as Beijing Science and Technology Program Grant Z181100008918010.

References

1. Ronneberger, O., Fischer, P., Brox, T.: U-Net: convolutional networks for biomedical image segmentation. In: Navab, N., Hornegger, J., Wells, W.M., Frangi, A.F. (eds.) MICCAI 2015. LNCS, vol. 9351, pp. 234–241. Springer, Cham (2015). https://doi.org/10.1007/978-3-319-24574-4_28

2. Myronenko, A.: 3D MRI brain tumor segmentation using autoencoder regularization. In: Crimi, A., Bakas, S., Kuijf, H., Keyvan, F., Reyes, M., van Walsum, T. (eds.) BrainLes 2018. LNCS, vol. 11384, pp. 311–320. Springer, Cham (2019). https://doi.org/10.1007/978-3-030-11726-9_28

3. Isensee, F., Kickingereder, P., Wick, W., Bendszus, M., Maier-Hein, Klaus H.: No new-net. In: Crimi, A., Bakas, S., Kuijf, H., Keyvan, F., Reyes, M., van Walsum, T. (eds.) BrainLes 2018. LNCS, vol. 11384, pp. 234–244. Springer, Cham (2019). https://doi.org/10.1007/978-3-030-11726-9_21

4. McKinley, R., Meier, R., Wiest, R.: Ensembles of densely-connected CNNs with label-uncertainty for brain tumor segmentation. In: Crimi, A., Bakas, S., Kuijf, H., Keyvan, F., Reyes, M., van Walsum, T. (eds.) BrainLes 2018. LNCS, vol. 11384, pp. 456–465. Springer, Cham (2019). https://doi.org/10.1007/978-3-030-11726-9_40

5. Lin, T.-Y., et al.: Focal loss for dense object detection. In: Proceedings of the IEEE International Conference on Computer Vision, pp. 2980–2988 (2017)

6. Goyal, P., et al.: Accurate, large minibatch SGD: Training imagenet in 1 h. arXiv preprint arXiv:1706.02677 (2017)

7. Bakas, S., et al.: Identifying the best machine learning algorithms for brain tumor segmentation, progression assessment, and overall survival prediction in the BRATS challenge. arXiv preprint arXiv:1811.02629 (2018)

8. Menze, B.H., et al.: The multimodal brain tumor image segmentation benchmark (BRATS) **34**, 1993–2024 (2014). https://doi.org/10.1109/tmi.2014.2377694

9. Bakas, S., et al.: Advancing the cancer genome atlas glioma MRI collections with expert segmentation labels and radiomic features. **4**, 170117 (2017). https://doi.org/10.1038/sdata.2017.117
10. Bakas, S., et al.: Segmentation labels and radiomic features for the pre-operative scans of the TCGA-GBM collection. **286** (2017). https://doi.org/10.7937/k9/tcia.2017.klxwjj1q
11. Bakas, S., et al.: Segmentation labels and radiomic features for the pre-operative scans of the TCGA-LGG collection. **286** (2017). https://doi.org/10.7937/k9/tcia.2017.gjq7r0ef
12. Simpson, A.L., et al.: A large annotated medical image dataset for the development and evaluation of segmentation algorithms. arXiv preprint arXiv:1902.09063 (2019)

Multi-resolution 3D CNN for MRI Brain Tumor Segmentation and Survival Prediction

Mehdi Amian[1]([✉]) and Mohammadreza Soltaninejad[2]

[1] Control and Intelligent Center of Excellence, School of Electrical
and Computer Engineering, University of Tehran, Tehran, Iran
mehdi.amian@ut.ac.ir
[2] School of Computer Science, University of Nottingham, Nottingham, UK
m.soltaninejad@nottingham.ac.uk

Abstract. In this study, an automated three dimensional (3D) deep segmentation approach for detecting gliomas in 3D pre-operative MRI scans is proposed. Then, a classification algorithm based on random forests, for survival prediction is presented. The objective is to segment the glioma area and produce segmentation labels for its different sub-regions, i.e. necrotic and the non-enhancing tumor core, the peritumoral edema, and enhancing tumor. The proposed deep architecture for the segmentation task encompasses two parallel streamlines with two different resolutions. One deep convolutional neural network is to learn local features of the input data while the other one is set to have a global observation on whole image. Deemed to be complementary, the outputs of each stream are then merged to provide an ensemble complete learning of the input image. The proposed network takes the whole image as input instead of patch-based approaches in order to consider the semantic features throughout the whole volume. The algorithm is trained on BraTS 2019 which included 335 training cases, and validated on 127 unseen cases from the validation dataset using a blind testing approach. The proposed method was also evaluated on the BraTS 2019 challenge test dataset of 166 cases. The results show that the proposed methods provide promising segmentations as well as survival prediction. The mean Dice overlap measures of automatic brain tumor segmentation for validation set were 0.86, 0.77 and 0.71 for the whole tumor, core and enhancing tumor, respectively. The corresponding results for the challenge test dataset were 0.82, 0.72, and 0.70, respectively. The overall accuracy of the proposed model for the survival prediction task is 55% for the validation and 49% for the test dataset.

Keywords: Convolutional neural network · U-Net · Deep learning · MRI · Brain tumor segmentation

1 Introduction

Brain tumors are caused by abnormal growth of the cells inside brain and have a wide variety of tumor types. They can be generally categorized into low-grade gliomas (LGG) or high-grade gliomas (HGG). Magnetic resonance imaging (MRI) plays an

© Springer Nature Switzerland AG 2020
A. Crimi and S. Bakas (Eds.): BrainLes 2019, LNCS 11992, pp. 221–230, 2020.
https://doi.org/10.1007/978-3-030-46640-4_21

important role regarding the clinical tasks related to brain tumors. Accurate segmentation of brain tumor may aid the measurement of tumor features to help diagnosis, treatment planning and survival prediction [1]. MR images can be generated using different acquisition protocols such as fluid attenuated inversion recovery (FLAIR), T1-weighted (with and without contrast agent), and T2-weighted to distinguish between different tumor sub-tissues.

Segmentation of human brain tumor in medical images is a vital and crucial task that traditionally is performed manually by physicians. The manual delineation practices are subjective and inherently prone to misinterpretation that can bring about sever and even fatal upcomings. So, developing a reliable and fast automated algorithm, undoubtedly, leads to much more accurate diagnosis, making a remarkable advance in long term in treatment planning for the patients. This becomes even more highlighted when having three-dimensional observation to images by the machine instead of a natural two-dimensional view of a human interpreter.

So far, many efforts addressed inventing such an automatic segmentation system. Undeniably, several big steps have been taken, yet there is a lot to be taken. On the other hand, thanks to emerging powerful computing processors as well as availability of big datasets, deep learning and particularly its recent advancements revolutionized many aspects of the technology by manifesting unprecedented amount of knowledge and learning about various data types, i.e. text, speech, and image.

Deep learning is widely being used in medical imaging domain in various ways such as denoising, finding biomarkers, pattern prediction, and detecting lesions and tumors. Applying deep learning techniques to multimodal MR images for tumor segmentation is naturally a challenging task due to high dimensionality of the input data, poor quality and problems of the image during capturing such as bias field, and after all designing an appropriate architecture for the specific objective.

Due to the recent advances in deep neural networks (DNN) in recognition of the patterns in the images, most of the recent tumor segmentations have focused on deep learning methods [2]. Fully convolutional networks (FCN) have been suggested for per-pixel classification with the advantage of end-to-end learning [3]. Despite the advantage of dense pixel classification, FCN-based methods suffer from the loss of spatial information, which occurs in the pooling layers, results in coarse segmentation [2]. U-Net [4] proposed using skip layers to tackle this problem and was suggested for fine medical image segmentation. Several methods have proposed using U-Net for brain tumor segmentation [5–7]. U-Net showed promising results in medical image segmentation tasks. So, several researches focused on U-Net modification to acquire even better outcomes. Cascaded U-Net is a successful example of such modifications as presented in [8] where the cascaded U-Net outperforms the standard version.

The present study on segmentation is inspired by two deep convolutional neural networks, i.e. U-Net and the one that is proposed in [9]. These networks are placed in parallel stream lines fed by original images. The U-Net is meant to capture the local features and make a fine learning of the data. On the other hand, the other pipeline is to maintain a coarse but global learning.

Furthermore, the proposed architecture is designed to take the whole image as the input for the network, rather than patch-based architectures which incorporate partial information from the images during training. This approach ensures that all the semantic features throughout the whole volume will be considered during training, which eliminates most of the false positives.

For the survival prediction task, a model based on the random forest (RF) [10] is presented. The output of the model is as the number of days. As an ensemble learning algorithm, RF is widely being used in both classification and regression purposes.

2 Materials and Methods

2.1 Dataset

The proposed network is trained using the Multimodal Brain Tumor Segmentation Challenge (BraTS) 2019 [11–15] training dataset which includes 259 HGG and 76 LGG patient cases. The dataset contains segmentation ground truth manually annotated by experts and provided on the Center for Biomedical Image Computing and Analytics (CBICA) portal. The network was evaluated using BRTAS 2019 validation dataset which includes 125 patient cases. For the task of post-operative survival prediction, 101 of the training patient cases were provided with the survival information. The network was also evaluated for the challenge test dataset which includes 166 patient cases for the segmentation task and 107 cases for the survival prediction task.

2.2 Segmentation

Our proposed segmentation method consists of two main pipelines steps with different resolution levels, i.e. original and low resolutions. The architecture of the proposed network is depicted in Fig. 1. The pre-processing stage consists of intensity normalization, histogram matching and bias filed correction. The intensities were normalized for each protocol by subtracting the average of intensities of the image and divided by their standard deviation. The histogram of each image was normalized and matched to a selected reference image, which is one of the patient cases. Bias field correction was performed using the toolbox provided in [16].

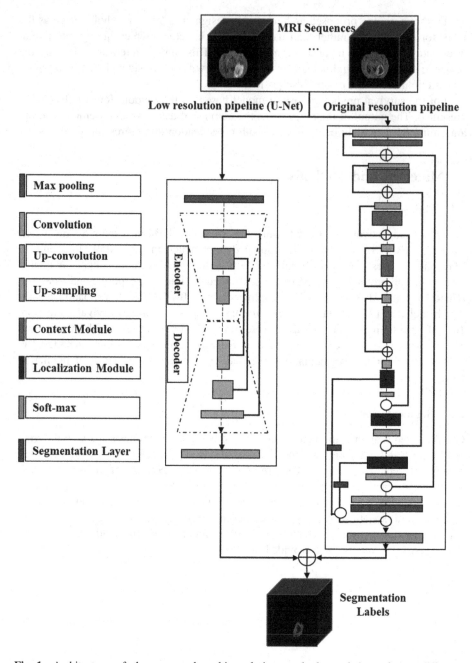

Fig. 1. Architecture of the proposed multi-resolution method consisting of two different pathways one with original resolution and the other with lower resolution and larger field of view.

2.3 Survival Prediction

The output masks forming the segmentation in the previous section are used for the task of survival prediction. Some spatial features are extracted for the whole tumor and each sub-tissue. Normalized volume size for the whole tumor, tumor core and enhanced tumor are calculated. The average intensity value for each sub-tissue is another feature. The feature vectors extracted for each volume of interest (VOI) were input to the random forests (RF). RF parameters, i.e. tree depth and the number of trees, were tuned by examining them on training datasets and evaluating the classification accuracy using 5-fold cross validation. The number of 30 trees with depth 10 provided an optimum generalization and accuracy. The RF classifier was used in regression mode which produced predictions as number of days. The schematic diagram of the proposed approach for the survival prediction is demonstrated in Fig. 2.

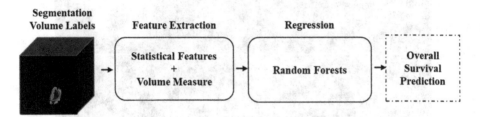

Fig. 2. The pipeline for overall survival prediction based on the volume labels extracted in the segmentation stage

3 Results

3.1 Segmentation Task

For the segmentation task, the proposed method was implemented using Keras Tensorflow backend on Nvidia GeForce GTX 1080 Ti GPU, RAM 11 GB, PC with CPU Intel Core i7 and RAM 16 GB with the operating system Linux. The U-Net [4] was modified and the method in [9] was implemented using [17]. The RF was implemented using MATLAB 2019a. The ground truth is provided for the training set, whilst for evaluation a blind testing system is utilized. The evaluation measures which are provided by the CBICA's Image Processing Portal, i.e. Dice score, sensitivity, specificity, Hausdorff distance, were used to compare the segmentation results with the gold standard (blind testing).

Table 1 presents the evaluation results obtained by applying the proposed segmentation method on BraTS 2019 validation dataset which was provided by CBICA blind testing system. Figure 3 shows segmentation results of the proposed multi-resolution approach for some cases of BraTS 2019 training dataset and the ground truth. Two modalities, i.e. FLAIR and T1-ce, are shown in Fig. 2 and the tumor sub-tissues are overlaid on T2 modality and depicted in axial, sagittal, and coronal views. Figure 4 represents segmentation results of the proposed multi-resolution method for two sample cases of BraTS 2019 validation dataset.

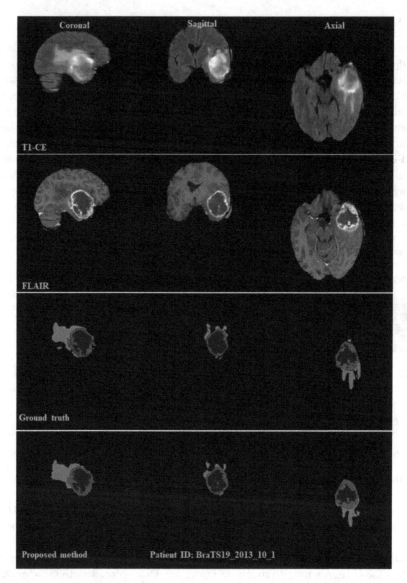

Fig. 3. Segmentation results for one sample training data using the proposed multi-resolution model, and comparison with the ground truth. Light blue: necrosis and on-enhancing, green: edema, red: enhancing. The Dice scores reported by the CBICA system for enhancing tumor, tumor core, and the whole tumor are as follows: BraTS19_2013_10_1: 0.82, 0.94, and 0.91. (Color figure online)

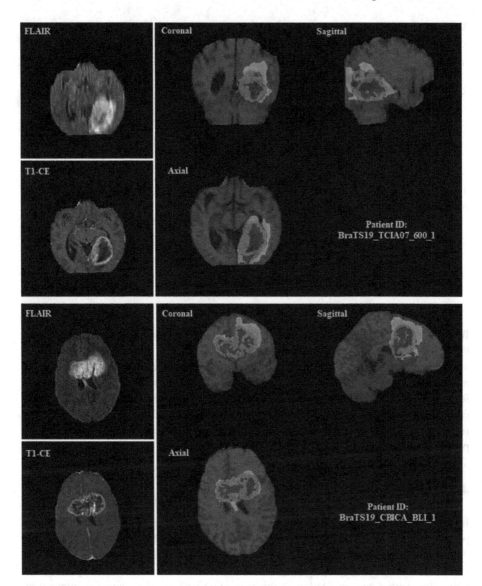

Fig. 4. Segmentation results for two validation data samples using the proposed multi-resolution model. Light blue: necrosis and on-enhancing, green: edema, red: enhancing. The Dice scores reported by the CBICA system for enhancing tumor, tumor core, and the whole tumor are as follows: Brats19_TCIA07_600_1: 0.86, 0.91, and 0.93; Brats19_CBICA_BLI_1: 0.84, 0.92, and 0.91. (Color figure online)

Table 1. Segmentation results for validation dataset provided by CBICA portal blind testing system. ET: enhancing tumor, WT: whole tumor, TC: tumor core.

Dataset		Dice			Sensitivity			Specificity			Hausdorff (95%)		
		ET	WT	TC	ET	WT	TC	ET	WT	TC	ET	WT	TC
Validation	Mean	0.71	0.86	0.77	0.69	0.85	0.76	1.00	0.99	1.00	6.92	8.42	11.55
	STD	0.25	0.09	0.18	0.25	0.11	0.18	0.00	0.01	0.01	11.87	14.21	20.04
Test	Mean	0.70	0.82	0.72	–	–	–	–	–	–	5.59	8.42	9.14
	STD	0.23	0.18	0.29	–	–	–	–	–	–	11.64	13.22	12.68

Table 2. The results of survival prediction for validation dataset provided by CBICA portal blind testing system.

Dataset	Accuracy	MSE	Median SE	STD SE	Spearman R
Validation	0.55	104253	43264	142579	0.26
Test	0.49	408632	69696	1219534	0.28

3.2 Survival Prediction

In order to validate the survival prediction method, 29 patient cases were specified by the CBICA portal, and the predictions were generated in terms of the number of survival days. Evaluation metrics for this task, were accuracy for classification mode and mean square error (MSE), median and standard deviation of SE, and Spearman R for regression mode. The classes of survival predictions were calculated based on three categories, short (less than 10 months), medium (between 10 to 15 months), and long (more than 15 months). The survival prediction results provided by CBICA system are presented in Table 2.

4 Conclusion

In the present study, a three-dimensional multi-resolution learning-based algorithm was proposed in which, instead of patching the image, the whole MR image is passed to the network. The low resolution path was inspired by the U-Net architecture which was modified to take a larger input receptive field and considered the whole input volume rather than partial patches. Although this procedure performed a coarser segmentation, the false positives were successfully eliminated, while the original resolution path produced fine segmentation boundaries. Fusion of these two resolution levels results in increasing accuracy, specificity and sensitivity compared to utilizing each single pipeline separately. The proposed algorithm reached to the Dice scores of 0.86, 0.77 and 0.71 for the whole tumor, core and enhancing tumor on the validation data, and 0.82, 0.72 and 0.70 on the test data set. The hand-crafted statistical and intensity-based features extracted from the segmentation masks are then applied to a random forest classifier for the task of survival prediction. The proposed method acquired MSE and classification accuracy of 104253 and 0.55, for the validation dataset. The corresponding results for the challenge dataset were 408632, and 0.49, respectively.

References

1. Gordillo, N., Montseny, E., Sobrevilla, P.: State of the art survey on MRI brain tumor segmentation. Magn. Reson. Imaging **31**, 1426–1438 (2013)
2. Soltaninejad, M., Zhang, L., Lambrou, T., Yang, G., Allinson, N., Ye, X.: MRI brain tumor segmentation and patient survival prediction using random forests and fully convolutional networks. In: Crimi, A., Bakas, S., Kuijf, H., Menze, B., Reyes, M. (eds.) BrainLes 2017. LNCS, vol. 10670, pp. 204–215. Springer, Cham (2018). https://doi.org/10.1007/978-3-319-75238-9_18
3. Long, J., Shelhamer, E., Darrell, T.: Fully convolutional networks for semantic segmentation. Presented at the Proceedings of the IEEE Conference on Computer Vision and Pattern Recognition (2015)
4. Ronneberger, O., Fischer, P., Brox, T.: U-Net: convolutional networks for biomedical image segmentation. In: Navab, N., Hornegger, J., Wells, W.M., Frangi, A.F. (eds.) MICCAI 2015. LNCS, vol. 9351, pp. 234–241. Springer, Cham (2015). https://doi.org/10.1007/978-3-319-24574-4_28
5. Feng, X., Tustison, N., Meyer, C.: Brain tumor segmentation using an ensemble of 3D U-Nets and overall survival prediction using radiomic features. In: Crimi, A., Bakas, S., Kuijf, H., Keyvan, F., Reyes, M., van Walsum, T. (eds.) BrainLes 2018. LNCS, vol. 11384, pp. 279–288. Springer, Cham (2019). https://doi.org/10.1007/978-3-030-11726-9_25
6. Chen, W., Liu, B., Peng, S., Sun, J., Qiao, X.: S3D-UNet: separable 3D U-Net for brain tumor segmentation. In: Crimi, A., Bakas, S., Kuijf, H., Keyvan, F., Reyes, M., van Walsum, T. (eds.) BrainLes 2018. LNCS, vol. 11384, pp. 358–368. Springer, Cham (2019). https://doi.org/10.1007/978-3-030-11726-9_32
7. Isensee, F., Kickingereder, P., Wick, W., Bendszus, M., Maier-Hein, K.H.: No new-net. In: Crimi, A., Bakas, S., Kuijf, H., Keyvan, F., Reyes, M., van Walsum, T. (eds.) BrainLes 2018. LNCS, vol. 11384, pp. 234–244. Springer, Cham (2019). https://doi.org/10.1007/978-3-030-11726-9_21
8. Lachinov, D., Vasiliev, E., Turlapov, V.: Glioma segmentation with cascaded UNet. In: Crimi, A., Bakas, S., Kuijf, H., Keyvan, F., Reyes, M., van Walsum, T. (eds.) BrainLes 2018. LNCS, vol. 11384, pp. 189–198. Springer, Cham (2019). https://doi.org/10.1007/978-3-030-11726-9_17
9. Isensee, F., Kickingereder, P., Wick, W., Bendszus, M., Maier-Hein, K.H.: Brain tumor segmentation and radiomics survival prediction: contribution to the BRATS 2017 challenge. In: Crimi, A., Bakas, S., Kuijf, H., Menze, B., Reyes, M. (eds.) BrainLes 2017. LNCS, vol. 10670, pp. 287–297. Springer, Cham (2018). https://doi.org/10.1007/978-3-319-75238-9_25
10. Breiman, L.: Random forests. Mach. Learn. **45**, 5–32 (2001)
11. Menze, B.H., et al.: The multimodal brain tumor image segmentation benchmark (BRATS). IEEE Trans. Med. Imaging **34**(10), 1993–2024 (2015). https://doi.org/10.1109/TMI.2014.2377694
12. Bakas, S., et al.: Advancing the cancer genome Atlas glioma MRI collections with expert segmentation labels and radiomic features. Nat. Sci. Data **4**, 170117 (2017). https://doi.org/10.1038/sdata.2017.117

13. Bakas, S., Reyes, M., Jakab, A., Bauer, S., Rempfler, M., Crimi, A., et al.: Identifying the best machine learning algorithms for brain tumor segmentation, progression assessment, and overall survival prediction in the BRATS challenge. arXiv preprint arXiv:1811.02629 (2018)

14. Bakas, S., et al.: Segmentation labels and radiomic features for the pre-operative scans of the TCGA-GBM collection. Cancer Imaging Archive (2017). https://doi.org/10.7937/K9/TCIA.2017.KLXWJJ1Q

15. Bakas, S., et al.: Segmentation labels and radiomic features for the pre-operative scans of the TCGA-LGG collection. Cancer Imaging Archive (2017). https://doi.org/10.7937/K9/TCIA.2017.GJQ7R0EF

16. Tustison, N.J., Avants, B.B., Cook, P.A., Yuanjie Zheng, E.A., Yushkevich, P.A., Gee, J.C.: N4ITK: improved N3 bias correction. IEEE Trans. Med. Imaging **29**, 1310–1320 (2010)

17. Keras 3D U-Net Convolution Neural Network (CNN) designed for medical image segmentation. https://github.com/ellisdg/3DU-NetCNN. Accessed 10 June 2019

Two-Stage Cascaded U-Net: 1st Place Solution to BraTS Challenge 2019 Segmentation Task

Zeyu Jiang[1], Changxing Ding[1(✉)], Minfeng Liu[2], and Dacheng Tao[3]

[1] School of Electronic and Information Engineering,
South China University of Technology, Guangzhou, China
chxding@scut.edu.cn
[2] Nanfang Hospital, Southern Medical University,
Guangzhou 510515, Guangdong, China
[3] UBTECH Sydney AI Centre, SIT, FEIT, University of Sydney, Sydney, Australia

Abstract. In this paper, we devise a novel two-stage cascaded U-Net to segment the substructures of brain tumors from coarse to fine. The network is trained end-to-end on the Multimodal Brain Tumor Segmentation Challenge (BraTS) 2019 training dataset. Experimental results on the testing set demonstrate that the proposed method achieved average Dice scores of 0.83267, 0.88796 and 0.83697, as well as Hausdorff distances (95%) of 2.65056, 4.61809 and 4.13071, for the enhancing tumor, whole tumor and tumor core, respectively. The approach won the 1st place in the BraTS 2019 challenge segmentation task, with more than 70 teams participating in the challenge.

Keywords: Deep learning · Brain tumor segmentation · U-Net

1 Introduction

Gliomas are the most common type of primary brain tumors. Automatic three-dimensional brain tumor segmentation can save doctors time and provide an appropriate method of additional tumor analysis and monitoring. Recently, deep learning approaches have consistently outperformed traditional brain tumor segmentation methods [6,10,17,20,24,27].

The multimodal brain tumor segmentation challenge (BraTS) is aimed at evaluating state-of-the-art methods for the segmentation of brain tumors [1–4,13]. The BraTS 2019 training dataset, which comprises 259 cases of high-grade gliomas (HGG) and 76 cases of low-grade gliomas (LGG), is manually annotated by both clinicians and board-certified radiologists. For each patient, a native pre-contrast (T1), a post-contrast T1-weighted (T1Gd), a T2-weighted (T2) and a T2 Fluid Attenuated Inversion Recovery (T2-FLAIR) are provided. An example image set is presented in Fig. 1. Each tumor is segmented into enhancing tumor, the peritumoral edema, and the necrotic and non-enhancing tumor core. A number of metrics (Dice score, Hausdorff distance (95%), sensitivity and specificity)

© Springer Nature Switzerland AG 2020
A. Crimi and S. Bakas (Eds.): BrainLes 2019, LNCS 11992, pp. 231–241, 2020.
https://doi.org/10.1007/978-3-030-46640-4_22

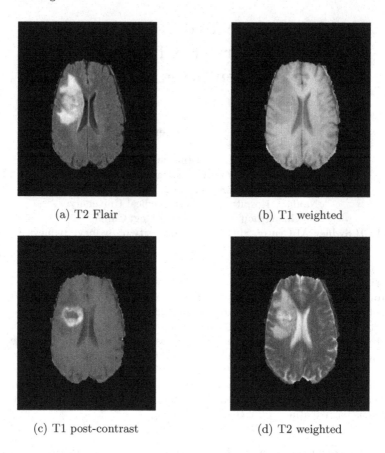

(a) T2 Flair
(b) T1 weighted

(c) T1 post-contrast
(d) T2 weighted

Fig. 1. Example of image modalities in the BraTS 2019 dataset.

are used to measure the segmentation performance of the algorithms proposed by participants.

In BraTS 2017, Kamnitsas et al. [9], who was the first-place winner of the challenge, proposed Ensembles of Multiple Models and Architectures (EMMA) for robust segmentation, which was achieved by combining several network architectures including DeepMedic [10], 3D U-Net [18] and 3D FCN [12]. These networks were trained with different optimization processes via diverse loss functions such as Dice loss [14] and cross-entropy loss. In BraTS 2018, Myronenko [15], who achieved the best performance on the testing dataset, utilized an asymmetrical U-Net with a larger encoder to extract image features, along with a smaller decoder to reconstruct the label. He fed a very large patch size ($160 \times 192 \times 128$ voxels) into the network, and also added a variational autoencoder (VAE) branch in order to regularize the shared encoder.

In this work, inspired by the cascaded strategy [19,22,25,26], we propose a novel two-stage cascaded U-Net. In the first stage, we use a variant of U-Net as the first stage network to train a coarse prediction. In the second stage, we increase the width of the network and use two decoders so as to boost performance. The second stage is added to refine the prediction map by concatenating a preliminary prediction map with the original input to utilize auto-context. We do not use any additional training data and only participate in the segmentation task in testing phase.

2   Methods

Myronenko [15] proposed an asymmetrical U-Net with a variational autoencoder branch [5,11]. In this paper, we take a variant of this approach as the basic segmentation architecture. We further propose a two-stage cascaded U-Net. The details are illustrated as follows.

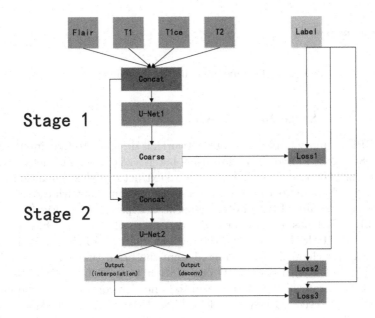

Fig. 2. Overview of the two-stage cascaded network.

2.1   Model Cascade

As can be seen in Fig. 2, in the first stage, multi-modal magnetic resonance images ($4 \times 128 \times 128 \times 128$) are passed into the first stage U-Net and predict a segmentation map roughly. The coarse segmentation map is fed together with the raw images into the second stage U-net. The second stage can provide a more accurate segmentation map with more network parameters. The two-stage cascaded network is trained in an end-to-end fashion.

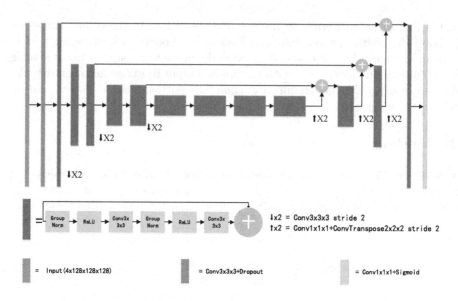

Fig. 3. The first stage network architecture.

2.2   The First Stage Network Architecture

Due to GPU memory limitations, our networks is designed to take input patches of size $128 \times 128 \times 128$ voxels and to use a batch size of one. The network architecture consists of a larger encoding path, to extract complex semantic features, and a smaller decoding path, to recover a segmentation map with the same input size. The architecture of the first stage network is presented in Fig. 3.

The 3D U-Net has an encoder and a decoder path, each of which have four spatial levels. At the beginning of the encoder, patches of size $128 \times 128 \times 128$ voxels with four channels are extracted from the brain tumor images as input, followed by an initial $3 \times 3 \times 3$ 3D convolution with 16 filters. We also use a dropout with a rate of 0.2 after the initial encoder convolution. The encoder part uses a pre-activated residual block [7,8]. Each of these blocks consists of two $3 \times 3 \times 3$ convolutions with Group Normalization [23] with group size of 8 and Rectified Linear Unit (ReLU) activation, followed by additive identity skip connection. The number of pre-activated residual blocks is 1, 2, 2, and 4 within each spatial level. Moreover, a convolution layer with a $3 \times 3 \times 3$ filter and a stride of 2 is used to reduce the resolution of the feature maps by 2 and simultaneously increase the number of feature channels by 2.

Unlike the encoder, the decoder structure uses a single pre-activated residual block for each spatial level. Before up-sampling, we use $1 \times 1 \times 1$ convolutions to reduce the number of features by a factor of 2. Compared with [15], we use a deconvolution with kernel size $2 \times 2 \times 2$ and a stride of 2 rather than trilinear interpolation in order to double the size of the spatial dimension. The network features shortcut connections between corresponding layers with the same resolution in the encoder and decoder by elementwise summation. At the end of the decoder, a $1 \times 1 \times 1$ convolution is used to decrease the number of output channels to three, followed by a sigmoid function. The detail of the structure is shown in Table 1.

2.3 The Second Stage Network Architecture

Different from the network in the first stage, we double the number of filters in the initial 3D convolution in order to increase the network width. What's more, we use two decoders. The structure of the two decoders is the same except that one uses a deconvolution and the other uses trilinear interpolation. The interpolation decoder is used only during training. Because the performance of the decoder used deconvolution is better than used trilinear interpolation and add a decoder used trilinear interpolation to regularize the shared encoder can improve the performance in our experiment. The architecture of the second stage network is presented in Fig. 4 and the detail of the structure is shown in Table 2.

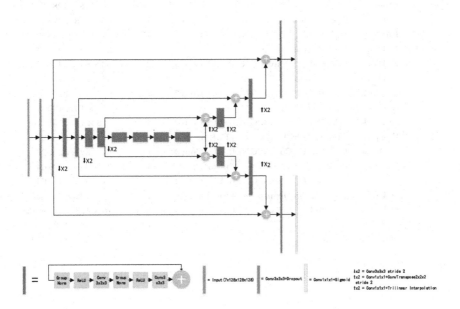

Fig. 4. The second stage network architecture.

2.4 Loss

The Dice Similarity Coefficient measures (DSC) the degree of overlap between the prediction map and ground truth. The DSC is calculated by Eq. 1, where S is the output of network, R is the ground truth label and $|\cdot|$ denotes the volume of the region.

$$DSC = \frac{2|S \cap R|}{|S| + |R|} \tag{1}$$

The soft Dice loss is designed as following:

$$\mathbf{L}_{dice} = \frac{2 * \sum S * R}{\sum S^2 + \sum R^2 + \epsilon} \tag{2}$$

Instead of learning the labels (e.g. enhancing tumor, edema, necrosis and non-enhancing), we directly optimize the three overlapping regions (whole tumor, tumor core and enhancing tumor) with the Dice loss, then simply add the Dice loss functions of each region together. We also add the loss of each stage together to arrive at the final loss.

Table 1. The first stage network structure, where + stands for additive identity skip connection, Conv3 - $3 \times 3 \times 3$ convolution, Conv1 - $1 \times 1 \times 1$ convolution, GN - group normalization, ConvTranspose - deconvolution with kernel size $2 \times 2 \times 2$.

		U-Net 1		
	Name	Details	Repeat	Size
	Input			$4 \times 128 \times 128 \times 128$
	InitConv	Conv3, Dropout	1	$16 \times 128 \times 128 \times 128$
	EnBlock1	GN, ReLU, Conv3, GN, ReLU, Conv3, +	1	$16 \times 128 \times 128 \times 128$
Encoder	EnDown1	Conv3 stride 2	1	$32 \times 64 \times 64 \times 64$
	EnBlock2	GN, ReLU, Conv3, GN, ReLU, Conv3, +	2	$32 \times 64 \times 64 \times 64$
	EnDown2	Conv3 stride 2	1	$64 \times 32 \times 32 \times 32$
	EnBlock3	GN, ReLU, Conv3, GN, ReLU, Conv3, +	2	$64 \times 32 \times 32 \times 32$
	EnDown3	Conv3 stride 2	1	$128 \times 16 \times 16 \times 16$
	EnBlock4	GN, ReLU, Conv3, GN, ReLU, Conv3, +	4	$128 \times 16 \times 16 \times 16$
	DeUp3	Conv1, ConvTranspose, +EnBlock3	1	$64 \times 32 \times 32 \times 32$
	DeBlock3	GN, ReLU, Conv3, GN, ReLU, Conv3, +	1	$64 \times 32 \times 32 \times 32$
	DeUp2	Conv1,ConvTranspose,+EnBlock2	1	$32 \times 64 \times 64 \times 64$
Decoder	DeBlock2	GN, ReLU, Conv3, GN, ReLU, Conv3, +	1	$32 \times 64 \times 64 \times 64$
	DeUp2	Conv1, ConvTranspose, +EnBlock1	1	$16 \times 128 \times 128 \times 128$
	DeBlock1	GN, ReLU, Conv3, GN, ReLU, Conv3, +	1	$16 \times 128 \times 128 \times 128$
	EndConv	Conv1	1	$3 \times 128 \times 128 \times 128$
	Sigmoid	Sigmoid	1	$3 \times 128 \times 128 \times 128$

3   Experiments

3.1   Data Pre-processing and Augmentation

Before feeding the data into the deep learning network, a preprocessing method is used to process the input data. Since the MRI intensity values are non-standardized, we apply intensity normalization to each MRI modality from each patient independently by subtracting the mean and dividing by the standard deviation of the brain region only.

Moreover, to prevent an overfitting issue from arising, we deploy three types of data augmentation. Firstly, we apply a random intensity shift between

Table 2. The second stage network structure, where $+$ stands for additive identity skip connection, Conv3 - $3 \times 3 \times 3$ convolution, Conv1 - $1 \times 1 \times 1$ convolution, GN - group normalization, ConvTranspose - deconvolution with kernel size $2 \times 2 \times 2$, Upsampling - trilinear interpolation, Decoder2 is used only during training.

	U-Net 2			
	Name	Details	Repeat	Size
	Input			$7 \times 128 \times 128 \times 128$
	InitConv	Conv3, Dropout	1	$32 \times 128 \times 128 \times 128$
	EnBlock1	GN, ReLU, Conv3, GN, ReLU, Conv3, +	1	$32 \times 128 \times 128 \times 128$
Encoder	EnDown1	Conv3 stride 2	1	$64 \times 64 \times 64 \times 64$
	EnBlock2	GN, ReLU, Conv3, GN, ReLU, Conv3, +	2	$64 \times 64 \times 64 \times 64$
	EnDown2	Conv3 stride 2	1	$128 \times 32 \times 32 \times 32$
	EnBlock3	GN, ReLU, Conv3, GN, ReLU, Conv3, +	2	$128 \times 32 \times 32 \times 32$
	EnDown3	Conv3 stride 2	1	$256 \times 16 \times 16 \times 16$
	EnBlock4	GN, ReLU, Conv3, GN, ReLU, Conv3, +	4	$256 \times 16 \times 16 \times 16$
	DeUp3	Conv1, ConTranspose, +EnBlock3	1	$128 \times 32 \times 32 \times 32$
	DeBlock3	GN, ReLU, Conv3, GN, ReLU, Conv3, +	1	$128 \times 32 \times 32 \times 32$
	DeUp2	Conv1, ConTranspose, +EnBlock2	1	$64 \times 64 \times 64 \times 64$
Decoder1	DeBlock2	GN, ReLU, Conv3, GN, ReLU, Conv3, +	1	$64 \times 64 \times 64 \times 64$
	DeUp2	Conv1, ConTranspose, +EnBlock1	1	$32 \times 128 \times 128 \times 128$
	DeBlock1	GN, ReLU, Conv3, GN, ReLU, Conv3, +	1	$32 \times 128 \times 128 \times 128$
	EndConv	Conv1	1	$3 \times 128 \times 128 \times 128$
	Sigmoid	Sigmoid	1	$3 \times 128 \times 128 \times 128$
	DeUp3_1	Conv1, Upsampling, +EnBlock3	1	$128 \times 32 \times 32 \times 32$
	DeBlock3_1	GN, ReLU, Conv3, GN, ReLU, Conv3, +	1	$128 \times 32 \times 32 \times 32$
	DeUp2_1	Conv1, Upsampling, +EnBlock2	1	$64 \times 64 \times 64 \times 64$
Decoder2	DeBlock2_1	GN, ReLU, Conv3, GN, ReLU, Conv3, +	1	$64 \times 64 \times 64 \times 64$
(Used only during training)				
	DeUp2_1	Conv1, Upsampling, +EnBlock1	1	$32 \times 128 \times 128 \times 128$
	DeBlock1_1	GN, ReLU, Conv3, GN, ReLU, Conv3, +	1	$32 \times 128 \times 128 \times 128$
	EndConv_1	Conv1	1	$3 \times 128 \times 128 \times 128$
	Sigmoid_1	Sigmoid	1	$3 \times 128 \times 128 \times 128$

$[-0.1 - 0.1]$ of the standard deviation of each channel, as well as a random scaling intensity of the input between scales $[0.9 - 1.1]$. Secondly, we train our network by randomly cropping the MRI data from $240 \times 240 \times 155$ voxels to $128 \times 128 \times 128$ voxels due to memory limitation. Finally, we use random flipping along each 3D axis with a probability of 50%.

3.2 Training Details

The implementation of our network is based on PyTorch 1.1.0 [16]. The maximum number of training iterations is set to 405 epochs with 5 epochs of linear warmup. We use Adam optimizer to update the weights of the network, with a batch size of 1 and an initial learning rate of $\alpha_0 = 1e - 4$ at the very beginning and decays it as following:

$$\alpha = \alpha_0 \times \left(1 - \frac{e}{N_e}\right)^{0.9} \tag{3}$$

where e is an epoch counter, and N_e is a total number of epochs. We regularize using an l2 weight decay of $1e - 5$. Training is performed on a Nvidia Titan V GPU with 12 Gb memory. However, our method requires slightly more than 12 Gb memory in our experiment. We utilize gradient checkpointing [21] by PyTorch to reduce the memory consumption.

3.3   Augmentation for Inference

At testing time, we segment the whole brain region at once instead of using a sliding window. The interpolation decoder is not used during the inference phase. To obtain a more robust prediction, we preserve eight weights of the model in the last time of the training progress for prediction. For each snapshot, the input images are used different flipping before being fed into the network. Finally, we average the output of the resulting eight segmentation probability maps.

3.4   Post-processing

We replace enhancing tumor with necrosis when the volume of predicted enhancing tumor is less than the threshold to post-process our segmentation results (The threshold is chosen for each experiment independently, depending on the performance of BraTS 2019 validation dataset).

4   Results

The variability of a single model can be quite high. We use total five networks from the 5-fold cross-validation as an ensemble to predict segmentation for BraTS 2019 validation dataset. Also, we use an ensemble of a set of 12 models, which are trained from scratch using the entire training dataset. The best single model is chosen from the set of 12 models.

We report the results of our approach on the BraTS 2019 validation dataset, which contains 125 cases with unknown glioma grade and unknown segmentation. All reported values are computed via the online evaluation platform (https://ipp.cbica.upenn.edu/) for evaluation of Dice score, sensitivity, specificity and Hausdorff distance (95%). Validation set results can be found in Table 3. The performance of the best single model is slightly better than ensemble of 5-fold cross-validation. The ensemble of 12 models results in a minor improvement compared with the best single model.

Testing set results are presented in Table 4. Our algorithm achieved the first place out of more than 70 participating teams.

Table 3. Mean Dice and Hausdorff measurements of the proposed segmentation method on BraTS 2019 validation set. DSC - dice similarity coefficient, HD95 - Hausdorff distance (95%), WT - whole tumor, TC - tumor core, ET - enhancing tumor core.

Method	DSC			HD95		
Validation	WT	TC	ET	WT	TC	ET
Ensemble of 5-fold	0.90797	0.85888	0.79667	4.35413	5.69195	3.12642
Best single model	0.90819	0.86321	0.80199	4.44375	5.86201	3.20551
Ensemble of 12 models	0.90941	0.86473	0.80211	4.26398	5.43931	3.14581

Table 4. Mean Dice and Hausdorff measurements of the proposed segmentation method on BraTS 2019 testing set. DSC - dice similarity coefficient, HD95 - Hausdorff distance (95%), WT - whole tumor, TC - tumor core, ET - enhancing tumor core.

Method	DSC			HD95		
Testing	WT	TC	ET	WT	TC	ET
Ensemble of 12 models	0.88796	0.83697	0.83267	4.61809	4.13071	2.65056

5 Conclusion

In this paper, we propose a two-stage cascaded U-Net. Our approach refines the prediction through a progressive cascaded network. Experiments on the BraTS 2019 validation set demonstrate that our method can obtain very competitive segmentation even though using single model. The testing results show that our proposed method can achieve excellent performance, winning the first position in the BraTS 2019 challenge segmentation task among 70+ participating teams.

Acknowledgements. Changxing Ding was supported in part by the National Natural Science Foundation of China (Grant No.: 61702193), Science and Technology Program of Guangzhou (Grant No.: 201804010272), and the Program for Guangdong Introducing Innovative and Entrepreneurial Teams (Grant No.: 2017ZT07X183). Dacheng Tao was supported by Australian Research Council Projects (FL-170100117, DP-180103424 and LP-150100671).

References

1. Bakas, S., et al.: Segmentation labels and radiomic features for the pre-operative scans of the TCGA-GBM collection. Cancer Imaging Archive (2017). https://doi.org/10.7937/K9/TCIA.2017.KLXWJJ1Q
2. Bakas, S., et al.: Segmentation labels and radiomic features for the pre-operative scans of the TCGA-LGG collection. Cancer Imaging Archive **286** (2017)
3. Bakas, S., et al.: Advancing the cancer genome atlas glioma MRI collections with expert segmentation labels and radiomic features. Sci. Data **4**, 170117 (2017). https://doi.org/10.1038/sdata.2017.117

4. Bakas, S., et al.: Identifying the best machine learning algorithms for brain tumor segmentation, progression assessment, and overall survival prediction in the brats challenge. arXiv preprint arXiv:1811.02629 (2018)

5. Doersch, C.: Tutorial on variational autoencoders. arXiv preprint arXiv:1606.05908 (2016)

6. Havaei, M., et al.: Brain tumor segmentation with deep neural networks. Med. Image Anal. **35**, 18–31 (2017)

7. He, K., Zhang, X., Ren, S., Sun, J.: Deep residual learning for image recognition. In: Proceedings of the IEEE Conference on Computer Vision and Pattern Recognition, pp. 770–778 (2016)

8. He, K., Zhang, X., Ren, S., Sun, J.: Identity mappings in deep residual networks. In: Leibe, B., Matas, J., Sebe, N., Welling, M. (eds.) ECCV 2016. LNCS, vol. 9908, pp. 630–645. Springer, Cham (2016). https://doi.org/10.1007/978-3-319-46493-0_38

9. Kamnitsas, K., et al.: Ensembles of multiple models and architectures for robust brain tumour segmentation. In: Crimi, A., Bakas, S., Kuijf, H., Menze, B., Reyes, M. (eds.) BrainLes 2017. LNCS, vol. 10670, pp. 450–462. Springer, Cham (2018). https://doi.org/10.1007/978-3-319-75238-9_38

10. Kamnitsas, K., et al.: Efficient multi-scale 3D CNN with fully connected CRF for accurate brain lesion segmentation. Med. Image Anal. **36**, 61–78 (2017)

11. Kingma, D.P., Welling, M.: Auto-encoding variational bayes. arXiv preprint arXiv:1312.6114 (2013)

12. Long, J., Shelhamer, E., Darrell, T.: Fully convolutional networks for semantic segmentation. In: Proceedings of the IEEE Conference on Computer Vision and Pattern Recognition, pp. 3431–3440 (2015)

13. Menze, B.H., et al.: The multimodal brain tumor image segmentation benchmark (BRATS). IEEE Trans. Med. Imaging **34**(10), 1993–2024 (2015). https://doi.org/10.1109/tmi.2014.2377694

14. Milletari, F., Navab, N., Ahmadi, S.A.: V-net: fully convolutional neural networks for volumetric medical image segmentation. In: 2016 Fourth International Conference on 3D Vision (3DV), pp. 565–571. IEEE (2016)

15. Myronenko, A.: 3D MRI brain tumor segmentation using autoencoder regularization. In: Crimi, A., Bakas, S., Kuijf, H., Keyvan, F., Reyes, M., van Walsum, T. (eds.) BrainLes 2018. LNCS, vol. 11384, pp. 311–320. Springer, Cham (2019). https://doi.org/10.1007/978-3-030-11726-9_28

16. Paszke, A., et al.: Automatic differentiation in PyTorch (2017)

17. Pereira, S., Pinto, A., Alves, V., Silva, C.A.: Brain tumor segmentation using convolutional neural networks in mri images. IEEE Trans. Med. Imaging **35**(5), 1240–1251 (2016)

18. Ronneberger, O., Fischer, P., Brox, T.: U-Net: convolutional networks for biomedical image segmentation. In: Navab, N., Hornegger, J., Wells, W.M., Frangi, A.F. (eds.) MICCAI 2015. LNCS, vol. 9351, pp. 234–241. Springer, Cham (2015). https://doi.org/10.1007/978-3-319-24574-4_28

19. Roth, H.R., et al.: A multi-scale pyramid of 3D fully convolutional networks for abdominal multi-organ segmentation. In: Frangi, A.F., Schnabel, J.A., Davatzikos, C., Alberola-López, C., Fichtinger, G. (eds.) MICCAI 2018. LNCS, vol. 11073, pp. 417–425. Springer, Cham (2018). https://doi.org/10.1007/978-3-030-00937-3_48

20. Shen, H., Wang, R., Zhang, J., McKenna, S.J.: Boundary-aware fully convolutional network for brain tumor segmentation. In: Descoteaux, M., Maier-Hein, L., Franz, A., Jannin, P., Collins, D.L., Duchesne, S. (eds.) MICCAI 2017. LNCS, vol. 10434, pp. 433–441. Springer, Cham (2017). https://doi.org/10.1007/978-3-319-66185-8_49

21. Siskind, J.M., Pearlmutter, B.A.: Divide-and-conquer checkpointing for arbitrary programs with no user annotation. Optim. Methods Softw. **33**(4–6), 1288–1330 (2018)
22. Tu, Z., Bai, X.: Auto-context and its application to high-level vision tasks and 3D brain image segmentation. IEEE Trans. Pattern Anal. Mach. Intell. **32**(10), 1744–1757 (2009)
23. Wu, Y., He, K.: Group normalization. In: Ferrari, V., Hebert, M., Sminchisescu, C., Weiss, Y. (eds.) ECCV 2018. LNCS, vol. 11217, pp. 3–19. Springer, Cham (2018). https://doi.org/10.1007/978-3-030-01261-8_1
24. Zhao, X., Wu, Y., Song, G., Li, Z., Zhang, Y., Fan, Y.: A deep learning model integrating FCNNS and CRFS for brain tumor segmentation. Med. Image Anal. **43**, 98–111 (2018)
25. Zhou, C., Chen, S., Ding, C., Tao, D.: Learning contextual and attentive information for brain tumor segmentation. In: Crimi, A., Bakas, S., Kuijf, H., Keyvan, F., Reyes, M., van Walsum, T. (eds.) BrainLes 2018. LNCS, vol. 11384, pp. 497–507. Springer, Cham (2019). https://doi.org/10.1007/978-3-030-11726-9_44
26. Zhou, C., Ding, C., Lu, Z., Wang, X., Tao, D.: One-pass multi-task convolutional neural networks for efficient brain tumor segmentation. In: Frangi, A.F., Schnabel, J.A., Davatzikos, C., Alberola-López, C., Fichtinger, G. (eds.) MICCAI 2018. LNCS, vol. 11072, pp. 637–645. Springer, Cham (2018). https://doi.org/10.1007/978-3-030-00931-1_73
27. Zhou, C., Ding, C., Wang, X., Lu, Z., Tao, D.: One-pass multi-task networks with cross-task guided attention for brain tumor segmentation. arXiv preprint arXiv:1906.01796 (2019)

Memory-Efficient Cascade 3D U-Net for Brain Tumor Segmentation

Xinchao Cheng[1], Zongkang Jiang[1], Qiule Sun[2], and Jianxin Zhang[1,3(✉)]

[1] Key Lab of Advanced Design and Intelligent Computing (Ministry of Education),
Dalian University, Dalian 116622, China
[2] School of Information and Communication Engineering,
Dalian University of Technology, Dalian 116024, China
[3] School of Computer Science and Engineering, Dalian Minzu University,
Dalian 116600, China
jxzhang0411@163.com

Abstract. Segmentation is a routine and crucial procedure for the treatment of brain tumors. Deep learning based brain tumor segmentation methods have achieved promising performance in recent years. However, to pursue high segmentation accuracy, most of them require too much memory and computation resources. Motivated by a recently proposed partially reversible U-Net architecture that pays more attention to memory footprint, we further present a novel Memory-Efficient Cascade 3D U-Net (MECU-Net) for brain tumor segmentation in this work, which can achieve comparable segmentation accuracy with less memory and computation consumption. More specifically, MECU-Net utilizes fewer down-sampling channels to reduce the utilization of memory and computation resources. To make up the accuracy loss, MECU-Net employs multi-scale feature fusion module to enhance the feature representation capability. Additionally, a light-weight cascade model, which resolves the problem of small target segmentation accuracy caused by model compression to some extent, is further introduced into the segmentation network. Finally, edge loss and weighted dice loss are combined to refine the brain tumor segmentation results. Experiment results on BraTS 2019 validation set illuminate that MECU-Net can achieve average Dice coefficients of 0.902, 0.824 and 0.777 on the whole tumor, tumor core and enhancing tumor, respectively.

Keywords: Deep learning · Brian tumor segmentation ·
Memory-efficient U-Net · Cascade strategy

1 Introduction

Malignant brain tumors, especially gliomas, belong to the aggressive and dangerous disease that leads to death worldwide [1]. Automatic tumor segmentation using computer technology plays an important role in assisting the diagnosis and treatment of brain tumors. However, due to the huge difference in intensity

© Springer Nature Switzerland AG 2020
A. Crimi and S. Bakas (Eds.): BrainLes 2019, LNCS 11992, pp. 242–253, 2020.
https://doi.org/10.1007/978-3-030-46640-4_23

ranges, as well as the various size, shape, and location of tumors, it is a challenging task to automatically achieve satisfied tumor segmentation results from brain images.

Recently, with the great success of deep learning in the medical image analysis community, deep learning based brain tumor segmentation methods have also achieved promising performance improvement. Generally speaking, current deep learning based brain segmentation models mainly consist of patch-wise based and fully convolution based segmentation networks. Patch-wise based segmentation networks utilize the idea of small-scale image patches classification to segment the brain tumor [2,3], successfully augmenting the sample size of brain images for the training of deep networks. In order to make better use of 3D features of MRI images, Kamnitsas et al. [4] proposed an efficient multi-scale brain tumor segmentation network model based on 3D convolutional neural network and fully connected conditional random fields (CRFs), which achieves the state-of-the-art performance on several public databases. However, patch-wise architectures lack spatial continuity and require huge storage space, leading to lower efficiency. Fully convolution based segmentation networks segment the whole brain tumor by the way of pixel classification, i.e., pixel-by-pixel prediction, which can largely improve the brain tumor segmentation efficiency. The initial fully convolutional networks (FCN) [5] is proposed for the nature image segmentation by Long et al., and it is quickly introduced to solve the medical image segmentation problems, including brain tumor segmentation. On this basis, a variant of FCN models, named U-Net [6] is further put forward, which utilizes the skip connection to reduce the loss of feature information and improve the ability to determine global location. Due to its light-weight and high-efficiency advantages, U-Net quickly becomes the priority choice for the brain segmentation task. In addition, since 2D U-Net approaches ignore 3D MRI images spatial context information, Cieck et al. proposed a 3D U-Net [7] model to achieve higher segmentation precision. To further improve its segmentation performance, a variety of more powerful modules (such as residual module and dense connection module [8]) and advanced strategies (such as multi-scale fusion cascade ideology [9,10]) are injected into the baseline model [7,11,12], which have largely promoted the development of brain tumor segmentation methods.

However, most advanced 3D depth models consume large amounts of memory and computation resources to achieve high segmentation accuracy, and they can only be performed on the computer/server equipped with high-performance graphic cards (at least 16G). More recently, to reduce the memory footprint of the existing advanced networks, Brügger et al. presented a novel partially reversible U-Net [13], which can be performed on a computer equipped with graphics card of 12GB capacity while achieves competitive accuracy with the state-of-the-art networks. Motivated by the partially reversible U-Net, in this work, we further present a novel Memory-Efficient Cascade 3D U-Net (MECU-Net) for brain tumor segmentation, which can be run on the smaller graphics card of 8G capacity, and evaluate the model in the brain tumor segmentation challenge (BraTS) [14]. Experiment results show that our MECU-Net can achieve comparable segmentation accuracy while requiring less memory and computation consumption. Our main contributions are concluded as follows: (1) We propose a novel Memory-Efficient Cascade 3D U-Net (MECU-Net) for brain tumor segmentation. MECU-Net largely reduces the number of down-sampling channels, i.e., from 60, 120 to 20, 48, which makes it suitable for a smaller graphics card of 8G capacity. (2) To make up the accuracy loss caused by the down-sampling channel reduction, cascade strategy, and multi-scale information fusion mechanism, as well as the combination of edge loss and weighted dice loss, are put forward to alleviate the oscillation problem in model training, and these strategies effectively improve the segmentation result for the small tumors. (3) Experiment results on BraTS 2019 dataset illuminate that MECU-Net can achieve competitive performance in brain tumor segmentation.

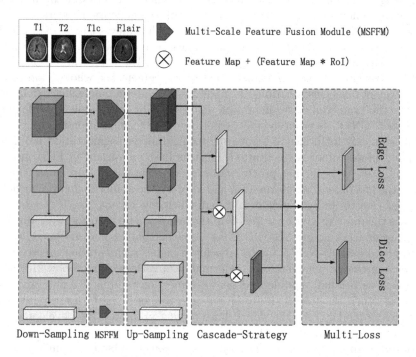

Fig. 1. Overview of the MECU-Net model. Our MECU-Net is mainly comprised of down-sampling, multi-scale feature fusion module (MSFFM), up-sampling, cascade strategy, and multi-loss module. It employs reversible convolution to put entire MRI images into the model, followed by several dilated convolutions with different dilation rates to extract various receptive field features for fusion. After the up-sampling stage, a light-weight cascade strategy is given to improve the segmentation accuracy of small targets. Finally, a combination of weighted dice loss and edge loss is employed for refining the tumor segmentation results.

2   Memory-Efficient Cascade 3D U-Net Model

Here, we mainly describe the details of the given Memory-Efficient Cascade 3D U-Net (MECU-Net) model. Our MECU-Net model is based on Robin's Partial Reversible U-Net [13], i.e., a recently proposed memory-efficient U-Net, which inherits and develops the reversible residual network [15]. MECU-Net largely reduces the number of down-sampling channels to make it suitable for the smaller graphics card of 8G capacity. To make up the accuracy loss caused by the down-sampling channel reduction, multi-scale information fusion mechanism, and cascade strategy, as well as the combination of edge loss and weighted dice loss, are introduced to alleviate the oscillation problem in model training, and these strategies effectively improve the segmentation result for the small tumors. The overview of our MECU-Net model is shown in Fig. 1, which mainly consists of down-sampling, multi-scale feature fusion, up-sampling, cascade strategy, and multi-loss module. The maximum memory requirement for the initial Partial Reversible U-Net model exists in the large-size feature map generated by the first and second layers of the down-sampling module. Therefore, we reduce the number of channels in the first two layers during the down-sampling stage, further compressing the parameters of the model. In this work, the channel number of the first down-sampling layer is reduced from 60 to 24, and that of second layers shrinks from 120 to 48. Then, it will be suitable for a smaller graphics card of 8G capacity. However, the compressed model has insufficient resolution ability and edge segmentation ability for small targets. Therefore, to make up the accuracy loss caused by the down-sampling channel reduction, we further introduce multi-scale information fusion mechanism and cascade strategy, as well as the combination of edge loss and weighted dice loss, to enhance the segmentation capacity for the small tumors, which also alleviates the oscillation problem of model training. The following three subsections will describe the details of the multi-scale information fusion, cascade strategy, and multi-loss module.

2.1   Multi-scale Feature Fusion Module

In the compressed U-Net feature extraction process, the feature map is gradually reduced by convolution operation, and the classification information of small targets is gradually diluted, resulting in the loss of the relevant classification information of small targets in the bottom feature map. Meanwhile, in the up-sampling process of U-Net, although the bottom information is gradually enlarged to the original size step by step, the loss of information of small targets can not be restored in the upper feature map. Traditional U-Net attempts to compensate for the loss of relevant information by directly superimposing the same size of the original atlas, but it does not extract the classification information of the large receptive field of small targets well. As the number of down-sampling channels decreases, the accuracy of small target segmentation of the compressed model decreases significantly. Therefore, we propose a multi-scale dilated convolution operation for the original features (as shown in Fig. 2). Under the same conditions, not only the classification information transmitted by the

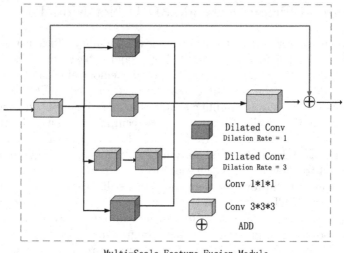

Multi-Scale Feature Fusion Module

Fig. 2. Overview of the multi-scale feature fusion module. The initial information is processed by multi-scale dilated convolution to obtain various information from multiple receptive fields. The fusion strategy consists of convolutions with various dilation rates and shortcut connections to provide richer classification information for each pixel, which makes up for the deficiency of lacking classification information for small tumors during the up-sampling process.

underlying features, but also the diversified classification information obtained by multi-scale receptive field convolution for the same layer features, together with the original feature map, are provided for the classification of small targets. The combined feature map provides enough feature information to improve the classification accuracy of each pixel in the small target area, thus improving the segmentation accuracy of the small target.

2.2 Cascade Strategy

For the brain tumor images, three levels of segmentation labels have a strong correlation with each other. The whole label, core label, and enhancing label are mosaic in turn. It can be intuitively seen that the segmentation results of the former stage have important guiding significance for the next stage. However, despite its promising performance, traditional cascade strategy not only introduces enormous model complexity and computational cost but also neglects the relevance among tasks. Therefore, we propose a light-weight cascade operation for the MECU-Net. We divide the segmentation operation into three parts and integrates the three tasks into one network in order. Using the segmentation results of the previous stage, we enhance the feature map and realize the attention mechanism by focusing on the target area. Moreover, each task owns an independent information extract block, a classification, and a loss layer, respectively. At the same time, most of the other parameters are shared to reduce the complexity of the model. Through a small amount of extra computation, the progressive attention of the model to small targets is realized, and the segmentation ability of the model to small targets is further improved.

2.3   Edge Loss and Weighted Dice Loss

Dice loss is effective for the brain tumor segmentation task. However, Dice loss is too sensitive to small target errors during the model training process. In the case where the pixel segmentation errors are the same, the fluctuation caused by Dice loss of large targets is much smaller than that caused by small targets. Although this phenomenon is conducive to the model's attention to small targets, the training of the model results will lead to instability of model loss and high volatility of training convergence. Especially in the later period of the model training, a slight mistake will lead to sharp fluctuations of the loss function, causing the model to oscillate during the training process. Therefore, according to the focal loss, we first add the weight information related to the total number of target pixels in the Dice loss. Meanwhile, we utilize the weighted dice loss function to stabilize the later process of the model training. In addition, the accuracy of correlation can be compensated by supplementary supervision of edge pixels in three segmentation regions, which could partially resolve the problem of attention loss of small targets. Inspired by this, we realize the 3D edge loss to focus on edge information of small targets, further improving the segmentation accuracy. Based on the above ideas, the multi-loss function formula can be given as follows:

$$L = \lambda_1 \cdot L_{dice} + \lambda_2 \cdot L_{edge} + \lambda_3 \cdot L_{mask} \tag{1}$$

where L_{mask} means the average binary cross-entropy loss, L_{dice} represents the added weighted Dice loss to optimize segmentation branch, L_{edge} represents the loss which focuses on the edge information, respectively. λ_i ($1 \leq i \leq 3$) is the hyper-parameter that controls the importance of each loss.

3   Experiments and Results

3.1   Experiment Dataset

The brain tumor MRI dataset adopted in this work is BraTS 2019 dataset, which is used for a brain tumor image segmentation challenge to compare the state-of-the-art methods at the MICCAI conference. The BraTS 2019 training dataset consists of 259 cases of high-grade gliomas (HGG) and 76 cases of low-grade gliomas (LGG), while the validation dataset includes 125 cases with hidden ground-truth. There are four sequence modalities, i.e., Flair, T1, T1ce and T2, which have already been co-registered, for each sample. According to the manual segmentation results given by the experts [16–18], each ground-truth for brain tumors contains four different regions, which are marked with three labels: label 1 for necrosis and non-enhancing, label 2 for edema, and label 4 for enhancing tumor. However, we evaluated the following three categories: whole tumor (all three labels), tumor core (label 1, 4), and the enhancing tumor (label 4) to construct segmentation results.

3.2 Experiment Settings

For pre-processing, we normalize each brain image individually with zero mean and unit variance (based on non-zero voxels only) to avoid the initial bias of the network towards one modality. Meanwhile, all of the images are cropped into the same size of $160 \times 192 \times 160$ pixels, which not only satisfies the limitation of machine performance but also ensures that each image contains the whole tumor area. Additionally, data enhancement is implemented by using diverse data augmentation strategies such as random rotation, random scaling, random elastic deformations, random flips, and small intensity shift. Due to the particularity of brain images, we keep the Z-axis constant, rotating, scaling, and elastically deforming only on the planes which are perpendicular to the z-axis. In our experiments, we utilize the Adam optimizer with an initial learning rate of 0.0001 is utilized. The learning rate is decreased by a factor of 5 after 35, 55 and 75 epochs, and the batch size is set to 1. Our model is implemented using the PyTorch deep learning framework on an NVIDIA GeForce GTX 1080Ti GPU with 11 GB of memory.

3.3 Evaluation Metrics

To evaluate the memory-efficient cascade 3D U-Net model, we utilize the official evaluation metrics given by BraTS 2019, including Dice score, Sensitivity, Specificity and Hausdorff distance defined as follows:

$$Dice = \frac{2TP}{FP + 2TP + FN} \tag{2}$$

$$Sensitivity = \frac{TP}{TP + FN} \tag{3}$$

$$Specificity = \frac{TN}{TN + FP} \tag{4}$$

$$Haus(T, P) = max\left\{sup_{t \in T} \inf_{p \in P} d(t, p), sup_{p \in P} \inf_{t \in T} d(t, p)\right\} \tag{5}$$

where True Positive (TP), False Positive (FP), True Negative (TN) and False Negative (FN) denote the number of false negative, true negative, true positive, and false positive voxels. t and p denote the pixels in the ground-truth regions T and the predicted regions P, respectively. $d(t, p)$ is the function that computes the distance between points t and p. Dice, Sensitivity and Specificity measure voxel-wise overlap between the predicted results and the ground-truth. The Hausdorff distance calculates the distance between predicted segmented regions and the ground-truth regions. Besides, Hausdorff95 is a metric of Hausdorff distance to measure the 95% quantile of the surface distance. Since Dice is the overall evaluation metric for the entire BraTS challenges, we adopt it as the key metric for evaluation consistently across all the challenges.

3.4   Experiment Results

We first train and test our proposed memory-efficient cascade 3D U-Net (MECU-Net) approach on the BraTS 2019 training dataset. Then, our model is evaluated on BraTS 2019 validation dataset and proves the effectiveness of the method. Finally, we carry out a series of ablation studies to demonstrate the validity of each module proposed in this paper.

Results on BraTS 2019 Training Dataset. We employ a random data split of 80% VS. 20% from the BraTS 2019 training dataset for training and testing, respectively. Therefore, 268 cases are used for training and the remained 67 cases are considered as test data, and the evaluation results of MECU-Net can be reported in Table 1. In addition to dice, we also given three other evaluation metrics, i.e., Sensitivity, Specificity, and Hausdorff95, required by the competition. As shown in Table 1, MECU-Net can achieve the dice score of 0.890, 0.811 and 0.765 on the whole tumor, core tumor and enhancing tumor, respectively.

Table 1. Evaluation results on BraTS 2019 training dataset

Tumor type	Dice	Sensitivity	Specificity	Hausdorff95
Enhancing	0.765	0.769	0.997	5.199
Whole	0.890	0.8980	0.994	5.381
Core	0.811	0.816	0.994	7.243

Additionally, we also analyze the segmentation results of the model and ground-truth by visual inspection, which is shown in Fig. 3. The six columns from left to right of image show not only one axial slice of MRI acquired in Flair, T1, T1ce and T2 modality which are used as inputs of model, but also ground truth (GT) and the prediction labels, respectively. As can be seen, the segmentation results of our proposed model are sensibly similar to Ground Truth. While labeling the pixel very well in the area with enhancing tag, the model can judge the pixels correctly in the area without enhancing label, thereby reducing False Positive (FP).

Results on BraTS 2019 Validation Dataset. In order to participate in the BraTS 2019 competition, all samples of the training dataset are employed for model training in this experiment. The evaluation results of MECU-Net achieved on the BraTS 2019 Validation dataset are tabulated in Table 2. Quantitatively, we achieve Dice scores of 0.9018, 0.8244 and 0.7765 for the whole tumor, core tumor and enhancing tumor, respectively. In addition, the table also lists the mean, standard deviation, median and 25th and 75th percentile of each metric. The validation dataset results are automatically generated by the evaluation mechanism provided by the official BraTS 2019 online website. It should be noted

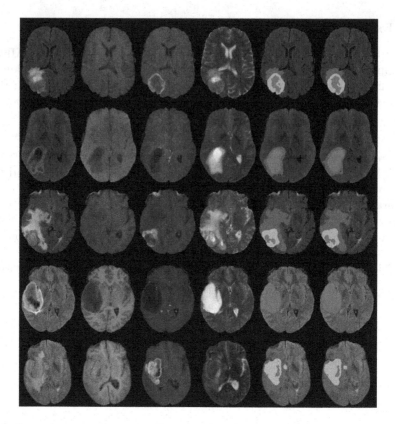

Fig. 3. Example segmentation results on the BraTS 2019 Training Dataset. From left to right, show the axial slice of MRI images in Flair, T1, T1ce and T2 modality, ground truth (GT) and the prediction labels. Among, enhancing tumor (yellow), edema (green) and necrotic and non-enhancing tumor (red). (Color figure online)

that a slight increase in performance on the validation dataset than the training dataset. The reason for this performance can be explained by the increasing number of training subjects. As the number of training subjects increases, the feature extraction ability of the model is improved, which improved the segmentation accuracy. Another possible reason is that our artificial split of training dataset brings greater data distribution differences than that between the validation dataset and the training dataset.

Ablation Studies on BraTS 2019 Validation Dataset. We also perform the ablation experiments on BraTS 2019 Validation Dataset to verify the effectiveness of the three embedded modules, whose results can be shown in Table 3. Here, the Partial Reversible U-Net is taken as the baseline model, which reduces the number of down-sampling channels at the first and second stages. On this basis, we further perform an experiment with improved loss function, and this

Table 2. Evaluation results on BraTS 2019 validation dataset

Metrics	Tumor type	Mean	StdDev	Median	25 quantile	75 quantile
Dice	Enhancing	0.777	0.257	0.862	0.793	0.906
	Whole	0.902	0.065	0.918	0.890	0.943
	Core	0.824	0.175	0.892	0.781	0.935
Sensitivity	Enhancing	0.779	0.269	0.860	0.758	0.933
	Whole	0.908	0.076	0.930	0.885	0.957
	Core	0.837	0.173	0.893	0.789	0.952
Specificity	Enhancing	0.998	0.003	0.999	0.998	0.999
	Whole	0.995	0.005	0.996	0.993	0.998
	Core	0.997	0.005	0.999	0.996	0.999
Hausdorff95	Enhancing	5.282	9.951	2.236	1.414	3.535
	Whole	5.412	9.397	3.162	2.236	5.000
	Core	7.263	11.849	3.606	1.799	6.633

Table 3. Ablation experiment results on BraTS 2019 validation dataset

Method	Dice		
	Enhancing	Whole	Core
Partial Reversible U-Net (Baseline)	0.757	0.895	0.808
Baseline + Combined loss	0.771	0.898	0.811
Baseline + Multi-scale Feature Fusion Module	0.772	0.900	0.817
Baseline + Cascade Strategy	0.766	0.900	0.821
MECU-Net	**0.777**	**0.902**	**0.824**

model is represented as Baseline + Combined loss, where combined loss stands for the combination of edge-loss and weighted dice loss. Note that the combined loss module gains 1.4% segmentation accuracy over the baseline model on the enhancing tumor. Multi-Scale Feature Fusion Module (MSFFM) represents a module that can extract different scale feature information from lower features through several convolution kernels with different sizes. We adopt the module before the fusion operation of each original feature and the upper sample feature. The experiments prove that the Baseline+MSFFM model outperforms the baseline with 1.5%, 0.5% and 0.9% on enhancing tumor, whole tumor, and core tumor, respectively, showing its well effectiveness for brain tumor segmentation task. Moreover, Cascade Strategy refers to the model which transfers the segmentation result of the whole label to the next task of enhancing label classification. This module achieves an average dice score of 0.9% over the baseline model. In particular, this module gains 1.3% accuracy improvement on the core tumor. Finally, based on the combination of the three modules, the MECU-Net obtains the optimal results on all of the three tumors, which outperforms the baseline module with a large margin.

4 Conclusion

In this work, we propose a novel Memory-Efficient Cascade 3D U-Net (MECU-Net) for automatic brain tumor segmentation. Our pipeline addresses some challenges of large memory requirement of traditional high-precision segmentation model and small target segmentation accuracy loss caused by model compression. By introducing the multi-scale information fusion mechanism, cascade strategy, as well as the combination of edge loss and weighted dice loss, MECU-Net can achieve comparable segmentation accuracy with less memory footprint.

Acknowledgements. This work was supported in part by the National Natural Science Foundation of China under Grant 61972062, the Program for Changjiang Scholars and Innovative Research Team in University under Grant IRT_15R07, the National Key R&D Program of China under Grant 2018YFC0910506, the Natural Science Foundation of Liaoning Province under Grant 2019-MS-011, the Key R&D Program of Liaoning Province under Grant 2019JH2/10100030, the High-level Talent Innovation Support Program of Dalian City under Grant 2016RQ078 and the Liaoning BaiQianWan Talents Program.

References

1. Zeng, H., et al.: Changing cancer survival in China during 2003–15: a pooled analysis of 17 populationbased cancer registries. Lancet Glob. Health **6**(5), e555–e567 (2018)
2. Pereira, S., et al.: Brain tumor segmentation using convolutional neural networks in MRI images. IEEE Trans. Med. Imaging **35**(5), 1240–1251 (2016)
3. Urban, G., et al.: Multi-modal brain tumor segmentation using deep convolutional neural networks. In: Proceedings of the Winning Contribution, MICCAI BraTS (Brain Tumor Segmentation) Challenge, pp. 31–35 (2014)
4. Kamnitsas, K., et al.: Efficient multi-scale 3D CNN with fully connected CRF for accurate brain lesion segmentation. Med. Image Anal. **36**, 61–78 (2017). www.sciencedirect.com/science/article/pii/S1361841516301839
5. Long, J., Shelhamer, E., Darrell, T.: Fully convolutional networks for semantic segmentation. In: Proceedings of the IEEE Conference on Computer Vision and Pattern Recognition, pp. 3431–3440 (2015)
6. Ronneberger, O., Fischer, P., Brox, T.: U-Net: convolutional networks for biomedical image segmentation. In: Navab, N., Hornegger, J., Wells, W.M., Frangi, A.F. (eds.) MICCAI 2015. LNCS, vol. 9351, pp. 234–241. Springer, Cham (2015). https://doi.org/10.1007/978-3-319-24574-4_28
7. Çiçek, Ö., Abdulkadir, A., Lienkamp, S.S., Brox, T., Ronneberger, O.: 3D U-Net: learning dense volumetric segmentation from sparse annotation. In: Ourselin, S., Joskowicz, L., Sabuncu, M.R., Unal, G., Wells, W. (eds.) MICCAI 2016. LNCS, vol. 9901, pp. 424–432. Springer, Cham (2016). https://doi.org/10.1007/978-3-319-46723-8_49

8. Dolz, J., Gopinath, K., Yuan, J., Lombaert, H., Desrosiers, C., Ayed, I.B.: HyperDense-Net: a hyper-densely connected CNN for multi-modal image segmentation. arXiv:180402967 (2018)

9. Wang, G., Li, W., Ourselin, S., Vercauteren, T.: Automatic brain tumor segmentation using cascaded anisotropic convolutional neural networks. In: Crimi, A., Bakas, S., Kuijf, H., Menze, B., Reyes, M. (eds.) BrainLes 2017. LNCS, vol. 10670, pp. 178–190. Springer, Cham (2018). https://doi.org/10.1007/978-3-319-75238-9_16

10. Zhou, C., Ding, C., Lu, Z., Wang, X., Tao, D.: One-pass multi-task convolutional neural networks for efficient brain tumor segmentation. In: Frangi, A.F., Schnabel, J.A., Davatzikos, C., Alberola-López, C., Fichtinger, G. (eds.) MICCAI 2018. LNCS, vol. 11072, pp. 637–645. Springer, Cham (2018). https://doi.org/10.1007/978-3-030-00931-1_73

11. Dong, H., Yang, G., Liu, F., Mo, Y., Guo, Y.: Automatic brain tumor detection and segmentation using U-Net based fully convolutional networks. In: Valdés Hernández, M., González-Castro, V. (eds.) MIUA 2017. CCIS, vol. 723, pp. 506–517. Springer, Cham (2017). https://doi.org/10.1007/978-3-319-60964-5_44

12. Bakas, S., et al.: Identifying the best machine learning algorithms for brain tumor segmentation, progression assessment, and overall survival prediction in the BRATS challenge, arXiv preprint arXiv:1811.02629 (2018)

13. Brügger, R., Baumgartner, C.F., Konukoglu, E.: A partially reversible U-Net for memory-efficient volumetric image segmentation. https://arxiv.org/abs/1906.06148

14. Menze, B.H., et al.: The multimodal brain tumor image segmentation benchmark (BRATS). IEEE Trans. Med. Imaging **34**(10), 1993–2024 (2015). https://doi.org/10.1109/TMI.2014.2377694

15. Gomez, A.N., Ren, M., Urtasun, R., Grosse, R.B.: The reversible residual network: backpropagation without storing activations. In: Advances in Neural Information Processing Systems, vol. 30, pp. 2214–2224. Curran Associates, Inc. (2017)

16. Bakas, S., et al.: Advancing the Cancer Genome Atlas glioma MRI collections with expert segmentation labels and radiomic features. Nat. Sci. Data **4**, 170117 (2017). https://doi.org/10.1038/sdata.2017.117

17. Bakas, S., et al.: Segmentation labels and radiomic features for the pre-operative scans of the TCGA-GBM collection. Cancer Imaging Arch. (2017). https://doi.org/10.7937/K9/TCIA.2017.KLXWJJ1Q

18. Bakas, S., et al.: Segmentation labels and radiomic features for the pre-operative scans of the TCGA-LGG collection. Cancer Imaging Arch. (2017). https://doi.org/10.7937/K9/TCIA.2017.GJQ7R0EF

A Baseline for Predicting Glioblastoma Patient Survival Time with *Classical* Statistical Models and *Primitive* Features Ignoring Image Information

Florian Kofler[1(✉)], Johannes C. Paetzold[1(✉)], Ivan Ezhov[1], Suprosanna Shit[1], Daniel Krahulec[2], Jan S. Kirschke[1], Claus Zimmer[1], Benedikt Wiestler[1], and Bjoern H. Menze[1]

[1] Technical University Munich, 81675 Munich, Germany
{florian.kofler,johannes.paetzold}@tum.de
[2] Philips Healthcare, MR R&D Clinical Science, 5684 PC Best, The Netherlands

Abstract. Gliomas are the most prevalent primary malignant brain tumors in adults. Until now an accurate and reliable method to predict patient survival time based on medical imaging and meta-information has not been developed [3]. Therefore, the survival time prediction task was introduced to the Multimodal Brain Tumor Segmentation Challenge (BraTS) to facilitate research in survival time prediction.

Here we present our submissions to the BraTS survival challenge based on classical statistical models to which we feed the provided metadata as features. We intentionally ignore the available image information to explore how patient survival can be predicted purely by metadata. We achieve our best accuracy on the validation set using a simple median regression model taking only patient age into account. We suggest using our model as a baseline to benchmark the added predictive value of sophisticated features for survival time prediction.

Keywords: Survival time prediction · Glioma · Glioblastoma · HGG · LGG · Brain tumor · Benchmark · BraTS · Medical imaging · MRI

1 Introduction

Accurate estimation of a patient's prognosis is at the heart of clinical decision-making, both for clinical trials as well as daily clinical care.

Survival time prediction and statistics are frequently requested not only by terminally ill patients, but also by the general public. Survival time prognosis is considered as one of the most important factors in palliative medicine for three major reasons [22]:

B. Wiestler and B.H. Menze—Contributed equally as senior authors.

© Springer Nature Switzerland AG 2020
A. Crimi and S. Bakas (Eds.): BrainLes 2019, LNCS 11992, pp. 254–261, 2020.
https://doi.org/10.1007/978-3-030-46640-4_24

1. Necessity for medical law and insurance decisions, e.g. in the United States of America two independent doctors have to agree on a survival prognosis to decide on hospice eligibility.
2. Survival prognosis is critical for medical decision making, which weights the risks of medical procedures against expected benefits. For instance, in pain management, it can be beneficial to deliver addictive and potentially harmful doses of antidepressants and neurolytic agents to patients with short life expectancy.
3. Lifetime prognosis enables doctors to assist patients in making critical life decisions [18].

Considering the relevance of reliable survival time predictions, it is particularly striking how statistics reveal that clinicians are often unsuccessful in predicting patient survival [6]. Many studies have shown this issue, e.g. [17] found that about 50% of survival predictions for patients with lung cancer are erroneous. Specifically, the patients did not survive half of the predicted time frame or survived more than double the predicted time. Most clinicians' predictions of survival time are overly optimistic [6]. An important finding is that, the longer the patient-doctor relationship exists, the larger the optimism bias is within the doctor's survival prognosis [4]. This indicates that human subjectivity is a major source of error, besides the difficulties for clinicians to integrate prognostic information from multiple sources (e.g. demographic, genomic or imaging information).

These inconsistencies and relevant bias in clinicians survival prediction demand more quantitative approaches such as statistical or learning based models to assist in creating more realistic survival predictions. This has been empirically studied in the literature for various terminal diseases, e.g. by Henderson et al. for patients with lung cell cancer, using statistical models [8]. Recently, learning based methods exploring image information have proven to outperform medical doctors in survival time predictions for a multitude of diseases [10,13].

The Brain Tumor Segmentation Challenge (BraTS) focuses on a specific type of brain neoplasms called gliomas. Gliomas are one of the most prevalent brain tumors in adults and can be roughly distinguished in two major classes: aggressive high-grade gliomas, and low-grade gliomas. The life expectancy of a patient with a high-grade glioma has a median remaining life span of fewer than two years, while for low-grade gliomas, it is more than five years [16]. The survival prediction task was introduced to the BraTS challenge to crowd-source the development of an accurate and generalizable prediction model [1–3,15].

The BraTS dataset was acquired at multiple clinical centers, therefore presenting several real-world challenges. For instance, scans are often acquired using different imaging protocols, and follow-up scans are acquired at varying time points. These inconsistencies, among others, pose severe problems to clinicians as well as automated diagnostic approaches.

In the BraTS survival challenge, the images as well as corresponding metadata are given to the participants to predict patient survival. Most contributions to the challenge explore image information using learning models, e.g. U-Net [3]. However, in the BraTS2018 survival challenge, a simple linear regression considering only patient age and simple tumor region sizes as features achieved third place [19,20]. This could be attributed to a lack of larger and diverse datasets, which could be resolved in future challenges by extending the data across clinics or using recently successful generative approaches [7,14]. Another methodological reason could be the insufficient structure of extracted imaging features and contradicting feature interpretation.

Inspired by Weninger et al. [19], we systematically explored how far one can get using only metadata for survival time prediction. We intentionally disregard image information and instead explore a multitude of classical statistical models and metadata based features.

2   Methods

2.1   Models

As a baseline, we implemented simple ordinary least squares (OLS) linear models [5]. Additionally, we fitted linear model with three orthogonal polynomials [9] and quantile regression models [11]. We computed p-values and confidence intervals for the model coefficients and evaluated the goodness of fit of the models by adjusted R^2 for the linear models and the quantile models by V, as suggested by Koenker respectively [12].

2.2   Features

We deliberately ignored image information and instead focused on primitive features extracted from the patients' metadata. Besides the patients' age we included resection status [21] and the clinical institution (extracted from the patient ID e.g. "CBICA") as predictors for our models. The clinical institution feature differentiates regional factors such as access to healthcare, different population etc. that might affect survival time.

2.3   Dataset

As the test set includes only patients with gross total resection (GTR), we evaluated our models' performance on the GTR subset. Additionally, we also took patients with only partial tumor resections into account to find out whether we can retrieve additional information from these cases.

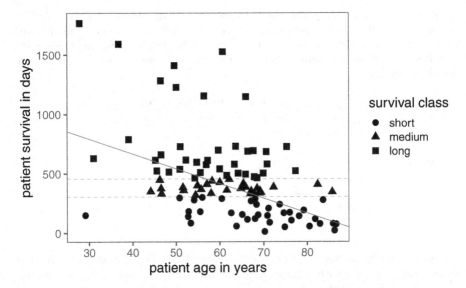

Fig. 1. Scatterplot of patient age versus survival time (*Pearson r*: −0.486), representing the space we fit our models to in the spirit of this XKCD comic [https://xkcd.com/2048/]. The dashed lines represent the thresholds to distinguish between short, medium and long-term survivors. The solid cyan line illustrates predictions of our proposed median regression model based on patient age.

3 Results

We designed our models on the training dataset considering measures of goodness of fit and p-values for model coefficients. Promising model configurations were evaluated on the validation dataset via the CBICA's Image Processing Portal (IPP).

3.1 Evaluation on the Training Set

The simple ordinary least squares (OLS) model outperformed the polynomial models on the training set, as reflected by higher values of adjusted R^2, see Table 1. Resection status and the clinical institution failed to add significant predictive value. These findings were also reflected in the analysis for the quantile models. For all models we achieved a much better fit on the subset of patients with gross total resection.

3.2 Evaluation on the Validation Set

Next, we evaluated on the validation set of 29 patients using the CBICA's Image Processing Portal (IPP). For the BraTS survival challenge, the predicted survival times are mainly evaluated by the accuracy of the survival prediction and secondarily by the metrics denoted in Table 2. Accuracy is defined as classifying

Table 1. Result table comparing the goodness of fit on the training set. We calculate the adjusted coefficient of determination R^2 for the OLS and polynomial models and V for the median model. The quantile model cannot be directly compared to the other models as the considered coefficients of determination (R^2 and V) are not the same as introduced by [12].

Model		R^2 all data	R^2 GTR only	V all data	V GTR only
OLS model	age	0.129	0.229	-	-
	age, resec., inst	0.147	-	-	-
	age, inst	0.129	0.243	-	-
Polyn. model	age	0.122	0.220	-	-
Median model	age	-	-	0.068	0.114
	age, resec., inst	-	-	0.099	0.121
	age, inst	-	-	0.072	0.114

patients correctly in one of three survival time bins. Three bins are defined as short-term survivors with a remaining survival time of fewer than ten months, mid-term survivors with a remaining survival time between ten and 15 months and long-term survivors with more than 15 months of remaining survival time. A glance at the scatterplot Fig. 1 reveals that these bins cannot be derived intuitively from the data and the accuracy-based challenge scoring might potentially lead to the paradox situation where a better fitting model performs worse in the classification-based challenge.

On the validation set, we find that the quantile models using only age as predictors achieve the best accuracy (0.552). We attribute this to the median models' decreased susceptibility to outliers, especially given the low number of patients in the training and validation dataset. However, the metrics for the survival time predictions in days are differing, for example, the polynomial model using age only as a predictor has the lowest mean squared error and the Spearman R is identical for five different solutions, see Table 2.

Given that we achieved a much better fit on the GTR subset for the training set and because features other than age fail to add predictive value reliably, we selected a median model trained solely on the GTR subset and taking only age as an input for evaluation on the test set. A positive side effect of this approach is the simple deployment in clinical and scientific practice. The predictions of this median model are illustrated in scatterplot 1.

Comparison to Other Challenge Participants. During the course of the challenge, we also compared our best performing model to the other participants on the validation set, knowing that most participating teams also consider image information. We monitored the leader board during the validation phase and found that our approach with an accuracy of 0.552 is within the best third of submissions. When comparing the metrics for fitting days of survival time, e.g. MSE, to the other submissions with equal accuracy, we found that our model shows solid performance. Overall, the total accuracy of our survival time

Table 2. Result table for the performance of our models on the validation set. Scores as calculated in the BRATS survival challenge leader board. Here the features used are encoded as age, resection status (resec.) and institution (inst.). The evaluation metrics for each submission (Subm.) are the accuracy, the mean squared error (MSE), median squared error (medianSE), standard squared error (stdSE) and SpearmanR.

Model		Accuracy	MSE	medianSE	stdSE	SpearmanR
OLS model	age; GTR only	0.448	90127.4	36773.6	123765.8	0.265
	age, inst.; GTR only	0.345	111571.2	40332.8	175070.8	0.165
	age, inst., resec.; all	0.310	105081.9	35523.3	161929.7	0.155
Polyn. model	age; all	0.448	90383.3	30953.2	131065.2	0.265
	age; GTR only	0.448	88113.3	32745.4	136508.8	0.265
Median model	age; all	**0.552**	101877.8	26958.2	116475.5	0.265
	age; GTR only	**0.552**	93572.3	30927.6	139847.1	0.265
	age, inst., resec.; all	0.483	96845.3	44466.3	155227.5	0.263
	age, inst.; GTR only	0.276	118450.0	54195.3	188132.4	0.184

predictions, but also of the best performing survival prediction, leaves much room for improvement. Even the best performing algorithms fail in more than one third of predictions. For perspective it is interesting to consider that last year the top performing algorithm used a U-Net to extract advanced image features. This shows that even the state-of-the-art in machine learning applied to this problem does not achieve a reliable survival prediction [3].

3.3   Evaluation on the Test Set

Finally, we consider our scores on the test set of 107 patients. We find a slight drop in performance with an accuracy of 0.486, a MSE of 419660.8, a medianSE 53177.5, a stdSE 1255102.9 and a SpearmanR of 0.358. Accuracy and medianSE are similar to our performance on the validation set. While the accuracy and medianSE scores remain comparable to the performance on the training and validation set, the outlier sensitive MSE and stdSE are substantially worse. This suggests that our drop in performance is mostly driven by statistical outliers.

4   Conclusion

We implemented simple OLS models, polynomial models and median regression models and experimented with different metadata-based predictor variables intentionally disregarding all image features. A simple median regression using only patient age as an input performed best to predict survival time for glioblastoma patients with gross total resection. Our model can serve as a baseline to evaluate the predictive value of sophisticated features.

Acknowledgments. Bjoern Menze, Benedikt Wiestler and Florian Kofler are supported through the SFB 824, subproject B12.

Supported by Deutsche Forschungsgemeinschaft (DFG) through TUM International Graduate School of Science and Engineering (IGSSE), GSC 81.

With the support of the Technical University of Munich – Institute for Advanced Study, funded by the German Excellence Initiative.

Johannes C. Paetzold is supported by the Graduate School of Bioengineering, Technical University of Munich.

Daniel Krahulec is supported by MR R&D Clinical Science, Philips Healthcare.

References

1. Bakas, S., et al.: Segmentation labels and radiomic features for the pre-operative scans of the TCGA-GBM collection. Cancer Imaging Arch. **286** (2017)
2. Bakas, S., et al.: Advancing the Cancer Genome Atlas glioma MRI collections with expert segmentation labels and radiomic features. Sci. Data **4**, 170117 (2017)
3. Bakas, S., et al.: Identifying the best machine learning algorithms for brain tumor segmentation, progression assessment, and overall survival prediction in the brats challenge. arXiv preprint arXiv:1811.02629 (2018)
4. Christakis, N.A., Smith, J.L., Parkes, C.M., Lamont, E.B.: Extent and determinants of error in doctors' prognoses in terminally ill patients: prospective cohort studycommentary: why do doctors overestimate? commentary: prognoses should be based on proved indices not intuition. BMJ **320**(7233), 469–473 (2000)
5. Everitt, B.: Book reviews: Chambers JM, Hastie TJ eds 1992: Statistical models in S. California: Wadsworth and Brooks/Cole. ISBN 0 534 16765-9. Stat. Methods Med. Res. **1**(2), 220–221 (1992). https://doi.org/10.1177/096228029200100208
6. Glare, P., et al.: A systematic review of physicians' survival predictions in terminally ill cancer patients. BMJ **327**(7408), 195 (2003)
7. Han, C., et al.: GAN-based synthetic brain MR image generation. In: 2018 IEEE 15th International Symposium on Biomedical Imaging (ISBI 2018), pp. 734–738. IEEE (2018)
8. Henderson, R., Keiding, N.: Individual survival time prediction using statistical models. J. Med. Ethics **31**(12), 703–706 (2005). https://doi.org/10.1136/jme.2005. 012427, https://jme.bmj.com/content/31/12/703
9. Kennedy, W.J., Gentle, J.E.: Statistical Computing. Routledge, Abingdon (2018)
10. Kim, D.W., Lee, S., Kwon, S., Nam, W., Cha, I.H., Kim, H.J.: Deep learning-based survival prediction of oral cancer patients. Sci. Rep. **9**(1), 6994 (2019)
11. Koenker, R., Bassett Jr, G.: Regression quantiles. Econ.: J. Econ. Soc. **46**, 33–50 (1978). https://www.jstor.org/stable/1913643?seq=1
12. Koenker, R., Machado, J.A.F.: Goodness of fit and related inference processes for quantile regression. J. Am. Stat. Assoc. **94**(448), 1296–1310 (1999). https://doi. org/10.1080/01621459.1999.10473882
13. Lao, J., et al.: A deep learning-based radiomics model for prediction of survival in glioblastoma multiforme. Sci. Rep. **7**(1), 10353 (2017)
14. Li, H., et al.: DiamondGAN: unified multi-modal generative adversarial networks for MRI sequences synthesis. In: Shen, D., et al. (eds.) MICCAI 2019. LNCS, vol. 11767, pp. 795–803. Springer, Cham (2019). https://doi.org/10.1007/978-3-030-32251-9_87
15. Menze, B.H., et al.: The multimodal brain tumor image segmentation benchmark (BRATS). IEEE Trans. Med. Imaging **34**(10), 1993–2024 (2015). https://doi.org/10.1109/TMI.2014.2377694

16. Ohgaki, H., Kleihues, P.: Population-based studies on incidence, survival rates, and genetic alterations in astrocytic and oligodendroglial gliomas. J. Neuropathol. Exp. Neurol. **64**(6), 479–489 (2005)
17. Parkes, C.M.: Commentary: prognoses should be based on proved indices not intuition. Br. Med. J. **320**, 473–473 (2000)
18. Steinhauser, K.E., Clipp, E.C., McNeilly, M., Christakis, N.A., McIntyre, L.M., Tulsky, J.A.: In search of a good death: observations of patients, families, and providers. Ann. Intern. Med. **132**(10), 825–832 (2000)
19. Weninger, L., Haarburger, C., Merhof, D.: Robustness of radiomics for survival prediction of brain tumor patients depending on resection status. Front. Comput. Neurosci. **13**, 73 (2019)
20. Weninger, L., Rippel, O., Koppers, S., Merhof, D.: Segmentation of brain tumors and patient survival prediction: methods for the BraTS 2018 challenge. In: Crimi, A., Bakas, S., Kuijf, H., Keyvan, F., Reyes, M., van Walsum, T. (eds.) BrainLes 2018. LNCS, vol. 11384, pp. 3–12. Springer, Cham (2019). https://doi.org/10.1007/978-3-030-11726-9_1
21. Yang, K., Nath, S., Koziarz, A., Badhiwala, J.H., Ghayur, H., Sourour, M., Catana, D., Nassiri, F., Alotaibi, M.B., Kameda-Smith, M., et al.: Biopsy versus subtotal versus gross total resection in patients with low-grade glioma: a systematic review and meta-analysis. World Neurosurg. **120**, e762–e775 (2018)
22. Youngner, S.J., Arnold, R.M.: The Oxford Handbook of Ethics at the End of Life. Oxford University Press, Oxford (2016)

Brain Tumor Segmentation and Survival Prediction Using 3D Attention UNet

Mobarakol Islam[1,2]($\boxtimes$), V. S. Vibashan[2,3], V. Jeya Maria Jose[2,3],
Navodini Wijethilake[2,4], Uppal Utkarsh[2,5], and Hongliang Ren[2]($\boxtimes$)

[1] NUS Graduate School for Integrative Sciences and Engineering,
NUS, Singapore, Singapore
mobarakol@u.nus.edu
[2] Department of Biomedical Engineering, National University of Singapore,
Singapore, Singapore
ren@nus.edu.sg
[3] Department of Instrumentation and Control Engineering, NIT,
Tiruchirappalli, India
[4] Department of Electronics and Telecommunications, University of Moratuwa,
Moratuwa, Sri Lanka
[5] Department of Electrical Engineering, Punjab Engineering College,
Chandigarh, India

Abstract. In this work, we develop an attention convolutional neural network (CNN) to segment brain tumors from Magnetic Resonance Images (MRI). Further, we predict the survival rate using various machine learning methods. We adopt a 3D UNet architecture and integrate channel and spatial attention with the decoder network to perform segmentation. For survival prediction, we extract some novel radiomic features based on geometry, location, the shape of the segmented tumor and combine them with clinical information to estimate the survival duration for each patient. We also perform extensive experiments to show the effect of each feature for overall survival (OS) prediction. The experimental results infer that radiomic features such as histogram, location, and shape of the necrosis region and clinical features like age are the most critical parameters to estimate the OS.

Keywords: Glioma · Tumor segmentation · Survival estimation · Attention · Regression

1 Introduction

Gliomas develop from glial cells, are the most common brain tumor with the highest mortality rate. The mean occurrence of gliomas is close to 190,000 cases annually in worldwide [4]. The average survival time of the glioma patients remains at approximately 12 months [6], and nearly 90% of patients are dead after 24 months of surgical resection [15]. Early detection, automatic delineation,

© Springer Nature Switzerland AG 2020
A. Crimi and S. Bakas (Eds.): BrainLes 2019, LNCS 11992, pp. 262–272, 2020.
https://doi.org/10.1007/978-3-030-46640-4_25

and volume estimation are vital tasks for survival prediction and treatment planning. However, gliomas are often difficult to localize and delineate with conventional manual segmentation due to their high variation of shape, location, and appearance. In addition, close supervision from a human expert is required to manually annotate the segmentation of tumor tissue, which is time-consuming and tedious. Automatic segmentation and survival rate prediction models will help the diagnosis and treatment to be much accurate and faster.

In recent years, deep learning has dominated most of the tasks like segmentation [9,11,12], tracking [8,10], and classification [13] in medical image analysis. Many studies are for brain tumor segmentation, and survival prediction utilizes deep learning techniques, especially convolutional neural network (CNN). In this paper, we design a 3D attention based UNet [19] for brain tumor segmentation from MR images. To predict the survival days for each patient, we extract shape and geometrical features and combine them with clinical features and train to analyze the performance of various regression techniques like Support Vector Machine (SVM), Artificial Neural Network (ANN), Random Forest and XGBoost.

2   Methods

2.1   Segmentation

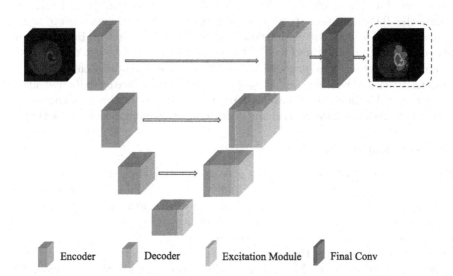

Fig. 1. Our proposed segmentation architecture 3D attention UNet by composing of sequential channel and spatial attention mechanism.

We adopt the UNet [19] architecture and convert it to 3D and integrate the 3D attention module with the decoder blocks. Further, we propose a 3D attention model with decoder blocks to enhance segmentation prediction [7]. Our proposed attention module consists of a channel and spatial attention in parallel with skip connection. Nonetheless, fusing parallelly exciting features may create inconsistency in feature learning. Integrating skip connection reduces this redundancy and sparsity of the network, as illustrated in Fig. 2. The overall architecture is illustrated in Fig. 1.

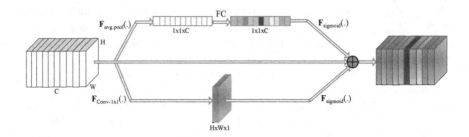

Fig. 2. Visual representation of the 3D spatial and channel attention with skip connection.

3D Skip Attention Unit. Spatial and channel attention enhances the quality of encoding throughout its feature hierarchy. Therefore we introduce 3D attention units to generate 3D spatial and channel attention by exploiting 3D inter-spatial and inter-channel feature relationships (as illustrated in Fig. 2). To obtain the 3D attention map, we first perform a $1 \times 1 \times C$ convolution to aggregate all spatial feature correlations into the $H \times W \times 1$ dimension. In parallel, we perform average pooling and feed it to the neural network to get the $1 \times 1 \times C$ channel correlation. The encoded 3D attention map encodes rich spatial and channel attention. Further, we fuse skip-connection to reduce sparsity and singularity caused by these parallel excitations. Moreover, integrating skip connection makes the learning more generic and enhancing the segmentation prediction.

2.2 Survival Prediction

Feature Extraction. Features that give information about the geometry, fractal nature of the tumor hold an important role in the number of days of survival as in our previous work [9]. The combination of features used in [9] produces the best accuracy for BraTS 2018 overall survival (OS) prediction task for the validation task. However, due to over-fitting of the data on the regression model, the method failed during the BraTS 2019 test phase. Therefore, the same combination of features is used in this work improvising the learning methods. The first axis, second axis, and third axis coordinates and lengths are extracted as geometrical features. In addition, centroid coordinates, eigenvalues, meridional and equatorial eccentricity, fractal dimensions, histogram features of the image including entropy, skewness, and kurtosis are also extracted for necrosis, tumor core, and whole tumor. All the features are normalized to 0–1 range to avoid the magnitude differences.

Feature Selection. To optimize the regression model prediction, we need to feed the model with the most decisive features for survival prediction. Thus, we explore recursive feature elimination (RFE) for feature ranking. The core idea of this method is to obtain the most significant features. The number of features is increased one by one to find the optimum number of features, which involves mostly for the overall survival (OS) prediction task.

Regression Model. We utilize the state-of-art XGBoost regression model [5] on the selected features, to predict the overall survival (OS). We tune the hyperparameters such as maximum tree depth, learning rate, the degree of verbosity, L1, and L2 regularization terms on weights to obtain the best performing model. As L1 and L2 terms control the sparsity and over-fitting, the utilization of regularization terms is an advantage in regression tasks. We also apply several other machine learning tools that are used commonly for regression tasks. For example, multi-layer perceptron (MLP), support vector machine (SVM) [21] and random forest (RF) [14].

3 Experiments

3.1 Dataset

Brain tumor dataset of BraTS 2019 [1–3,16] is used to conduct all the experiments in work. The train set of BraTS 2019 consists of 335 cases with high and low-grade glioma of 259 and 76, respectively. There are 125 and 166 cases in the validation and test set, respectively. Each case contains MRI images of 4 modalities - a) native (T1) b) post-contrast T1-weighted (T1Gd), c) T2-weighted (T2), and d) T2 Fluid Attenuated Inversion Recovery (T2-FLAIR). The voxel size of the modality is $240 \times 240 \times 155$. There is also a segmentation annotation in the train set where 3 regions are labels as 1, 3, and 4 pixels values. The annotated labels denote the necrotic and non-enhancing tumor core (NCR/NET: 1), the peritumoral edema (ED: 2), and GD-enhancing tumor (ET: 4).

3.2 Implementation Details

Our model is trained using Pytorch [18] deep learning framework. The learning rate and weight decay are adopted as 0.00015 and 0.005, respectively. We use the ADAM optimizer to train the model. Two NVIDIA GTX 1080 Ti 12 GB GPUs are exploited to conduct all the experiments in this work.

As a model input, we use the 3D voxel in 4 available modalities by cropping the brain region. The study [17] utilizes the 3D data of $128 \times 128 \times 128$ to fit the GPU memory and achieves the best accuracy in BraTS 2018 challenge. We apply a random crop of $128 \times 192 \times 192$ and mean normalization inside the data loader to prepare our model input.

4   Results

To evaluate our model prediction, we submit the model prediction into the BraTS 2019 portal and obtain several measurement metrics such as Dice, Hausdorff, Sensitivity, and Specificity. The performances of the BraTS 2019 validation set are demonstrated in Table 1. The visualization of the validation set prediction is

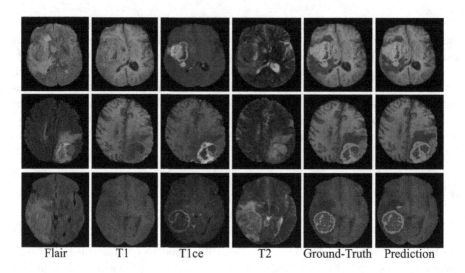

Flair   T1   T1ce   T2   Ground-Truth   Prediction

Fig. 3. Flair, T1, T1ce, and T2 modalities of the brain tumor visualized with the Ground-Truth and Predicted segmentation of tumor sub-regions for BraTS 2019 cross-validation dataset. Red label: Necrosis, yellow label: Edema and Green label: Edema. (Color figure online)

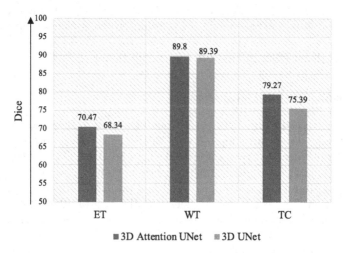

Fig. 4. Performance comparison between our proposed model 3D attention UNet and original model 3D UNet.

illustrated in Fig. 3. The performance graph of our proposed 3D attention UNet and original 3D UNet is plotted in Fig. 4. It is clearly shown that 3D attention UNet outperforms the original model for all the regions such as ET, WT, and TC.

The quantitative results for the BraTS 2019 test set are showed in Table 2. In Fig. 5, we can infer the prediction of our model for the BraTS 2019 testing dataset.

Table 1. Dice, Hausdorff, Sensitivity, and Specificity metrics evaluation of BraTS 2019 validation set for segmentation task.

	Dice			Hausdorff			Sensitivity			Specificity		
	ET	WT	TC	ET	WT	TC	ET	WT	TC	ET	WT	TC
Mean	0.704	**0.898**	0.792	7.05	6.29	8.76	0.751	0.900	0.816	0.998	0.994	0.996
StdDev	0.311	0.070	0.190	13.09	11.14	13.95	0.284	0.086	0.191	0.003	0.005	0.007
Median	0.835	0.917	0.868	2.23	3.31	4.24	0.859	0.926	0.894	0.999	0.996	0.998

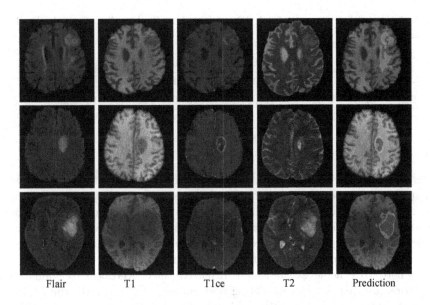

| Flair | T1 | T1ce | T2 | Prediction |

Fig. 5. Flair, T1, T1ce and T2 modalities of the brain tumor visualized with the Ground-Truth and Predicted segmentation of tumor sub-regions for BraTS 2019 testing dataset. The annotation color can be interpreted as red - necrosis, yellow - enhance tumor, and green - edema. (Color figure online)

Table 2. Dice and Hausdorff metrics evaluation of BraTS 2019 testing set for segmentation task.

	Dice			Hausdorff		
	ET	WT	TC	ET	WC	TC
Mean	0.7780	**0.8689**	0.7771	3.6730	7.3071	6.8196
StdDev	0.2111	0.1496	0.2873	6.1930	13.6302	11.3926
Median	0.8389	0.9130	0.8949	2.0000	3.6055	3.08114

Table 3. Performance comparison of SVM, XGBoost, MLP and Random Forest (RF) on validation set for overall survival prediction. MSE and stdSE denotes as the mean square error and standard deviation of the predicted survival days.

Method	Accuracy	MSE	MedianSE	stdSE
XGBoost [5]	**42.86%**	110012.835	38444.333	207273.871
MLP [20]	41.4%	**102839.036**	49823	138563.601
Random Forest [14]	35.6%	268310.586	58369.883	12603.182
SVM [21]	32.9%	107569.325	72686.271	106573.219

Table 4. Quantitative results for XGBoost based survival prediction on the BraTS19 validation and test dataset

Dataset	Cases	Accuracy	MSE	MedianSE	stdSE	SpearmanR
Valid	94	**48.3%**	127478.649	35101.147	211645.67	0.187
Test	107	38.3%	417633.26	68150.079	1215799.813	0.238

4.1 Survival Prediction

Several state-of-the-art regression models are used to estimate the survival rate in our study. There are 125 cases in the validation set, but only 29 anonymous cases are chosen to validate the model in BraTS 2019 evaluation portal. We have done 4-fold cross-validation to evaluate the regression model on the training dataset. Table 3 shows the performance comparison among all the models. We observe that XGBoost outperforms all other regression models with the highest accuracy where MLP achieves the lowest MSE. We select XGBoost to evaluate the validation and test set by considering performance. Table 4 shows the XGBoost OS performance on BraTS 2019 validation and test dataset.

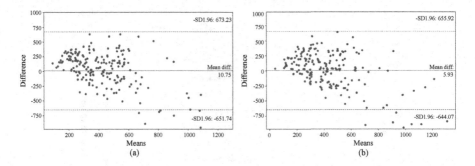

Fig. 6. Bland Altman plot obtained from the training cross-validation results of overall survival prediction model (a) Bland Altman plot obtained for all the extracted features. This gives a mean difference of 10.75 days. (b) Bland Altman plot obtained for the selected 14 features. This gives a mean difference of 5.95 days.

5   Discussion

The results infer that our 3D attention UNet produces better accuracy than the original 3D UNet. Especially, the prediction of tumor core boosts up in our model (as shown in Fig. 4), which is a very important region to define tumor prognosis. To estimate the OS, we exploit 4 different regression models where XGBoost outperforms in terms of accuracy. To design an efficient model, we select the 14 most important features and train the models. A Bland Altman plot in Fig. 6 (a and b) represents the distribution of regression output for all extracted features and 14 selected features. The mean difference between the ground truth and the predicted survival rate is almost half (5.93 days) for the selected features comparing to all features.

Figure 7 demonstrate the importance of the selected features for the model performance. SHAP (SHapley Additive exPlanations) analysis, based on game theory, is an approach to explain the output of tree ensemble methods such as XGBoost. The red color represents the high feature values, and blue represents the low values. The y-axis of the plot shows the 14 features selected for our experiments. We can infer that age has the highest contribution to model performance. In addition, the histogram of necrosis, eigenvalue, whole tumor volume, and 2nd axis length of the tumor voxel are some of the significant features that contributed to predicting OS. Figure 8 shows the regression plot of the ground truth and the prediction of the model.

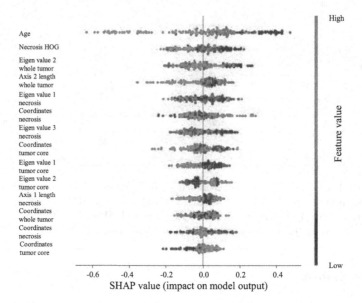

Fig. 7. Effect of the features for the outcome of the model. The red color represents the high feature values, and blue represents the low values to determine the significance of the features in model prediction. (Color figure online)

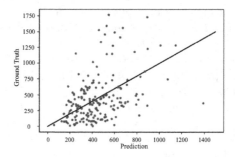

Fig. 8. Regression scatter plot for the predicted overall survival and ground truth overall survival in days.

6 Conclusion

In this paper, we present a segmentation and survival prediction model for automatic brain tumor prognosis using MRI. We adopt UNet and integrate the 3D attention technique into a novel way to capture the significant features in model learning. We also extract many novel geometric and shape features to estimate the survival days using the regression model. We observe that the location, shape, and size of the necrosis region is the most significant parameters in glioma prognosis estimation.

Acknowledgement. This work was supported by the Singapore Academic Research Fund under Grant R-397-000-297-114, and NMRC Bedside & Bench under grant R-397-000-245-511 awarded to Dr. Hongliang Ren.

References

1. Bakas, S., et al.: Segmentation labels and radiomic features for the pre-operative scans of the TCGA-GBM collection. Cancer Imaging Arch. **286** (2017)
2. Bakas, S., et al.: Advancing the Cancer Genome Atlas glioma MRI collections with expert segmentation labels and radiomic features. Sci. Data **4**, 170117 (2017)
3. Bakas, S., et al.: Identifying the best machine learning algorithms for brain tumor segmentation, progression assessment, and overall survival prediction in the brats challenge. arXiv preprint arXiv:1811.02629 (2018)
4. Castells, X., et al.: Automated brain tumor biopsy prediction using single-labeling cDNA microarrays-based gene expression profiling. Diagn. Mol. Pathol. **18**(4), 206–218 (2009)
5. Chen, T., Guestrin, C.: XGBoost: a scalable tree boosting system. In: Proceedings of the 22nd ACM SIGKDD International Conference on Knowledge Discovery and Data Mining, pp. 785–794. ACM (2016)
6. Furnari, F.B., et al.: Malignant astrocytic glioma: genetics, biology, and paths to treatment. Genes Dev. **21**(21), 2683–2710 (2007)
7. Hu, J., Shen, L., Sun, G.: Squeeze-and-excitation networks. In: Proceedings of the IEEE Conference on Computer Vision and Pattern Recognition, pp. 7132–7141 (2018)
8. Islam, M., Atputharuban, D.A., Ramesh, R., Ren, H.: Real-time instrument segmentation in robotic surgery using auxiliary supervised deep adversarial learning. IEEE Robot. Autom. Lett. **4**(2), 2188–2195 (2019)
9. Islam, M., Jose, V.J.M., Ren, H.: Glioma prognosis: segmentation of the tumor and survival prediction using shape, geometric and clinical information. In: Crimi, A., Bakas, S., Kuijf, H., Keyvan, F., Reyes, M., van Walsum, T. (eds.) BrainLes 2018. LNCS, vol. 11384, pp. 142–153. Springer, Cham (2019). https://doi.org/10. 1007/978-3-030-11726-9_13
10. Islam, M., Li, Y., Ren, H.: Learning where to look while tracking instruments in robot-assisted surgery. In: Shen, D., et al. (eds.) MICCAI 2019. LNCS, vol. 11768, pp. 412–420. Springer, Cham (2019). https://doi.org/10.1007/978-3-030-32254-0_46
11. Islam, M., Sanghani, P., See, A.A.Q., James, M.L., King, N.K.K., Ren, H.: ICHNet: Intracerebral Hemorrhage (ICH) segmentation using deep learning. In: Crimi, A., Bakas, S., Kuijf, H., Keyvan, F., Reyes, M., van Walsum, T. (eds.) BrainLes 2018. LNCS, vol. 11383, pp. 456–463. Springer, Cham (2019). https://doi.org/10.1007/ 978-3-030-11723-8_46
12. Islam, M., Vaidyanathan, N.R., Jose, V.J.M., Ren, H.: Ischemic stroke lesion segmentation using adversarial learning. In: Crimi, A., Bakas, S., Kuijf, H., Keyvan, F., Reyes, M., van Walsum, T. (eds.) BrainLes 2018. LNCS, vol. 11383, pp. 292–300. Springer, Cham (2019). https://doi.org/10.1007/978-3-030-11723-8_29
13. Li, Q., Cai, W., Wang, X., Zhou, Y., Feng, D.D., Chen, M.: Medical image classification with convolutional neural network. In: 2014 13th International Conference on Control Automation Robotics & Vision (ICARCV), pp. 844–848. IEEE (2014)
14. Liaw, A., Wiener, M., et al.: Classification and regression by randomforest. R News **2**(3), 18–22 (2002)

15. Louis, D.N., et al.: The 2007 who classification of tumours of the central nervous system. Acta Neuropathol. **114**(2), 97–109 (2007)
16. Menze, B.H., et al.: The multimodal brain tumor image segmentation benchmark (BRATS). IEEE Trans. Med. Imaging **34**(10), 1993 (2015)
17. Myronenko, A.: 3D MRI brain tumor segmentation using autoencoder regularization. In: Crimi, A., Bakas, S., Kuijf, H., Keyvan, F., Reyes, M., van Walsum, T. (eds.) BrainLes 2018. LNCS, vol. 11384, pp. 311–320. Springer, Cham (2019). https://doi.org/10.1007/978-3-030-11726-9_28
18. Paszke, A., et al.: Automatic differentiation in PyTorch (2017)
19. Ronneberger, O., Fischer, P., Brox, T.: U-Net: convolutional networks for biomedical image segmentation. In: Navab, N., Hornegger, J., Wells, W.M., Frangi, A.F. (eds.) MICCAI 2015. LNCS, vol. 9351, pp. 234–241. Springer, Cham (2015). https://doi.org/10.1007/978-3-319-24574-4_28
20. Ruck, D.W., Rogers, S.K., Kabrisky, M., Oxley, M.E., Suter, B.W.: The multilayer perceptron as an approximation to a bayes optimal discriminant function. IEEE Trans. Neural Netw. **1**(4), 296–298 (1990)
21. Suykens, J.A., Vandewalle, J.: Least squares support vector machine classifiers. Neural Process. Lett. **9**(3), 293–300 (1999)

Brain Tumor Segmentation Using Dense Channels 2D U-net and Multiple Feature Extraction Network

Wei Shi[1,2], Enshuai Pang[1,2], Qiang Wu[1,2(✉)], and Fengming Lin[1,2]

[1] School of Information Science and Engineering, Shandong University, Jinan, China
wuqiang@sdu.edu.cn
[2] Institute of Brain and Brain-Inspired Science, Shandong University, Jinan, China

Abstract. Semantic segmentation plays an important role in the prevention, diagnosis and treatment of brain glioma. In this paper, we propose a dense channels 2D U-net segmentation model with residual unit and feature pyramid unit. The main difference compared with other U-net models is that the number of bottom feature components is increased, so that the network can learn more abundant patterns. We also develop a multiple feature extraction network model to extract rich and diverse features, which is conducive to segmentation. Finally, we employ decision tree regression model to predict patient overall survival by the different texture, shape and first-order features extracted from BraTS 2019 dataset.

Keywords: Brain tumor segmentation · 2D U-net · Dense channels · Multiple feature extraction · Feature pyramid

1 Introduction

The most primary malignant brain tumor is gliomas with different histological sub-regions including edema, necrotic core, enhancing and non-enhancing tumor core [1–5]. Usually, multi-modal MRI scans are used to detect the various sub-regions of gliomas by different intensity distribution [1–5]. Therefore, how to segment the multi-modal MRI of gliomas automatically becomes an important clinical solution for tumor prevention, diagnosis, and treatment.

VGG [6], FCN [7] and U-net [8] are commonly used methods for medical image segmentation. U-net is the most frequently used method for medical image segmentation. The typical characteristic of U-net network is that its structure is U-symmetrical, the left side is the encoder, and the right side is the decoder [8]. Another characteristic is that each convolution layer of U-net network encoder will concatenate to the upsample layer of the corresponding decoder, so that the resulting feature map contains not only features of high-level, but also features of low-level, then it achieves the integration of features under different scales, by retaining the location information, to improve the segmentation performance [8].

A. Crimi and S. Bakas (Eds.): BrainLes 2019, LNCS 11992, pp. 273–283, 2020.
https://doi.org/10.1007/978-3-030-46640-4_26

In practice, the number of medical images training samples is less and the scanning position and number of slices will be different, so the 2D segmentation model is undoubtedly more flexible in the real application. In this paper, we propose a dense channels 2D U-net model, as an extension of 2D U-net integrating the idea of feature pyramid network [9,10] and residual network [11]. By increasing the number of channels, the segmentation performance is improved. In order to conserve computational resources and extract more diverse features, based on dense channels 2D U-net model, we develop a multiple feature extraction network to segment the multi-modal MRI scans of brain tumor. Finally, by the different texture, shape and first-order features extracted from BraTS 2019 dataset, we employ decision tree regression model to predict patient overall survival.

2   Segmentation Methods

2.1   Dense Channels 2D U-net Model

2D U-net model is the most commonly used model in medical image segmentation, and its segmentation accuracy is higher than other traditional models. 2D U-net model's unique skip connection structure enables it to obtain location information of low-level features, which is conducive to segmentation [8]. However, the segmentation performance of 2D U-net model can be improved because the details of tumor can not be accurately segmented [10]. We propose a dense channels 2D U-net model (DCU-net) which consists of three parts: encoder, decoder and feature pyramid structure to extract more features and improve the segmentation performance [9,10]. The structure of dense channels 2D U-net model is shown in Fig. 1.

The encoder is composed of five residual units (RU) [10] and four down-sampling units, which processes the image down sampling. The decoder is composed of four residual units and four skip connection units (SCU), which can get the location information of low-level features. The feature pyramid structure is composed of three up-sampling units and one convolution layer, which can obtain both low-level location information and high-level detail information.

Encoder. The data is sampled by encoder. The encoder is composed of five residual units (RU) [10] and four down-sampling units. The residual unit consists of two convolution operations with 3×3 convolution kernel size in series and one convolution operation with 1×1 convolution kernel size in parallel. The strides of the convolution operation is 1×1, and the activation function is RELU. The main function of residual unit is to extract image features and solve the problem of gradient disappearance of network [10]. From the top to bottom, the number of features extracted by each residual unit is 256, 256, 512, 512 and 512, respectively. Down-sampling operation by maxpool with 2×2 kernel size and 2×2 strides changes the image to half of the original size.

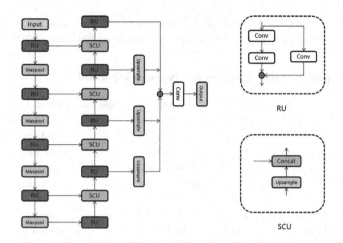

Fig. 1. Architecture of the dense channels 2D U-net model.

The input data is normalized artificially for each sample. The distribution of input data in each back layer of the network is always changing, because the updating of the training parameters in the front layer will lead to the changing distribution of input data in the back layer. In order to solve this problem, we use batch normalization operation after each convolution operation [10,12].

Decoder. The decoder is designed for restoring the data processed by the encoder to the original image size. The decoder consists of four residual units (RU) and four skip connection units (SCU). The structure of residual unit is the same as that of residual unit in encoder. From the bottom to top, the number of features extracted by each residual unit is 512, 512, 256 and 256, respectively. The skip connection unit consists of an upsample operation with 2×2 kernel size and 2×2 strides and a concatenate operation. The main function of the skip connection unit is to connect the encoder and decoder so that the decoder can obtain the position information of the pixels [10].

Feature Pyramid Structure. We employ the feature pyramid structure to achieve the combination of low-level and high-level features [9,10]. In traditional neural networks, low-level features focus on location information, while high-level features focus on semantic information [9,10]. We combine low-level features with high-level features to achieve more accurate segmentation of medical images. Traditional segmentation networks ignore the influence of low-level features, resulting in poor segmentation performance. In the feature pyramid structure, we select feature maps of different levels. First, we upsample feature maps of different levels to the same size as the original maps. Then, we adjust them by a convolution operation with kernel size and strides of 1×1 and RELU activation function. The number of features extracted by each convolution operation is 32. Finally, we add these feature maps of the same size together, so that we can get both high-level features and low-level features [9,10].

2.2 Multiple Feature Extraction Network Model

Due to the large number of feature components, dense channels 2D U-net model needs a lot of computational and storage resources. We propose a multiple feature extraction network model (MFEN). Compared with dense channels 2D U-net model, multiple feature extraction network model can extract more abundant and different features, and conserve computational and storage resources at the same time. The structure of multiple feature extraction network is shown in Fig. 2.

Multiple feature extraction network model consists of four parts: encoder, decoder, pre-segmentation structure and feature pyramid structure. The pre-segmentation structure consists of a traditional FCN [7] model and an attention unit. The encoder is composed of five multiple feature extraction units [16] and four down-sampling units. The decoder is composed of four multiple feature extraction units and four skip connection units. The feature pyramid structure is composed of three up-sampling units and one convolution layer.

Pre-segmentation Structure. The pre-segmentation structure makes a rough segmentation of the image. The segmentation result and the original image are sent to the subsequent structure for more accurate segmentation. The pre-segmentation structure consists of a traditional FCN [7] model and an attention unit. The attention unit (AU) multiplies the FCN model segmentation result with the original data, then adds the multiplied result with the original data, and finally concatenates the added result and the original data. The attention unit not only roughly segment the original data, but also retain the information of the original data, which contributes to the subsequent accurate segmentation.

Encoder. The overall structure of the encoder is similar to that of the encoder in dense channels 2D U-net model, but residual units are replaced by multiple feature extraction units (MFEU) [16]. The input data is firstly processed by five different parallel operation modules. The five operation modules are one convolution operation with 1×1 convolution kernel size, one convolution operation with 3×3 convolution kernel size, two convolution operations with 3×3 convolution kernel size in series, three convolution operations with 3×3 convolution kernel size in series and one maxpool operation with 3×3 kernel size [15,16]. These five operation modules can extract a large number of rich and more different feature components. The data processed by these five operation modules are concatenated, then one convolution operation with 1×1 convolution kernel size is used to reduce the number of features to conserve computational and storage resources. Finally, the original data is convoluted by one convolution operation with 1×1 convolution kernel size to get the same number of features, and then added with the reduced number of features. This operation is equivalent to a

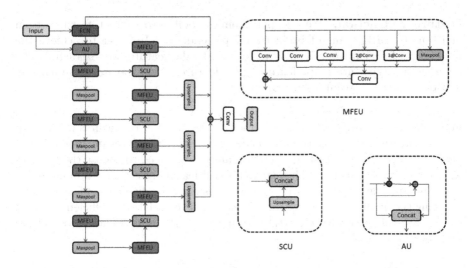

Fig. 2. Architecture of the multiple feature extraction network model.

residual unit, which can solve the problem of gradient disappearance of network. The strides of the all operations is 1 × 1, and the activation function is RELU. Batch normalization operation is used before the activation function. From the top to bottom, the number of features extracted by each multiple feature extraction unit is 32, 64, 128, 256 and 512, respectively.

Decoder. The structure of decoder is the same as that of dense channels 2D U-net model, and residual units are replaced by multiple feature extraction units. The structure of multiple feature extraction unit is exactly the same as that described in encoder. From the bottom to top, the number of features extracted by each multiple feature extraction unit is 256, 128, 64 and 32, respectively.

Feature Pyramid Structure. Compared with dense channels 2D U-net model, the structure of feature pyramid has more abundant information, because the structure of feature pyramid obtains the information from FCN model, which is more effective for segmentation results.

3   Prediction of Patient Overall Survival

3.1   Survival Prediction Methods

First, we separate the different tumor area from the four modalities, and then we take the whole brain tumor area as mask to extract features on four modalities, then take the edema area, the enhancing area, the non-enhancing and necrotic area as mask to extract features on Flair, T1ce, T1 MRI scans. In this way, we can get seven different tumor data in total. The following operations are carried out on these seven data.

Feature Extraction. We use pyradiomics toolkit to extract features from multi-modal MRI scans of brain tumor patients [13]. In order to achieve the prediction of patient overall survival task, 714 valid features are extracted. The main feature types are shape feature, glcm feature, glszm feature, gldm feature, first-order feature, glrlm feature and etc.

Feature Selection. Because we extract a large number of features in the process of feature extraction, the impact factors of these features on prediction of patient overall survival task are not the same, so we need to choose the features with larger impact factors. The significance of feature selection lies in eliminating some irrelevant and repetitive features, ensuring the diversity of features and reducing the number of features. In this paper, the correlation coefficients of each feature are calculated to select the features with strong correlation coefficients by f_regression [13].

Regression Framework. In this paper, we employ decision tree regression to complete the regression task [13]. The input is the training dataset, and the output is the regression tree. In the input space where the training dataset is located, each region is recursively divided into two sub-regions and the output value of each sub-region is determined. A binary decision tree is constructed.

The prediction of patient overall survival regression framework is shown in Fig. 3.

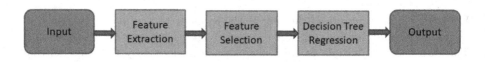

Fig. 3. Regression framework.

4 Experiments

4.1 Data

BraTS 2019 dataset is divided into training dataset, validation dataset and testing dataset. There are 335 samples in the training dataset, of which 259 are high-grade gliomas and 76 are low-grade gliomas. There are 125 samples in the validation dataset and 166 samples in the testing dataset, including both high-grade and low-grade gliomas. Each sample of training dataset, validation dataset and testing dataset contains four modalities with size of $155 \times 240 \times 240$, which are Flair, T1, T1ce and T2. The training dataset also contains ground truth with size of $155 \times 240 \times 240$, including the enhancing tumor (label 4), the edema tumor (label 2), and the necrotic and non-enhancing tumor (label 1) [1–5].

The competition does not provide the ground truth of the validation dataset and testing dataset. Four modal MRI scans and their ground truth are shown in Fig. 4.

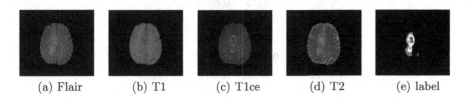

(a) Flair   (b) T1   (c) T1ce   (d) T2   (e) label

Fig. 4. Four modal MRI scans and their ground truth.

4.2 Preprocessing

The four modalities in the dataset provide the information of different regions of the same tumor. In order to segment different regions of the tumor accurately, the information provided by these four modalities must be used in the training process of the network. We firstly reshape the data of four modalities into a multi-channel data with the size of $155 \times 240 \times 240 \times 4$.

There are a lot of unlabeled data in the training dataset. These data do not contribute to the training of the network, but will waste a lot of computational resources, so we remove the slices of the label with all tags of zero and corresponding four modalities of the slices. Then, we normalize the each modality of whole dataset with mean value of zero and standard deviation of one, which improves the convergence speed and accuracy of the model. After these preprocessing steps, the size of the data eventually becomes $slices \times 240 \times 240 \times 4$.

4.3 Results

Segmentation of Gliomas. Using our segmentation methods, the training data processed in Sect. 4.2 are divided into five parts. Four parts are the training dataset and the other one is the validation dataset. We rotate half of the training dataset 90°, and rotate the other half of the training dataset −90°. The augmented training dataset is 2 times larger than the original one. Five experiments are performed. The results of the last epoch validation dataset of each experiment are averaged. The results are shown in Table 1.

In order to prove that our model is more effective, we designed a comparative experiment. We selected some of the most commonly used models for medical image segmentation tasks, including VGG [6], FCN [7], U-net [8] and HPU-net [10]. All models use the same processed dataset, and the results of 5-fold cross validation are shown in Table 1. Compared with other models, we can see that the models in this paper is better in three evaluation indicators.

Table 1. The 5-fold cross validation average performance on the training dataset.

	ET dice	WT dice	TC dice
VGG	0.6383	0.8649	0.6631
FCN	0.6343	0.8813	0.6865
U-net	0.7136	0.8936	0.7558
HPU-net	0.7626	0.8985	0.7959
DCU-net	**0.8266**	**0.9019**	**0.8394**
MFEN	**0.8155**	**0.9023**	**0.8284**

Table 2. The segmentation results of dense channels 2D U-net model on the validation dataset.

	Dice ET	Dice WT	Dice TC	Hausdorff95 ET	Hausdorff95 WT	Hausdorff95 TC
Mean	0.6714	0.8855	0.7712	12.9919	19.7379	17.3661
StdDev	0.3209	0.0876	0.2359	26.0555	29.5509	26.3002
Median	0.8207	0.9089	0.8700	2.4495	4.2426	5.8310
25quantile	0.6113	0.8809	0.6995	1.4142	2.8284	2.4495
75quantile	0.8911	0.9329	0.9254	6.6575	27.9106	15.2725

Table 3. The segmentation results of multiple feature extraction network model on the validation dataset.

	Dice ET	Dice WT	Dice TC	Hausdorff95 ET	Hausdorff95 WT	Hausdorff95 TC
Mean	0.6910	0.8867	0.7702	5.8884	21.1901	12.1920
StdDev	0.3079	0.0842	0.2484	10.8221	31.3611	19.2535
Median	0.8292	0.9098	0.8810	2.2361	4.1231	4.5826
25quantile	0.6275	0.8796	0.7108	1.4142	2.4495	2.2361
75quantile	0.8882	0.9367	0.9306	3.7417	28.7576	13.0384

Validation dataset is processed as follows. Firstly, multi-modal data is reshaped into multi-channel data. Secondly, each modality of the dataset was normalized to zero mean and one standard deviation. The segmentation results of dense channels 2D U-net model on the validation dataset are shown in Table 2. The segmentation results of multiple feature extraction network model on the validation dataset are shown in Table 3.

According to the segmentation results of the validation dataset, we finally use multiple feature extraction network model in the testing dataset. Testing dataset is processed exactly the same way as the validation dataset. The segmentation results of multiple feature extraction network model on the testing dataset are shown in Table 4.

Table 4. The segmentation results of multiple feature extraction network model on the testing dataset.

	Dice ET	Dice WT	Dice TC	Hausdorff95 ET	Hausdorff95 WT	Hausdorff95 TC
Mean	0.7599	0.8645	0.7974	5.3315	21.1373	14.4187
StdDev	0.2434	0.1313	0.2723	13.5876	26.7395	25.3849
Median	0.8336	0.9073	0.9079	1.4142	5.5649	3.1623
25quantile	0.7325	0.8478	0.8219	1.4142	2.4495	2.0000
75quantile	0.9016	0.9399	0.9451	2.8284	41.2851	10.3051

Prediction of Patient Overall Survival. In the training phase, we also used the 5-fold cross validation for prediction of patient overall survival task. In order to evaluate the performance of the classification framework, the overall survival data is divided into three classes: long-term data (>15 months), medium-term data (≥ 10 months and ≤ 15 months) and short-term data (<10 months) [14]. The prediction results are shown in Table 5.

Table 5. The results of prediction of patient overall survival on the training dataset. Horizontal column is a feature selection method and vertical column is a regression method.

	variance threshold	f_regression	mutual_info_regression
Linear regression	0.375	0.394	0.361
Decision tree regression	0.446	**0.493**	0.375
KNN	0.412	0.408	0.365
Random forest regression	0.422	0.455	0.408
Adaboost	0.388	0.351	0.350
GBRT	0.365	0.379	0.342
XGboost	0.412	0.388	0.398

In order to find a better prediction performance, we achieved variance threshold and mutual_info_regression methods besides f_regression in this paper [13]. In addition to the decision tree regression method, we also achieved linear regression, KNN, random forest regression, Adaboost, GBRT and XGboost regression framework in this paper [13]. From the comparison results in Table 5, we can see that the accuracy of model using f_regression and decision tree regression is better than other approaches. The results of patient overall survival task on the testing dataset are shown in Table 6.

Table 6. The results of prediction of patient overall survival on the testing dataset.

CasesExpected	CasesEvaluated	Accuracy	MSE	medianSE	stdSE	SpearmanR
107	107	0.486	488865.700	83521	1073912.00	0.229

5   Conclusion

We propose two image segmentation models based on extended 2D U-net model and feature pyramid structure. These models extract a large number of feature components in the process of segmentation, which is conducive to improving the accuracy of segmentation. We also propose a patient overall survival prediction framework, which uses the segmentation results of the above segmentation model and multi-modal MRI data to extract features, and then uses decision tree regression to predict patient overall survival based on these features. Our models provide a good performance on the testing dataset.

Acknowledgments. The work is supported by the Fundamental Research Funds of Shandong University (Grant No. 2017JC013), the Shandong Province Key Innovation Project (Grant No. 2017CXGC1504, 2017CXGC1502) and the Natural Science Foundation of Shandong Province (Grant No. ZR2019MH049).

References

1. Menze, B.H., Jakab, A., Bauer, S., Kalpathy-Cramer, J., Farahani, K., Kirby, J., et al.: The multimodal brain tumor image segmentation benchmark (BRATS). IEEE Trans. Med. Imaging **34**(10), 1993–2024 (2015)
2. Bakas, S., Akbari, H., Sotiras, A., Bilello, M., Rozycki, M., Kirby, J.S., et al.: Advancing the cancer genome atlas glioma MRI collections with expert segmentation labels and radiomic features. Nat. Sci. Data **4**, 170117 (2017)
3. Bakas, S., Reyes, M., Jakab, A., Bauer, S., Rempfler, M., Crimi, A., et al.: Identifying the best machine learning algorithms for brain tumor segmentation, progression assessment, and overall survival prediction in the BRATS challenge. arXiv preprint arXiv:1811.02629 (2018)
4. Bakas, S., Akbari, H., Sotiras, A., Bilello, M., Rozycki, M., Kirby, J., et al.: Segmentation labels and radiomic features for the pre-operative scans of the TCGA-GBM collection. The Cancer Imaging Archive (2017)
5. Bakas, S., Akbari, H., Sotiras, A., Bilello, M., Rozycki, M., Kirby, J., et al.: Segmentation labels and radiomic features for the pre-operative scans of the TCGA-LGG collection. The Cancer Imaging Archive (2017)
6. Simonyan, K., Zisserman, A.: Very deep convolutional networks for large-scale image recognition. arXiv preprint arXiv:1409.1556 (2014)
7. Long, J., Shelhamer, E., Darrell, T.: Fully convolutional networks for semantic segmentation. In: The IEEE Conference on Computer Vision and Pattern Recognition, pp. 3431–3440 (2015)
8. Ronneberger, O., Fischer, P., Brox, T.: U-Net: convolutional networks for biomedical image segmentation. In: Navab, N., Hornegger, J., Wells, W.M., Frangi, A.F. (eds.) MICCAI 2015. LNCS, vol. 9351, pp. 234–241. Springer, Cham (2015). https://doi.org/10.1007/978-3-319-24574-4_28

9. Tsung-Yi, L., et al.: Feature pyramid networks for object detection. In: Proceedings of the IEEE Conference on Computer Vision and Pattern Recognition, pp. 2117–2125 (2017)
10. Kong, X., Sun, G., Wu, Q., Liu, J., Lin, F.: Hybrid pyramid U-Net model for brain tumor segmentation. In: Shi, Z., Mercier-Laurent, E., Li, J. (eds.) IIP 2018. IAICT, vol. 538, pp. 346–355. Springer, Cham (2018). https://doi.org/10.1007/978-3-030-00828-4_35
11. He, K., et al.: Deep residual learning for image recognition. In: Proceedings of the IEEE Conference on Computer Vision and Pattern Recognition, pp. 770–778 (2016)
12. Ioffe, S., Szegedy, C.: Batch normalization: accelerating deep network training by reducing internal covariate shift. arXiv preprint arXiv:1502.03167 (2015)
13. Griethuysen, J.J.M., et al.: Computational radiomics system to decode the radiographic phenotype. Cancer Res. **77**(21), 104–107 (2017)
14. Multimodal Brain Tumor Segmentation Challenge 2019: Evaluation Framework. (2019-6-10) [2019-10-21]. https://www.med.upenn.edu/cbica/brats2019/evaluation.html
15. Szegedy, C., et al.: Going deeper with convolutions. In: Proceedings of the IEEE Conference on Computer Vision and Pattern Recognition, pp. 1–9 (2015)
16. Szegedy, C., et al.: Rethinking the inception architecture for computer vision. In: Proceedings of the IEEE Conference on Computer Vision and Pattern Recognition, pp. 2818–2826 (2016)

Brain Tumour Segmentation on MRI Images by Voxel Classification Using Neural Networks, and Patient Survival Prediction

Subin Sahayam[1]([⊠])(ID), Nanda H. Krishna[2](ID), and Umarani Jayaraman[1]

[1] Indian Institute of Information Technology Design and Manufacturing,
Kancheepuram, Chennai 600127, India
{coe18d001,umarani}@iiitdm.ac.in
[2] Sri Sivasubramaniya Nadar College of Engineering,
Kalavakkam, Chennai 603110, India
nanda17093@cse.ssn.edu.in

Abstract. In this paper, an algorithm for segmentation of brain tumours and the survival prediction of a patient in days has been proposed. The delineation of brain tumours from magnetic resonance imaging (MRI) by experts is a time-consuming process and is susceptible to human error. Recently, most methods in the literature have used convolution neural network architectures, its variants, and an ensemble of several models to achieve the state-of-the-art result. In this paper, we study a neural network architecture to classify voxels in 3D MRI brain images into their respective segment classes. The study focuses on class imbalance among tumour regions, and pre-processing. The method has been trained and tested on the BraTS2019 dataset. The average Dice score for the segmentation task in the validation set is 0.47, 0.43, and 0.23 for enhancing, whole, and core tumour regions, respectively. For the second task, linear regression has been used to predict the survival of a patient in days. It achieved an accuracy of 0.465 on the online evaluation engine for the training dataset.

Keywords: Voxel classification · Neural network · Brain tumour segmentation · Survival prediction · Linear regression

1 Introduction

Glial cells are support cells which surround the neurons in the brain. Tumours that originate from these cells are known as gliomas. High-grade gliomas (HGG) and low-grade gliomas (LGG) are two broad grades of tumours based on its growth rate (aggressiveness) and capability to infiltrate nearby tissues. The higher-grade is more deadly than its lower counterpart with the average survival for HGG falling between 12 to 14 months [11,18]. There are several MRI modalities which give information about the difference in tissue water (T1, T2,

© Springer Nature Switzerland AG 2020
A. Crimi and S. Bakas (Eds.): BrainLes 2019, LNCS 11992, pp. 284–294, 2020.
https://doi.org/10.1007/978-3-030-46640-4_27

FLAIR), water diffusion (DTI), and contrast-enhanced images (T1ce) [13]. This information helps in the diagnosis and treatment planning of diseases. Human experts will manually delineate the tumour, which is a time-consuming and human error-prone task. They use rough measures to estimate tumours, which results in inter-rater and intra-rater errors. Hence, a robust and accurate method is required to perform the segmentation task [5,13].

There are several challenges in the segmentation of brain tumours from MRI images. MRI images are 3D volumetric data in which each voxel location has an intensity value. The intensity changes across MRI modalities are used to identify a tumour and its sub-structures. For example, a set of voxels with the highest intensity value in T1 contrast-enhanced (T1ce) images are used to identify enhancing tumours. Bias field artefacts affect the intensity values in an image. It results in intensity inhomogeneity of voxels in the same region in a single frame, across a series of frames, and in images acquired from the same acquisition scanner at different intervals of time. The intensity in MRI images varies across acquisition scanners as various scanners use different magnetic field strength. High variability in size, shape and location of the tumour makes the segmentation process more challenging [2,13].

In the literature, convolution neural networks and its variants have dominated the segmentation task. These architectures use region-based segmentation approach to achieve good results [10]. U-Nets have performed well in several segmentation challenges across the image processing domain [7–10,12,14,16,17]. The winner of BraTS 2018 segmentation challenge, Andriy used an encoder-decoder CNN based network. A large encoder has been used to extract deep features. A decoder has been used to reconstruct the segmentation part. Variational auto-encoders have been used to reduce the problem of over-fitting [14]. Kamnitsas et al., the winner of BraTS 2017, proposed an ensemble model called EMMA (Ensembles of Multiple Models and Architectures) for robust segmentation. EMMA used the ensemble results of various segmentation models like DeepMedic [10], FCN [12] and U-Net [17] to produce a target segmented image [9]. Isensee et al. proposed a modified 3D U-Net. They claim that a well-tuned U-Net architecture can achieve state-of-the-art results. They validated their model in BraTS 2018 dataset. Their model achieved a Dice score of 78.62, 91.75, and 85.69 in enhancing, whole, and core tumour regions [8].

For survival prediction, Xue Feng et al. [7] with their linear regression model achieved an accuracy of 0.32 in the validation set. They used age, resection status, and 7 other extracted features from the tumour segmented image. Deep learning neural network-based models require a lot of data for training. The limited number of features and instances for the training phase overfits such models to the training data. Thus, deep learning neural network models failed to predict patient survival when compared to traditional machine learning algorithms like linear regression. A detailed review of BraTS 2018 segmentation and survival prediction tasks are given in [4].

In this paper, we aim to solve the class imbalance among brain tumours and it's sub-structures especially enhancing tumour region. For which we segmented them using only intensity values in the MRI image modalities without the usage of neighbourhood and temporal information. This is because balancing among classes is easier in lower dimensions (intensity values - zero dimension (0D)) than in neighbourhood (2D) and, (3D). We used a simple dense artificial neural network (ANN) to carry out the segmentation. From the sensitivity and specificity results, it is clear that the model can segment the tumour and its sub-structures but it also classified healthy regions as tumourous regions.

2   Proposed Segmentation Model

The training phase of the tumour segmentation task consists of image pre-processing, frame selection, voxel selection, and neural network architecture. The trained neural network model is used to classify a voxel into background, edema, core or enhancing tumour classes. Figure 1 shows the flow diagram for the segmentation task.

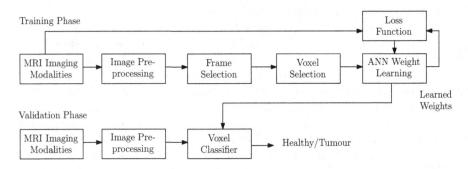

Fig. 1. The flow diagram of the proposed segmentation task. The trained weights from the training phase is used for voxel classification in the validation phase for each image. All classified voxels are then reconstructed into an MRI image

The training phase of the patient survival prediction task consists of feature extraction and fitting a linear regression model. The trained model is used to predict the survival of a patient in days.

2.1   Pre-processing

The intensity for 3D MRI images varies over different acquisition scanners used across several institutions [15]. So, the mean (M_o) of each MRI image (I_o) has been normalised to zero. Further, to normalise the distribution of intensities across MRI images, the voxels in zero-mean images has been divided by the standard deviation (Sd_o) of the original image. Intensity values in the resulting image have been scaled up by a factor of 100 as conversion from a floating-point value to an integer may result in loss of image intensity data. The resultant normalised image is denoted as I_n. The normalisation is given by,

$$\text{Normalized Image } I_n = ((I_o - M_o)/Sd_o) \cdot 100$$

Most of the voxel data belong to background or healthy tissue class. So, most of the voxels can be ignored. To achieve this, the nth percentile intensity value is calculated for each image and all voxels with intensity value less than the nth percentile have been changed to 0. The value of 'n' is fixed as 98 as it produced the best Dice score value.

2.2 Frame Selection

Each MRI image has 8,928,000 (240 × 240 × 155) voxels. Most of the voxels are non-important redundant background information represented by '0' intensity value in the ground truth MRI image. Hence, the use of all these voxels may result in large computation time in terms of hours to days. To avoid this, the sum for each frame in a ground truth MRI image along each of its three-axis (axial, coronal and sagittal) has been calculated. The frames with a sum of '0' have been dropped in all modalities (FLAIR, T1, T2 and T1ce). All the frames with a sum greater than '0' for a MRI image has been reconstructed and retained as a new MRI image. This process has been repeated for all the images in the training set. Thus, a new set of MRI images across all modalities with the selected frames has been generated. This new set of MRI images have been used for further processing.

2.3 Voxel Selection

The new set of MRI images generated in Sect. 2.2 have been used for voxel selection. All the voxels from the new T1, T2, T1ce and FLAIR MRI modalities have been selected. The ground truth MRI image consists of four labels namely, '0' for background, '1' for the core tumour area, '2' represents edema, and '4' enhancing tumour region. The class labels '0', '1', '2', and '4' consist of 60,981,446, 7,031,170, 17,438,190, and 5,561,228 voxels respectively for all 335 patients. Each instance for a target class intensity label at a given voxel (x,y,z) consists of intensity value from T1, T2, T1ce and FLAIR at that respective voxel (x,y,z). For example, let the voxel (x,y,z) contain intensity values 453, 515, 823, and 652 in T1, T2, T1ce, and FLAIR modalities. These four intensity values correspond to a single input instance. Since the enhancing intensity value is significantly higher than the remaining three intensities, the input instance is classified as label '4'. This is tabulated in Table 1. The total number of instances is the sum of all intensity class labels '0', '1', '2', and '4' which is a total of 91,012,034 instances. Among all the instances, 67% of the voxels belong to intensity class label '0'. Training on such target labels may bias any neural network towards the target class with the most number of instances [16]. Hence, first 'n' instances have been selected from each class label. 'n' is class intensity label with the least number of instances. Class intensity label '4' has the least number of instances with a total of 5,561,228 instances. Therefore, from classes '0', '1', and '4', first 5,561,228 instances have been selected.

2.4   Network Architecture and Training

A simple neural network architecture has been designed to perform voxel classification. The input instance consists of intensity value from T1, T2, T1ce, and FLAIR respectively for a corresponding voxel location. Therefore, the input layer for the neural network consists of 4 neurons. There are 4 target class intensity labels namely, background '0', core '1', edema '2', and enhancing '4'. Therefore, the output layer consists of 4 neurons with a single neuron for each class. Each target class intensity label is a single value. So, the dimensions of the target intensity value (1×1) and the neuron output layer don't match (1×4). Hence, each target intensity value is transformed into a one-hot vector of dimension 1×4. An input instance with a target class intensity value of '4' will be changed to 0,0,0,1 where the values represents the target class labels background '0', core '1', edema '2' and enhance '4' respectively. This is given in Table 1.

Table 1. Shows a sample input instance and it's respective target one-hot vector

Input instance				Target one-hot vector			
T1 (x, y, z)	T2 (x, y, z)	T1ce (x, y, z)	FLAIR (x, y, z)	Background(0)	Core(1)	Edema(2)	Enhance(4)
453	515	823	652	0	0	0	1

Softmax function has been used as the activation function in the output layer. The architecture has used a two fully connected hidden layer with 100 and 500 respectively neurons. Leaky ReLu has been used as the activation function with leakiness parameter $\alpha = 0.01$. This is shown in Table 2.

Table 2. Shows the artificial neural network (ANN) architecture of the proposed model.

	Type	Neurons	Activation fuction
Layer 1	Input	4	-
Layer 2	Hidden layer 1	100	Leaky ReLu
Layer 3	Hidden layer 2	500	Leaky ReLu
Layer 4	Output	4	Softmax

Adam has been used as the optimisation algorithm. Categorical cross-entropy has been used as the loss function since the problem is a multi-class classification problem. The small architecture and input size allowed the usage of a large batch size of 500000 instances per batch. There is a total of 22,244,911 instances in which 20% (4,448,983) of it has been used for validation and the remaining 80% (17795928) has been used for training. The model has been trained for 50 epochs as the accuracy stopped improving around 40 epoch. Keras [6] has been used to implement the mentioned architecture. The GPU provided by Google Colaboratory has been used to implement the architecture. Each epoch took 15 seconds and the entire training took 12.5 min.

2.5 Tumour Volume Segmentation

All the voxels from the 3D MRI modalities (T1, T2, T1ce, and FLAIR) has been flattened into a 1D array to perform tumour volume segmentation. Each voxel (x,y,z) across modalities correspond to a single input instance with the arrangement mentioned in Table 1. The input instances have been given to the trained model and the respective predicted output is stored into a 1D array. The 1D array has been reshaped into an MRI image of dimension (240, 240, 155) and saved as patient_ID.nii.gz.

3 Proposed Survival Prediction Model

The survival prediction of patients consists of two stages namely, feature extraction from a segmented tumour, and linear regression for prediction of survival of the patient in days. The flow diagram for the patient survival prediction task is shown in Fig. 2. The training data consist of segmented tumour (ground truth), a patients age, resection status, and the number of days the patient survived. Resection of the tumour has been done on 101 out of 335 patients. Patients (101 instances) for whom tumour resection has been done are considered for survival prediction. The ground truth segmentation labels are used for feature extraction in the training phase. Thus, the accuracy and robustness of prediction depends on the quality of tumour segmentation. The features extracted from the segmentations are:

1. Tumour volume (for 3 tumour classes) - obtained by counting the number of voxels belonging to each class.
2. Tumour centroid (for 3 tumour classes) - obtained by averaging the x, y, and z values of the voxels of each class.
3. Extent of the tumour (for 3 tumour classes) - obtained from iteratively checking the minimum and maximum x, y and z values of voxels for each class.
4. Area of projection of tumour - obtained by flattening the 3D scan to 2D (by adding values of all slices as matrices) and counting the number of non-zero values in the resultant matrix.

Along with these features, the non-radiomic features namely, resection status and age have been used for survival prediction. The volume feature has been calculated by counting the number of intensity value belonging to each tumour sub-class as discussed in [7]. Traditional machine learning models (excluding

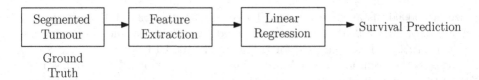

Fig. 2. The flow diagram for the proposed survical prediction system. Ground Truth segmentation labels were used for survival prediction task

neural network models) like linear regression performed better than the deep learning neural network-based models in BraTS 2018. Neural network models overfit to the training data when the number of instances and features for training are limited. They also tend to overfit if the training data does not capture all the principle features required for prediction [4]. Hence, a linear regression model has been used for the training phase.

4   Experimental Results

4.1   BraTS Dataset

The Brain Tumour Segmentation (BraTS) challenge aims to provide a publicly available dataset and a community benchmark. The BraTS 2019 challenge consist of MRI images acquired from 19 different institutions. It is collected through regular clinical evaluation of pre-operative scans for patients with glioblastoma multiforma (GBM/HGG), or low-grade glioma (LGG). Each voxel in the MRI image is normalised to 1 mm^3 dimensions and the images are skull stripped [1,3]. Each 3D MRI image has dimensions $240 \times 240 \times 155$. The patients' data is split into training and validation sets. The training set consists of 259 HGG and 76 LGG patients for a total of 335 patients. Each patient has a T1-weighted, T2-weighted, T2 Fluid Attenuated Inversion Recovery (T2-FLAIR), post-contrast T1 weighted image (T1Gd/T1ce), and the target ground truth segmentation. The ground truth consists of four labels namely, '0' for background, '1' for the core tumour area, '2' represents edema, and '4' enhancing tumour region. The MRI images are stored in the Neuroimaging Informatics Technology Initiative (NIfTI) format with '.nii.gz' extension. The validation set consist of 125 patients without ground truth segmented images [2,3].

4.2   Results of Segmentation

The segmentation for all the patients in the training (335 patients) and validation (125 patients) datasets has been obtained, as discussed in Sect. 2. The resultant MRI images have been evaluated on the online CBICA's Image Processing Portal (IPP). An example segmentation for a patient is shown in Fig. 3.

Segmentation of additional non-tumourous/healthy tissues along with tumourous tissues has been observed. This is because some regions of the healthy

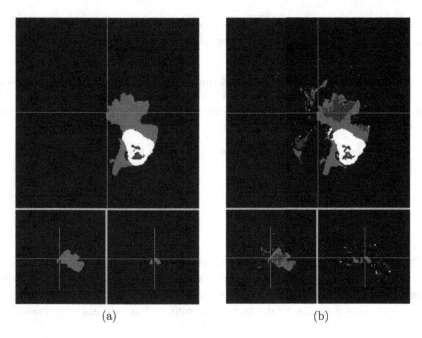

(a)   (b)

Fig. 3. A single axial, coronal and sagittal slice of segmented brain tumour. The light-grey area is the whole tumour region, dark-grey area is the core tumour area and, white area is the enhancing tumour area. a) A slice of the Ground truth image b) The same slice after segmentation using the proposed method

tissues exhibit tumorous hyperintensities. Also, some regions of the whole tumour (light grey area) have been misclassified as core tumour region (dark grey area). This may be due to bias field distortion resulting in inhomogeneity of voxels in the same region. Hence, bias field correction may be required to improve the signal-to-noise ratio as mentioned in [14].

The metrics used for evaluation are Dice score, sensitivity, specificity and 95 Hausdorff's distance. They have been used to evaluate the segmentation of edema, core, and enhancing tumour class intensity labels. The mean, standard deviation, and median for Dice score and Hausdorff distance for training and validation sets are given in Table 3, while Table 4 gives the sensitivity and specificity values.

From Table 3, it can be observed that the Dice and Hausdorff distance for enhancing tumour and whole tumour is better when voxels are suppressed. Tumour core segmentation results are better without suppression. On analysing the suppressed images before segmentation, it has been observed that tumour core segments are voxels from T2 images and the 98% tumour suppression captured more healthy tissues along with tumour core class. The above excess misclassifications can be removed with an efficient post-processing algorithm. An efficient pre-processing algorithm which can separate tumour sub-regions from each other in terms of intensity values can also reduce misclassification among tumour regions. It can also be noted that the model learned whole tumour regions from FLAIR and enhancing tumour regions from T1 image.

Table 3. Shows the mean Dice score, and Hausdorff-95 results of training and validation sets, where Dice represents dice score, Dist represents 95 Hausdroff distance, ET represents enhancing tumour, TC represents tumour core and, WT represents whole tumour

		Dice_ET	Dice_WT	Dice_TC	Dist_ET	Dist_WT	Dist_TC
Training	No suppression	0.33937	0.43678	0.32676	81.20324	70.06155	81.82787
	98% suppression	0.50729	0.42871	0.24424	57.53872	69.3061	82.73122
Validation	No suppression	0.32712	0.44418	0.30087	85.59015	74.91022	85.05087
	98% suppression	0.4736	0.43053	0.23625	59.6719	72.92901	84.63

From Table 4, high sensitivity shows that the model is fine-tuned to the presence of a tumour. Also, the high specificity value shows that the model can distinguish between various tumour regions accurately. Also, the sensitivity decreases with suppression meaning that some tumourous regions are also suppressed.

Table 4. Shows the mean sensitivity, and specificity results of the training and validation sets, where Sen represents sensitivity, Spe represents specificity, ET represents enhancing tumour, TC represents tumour core and, WT represents whole tumour

		Sen_ET	Sen_WT	Sen_TC	Spe_ET	Spe_WT	Spe_TC
Training	No suppression	0.80658	0.90346	0.70311	0.9688	0.84367	0.93808
	98% suppression	0.65864	0.71218	0.69739	0.99399	0.90164	0.90263
Validation	No suppression	0.7505	0.8948	0.62438	0.97241	0.86056	0.94401
	98% suppression	0.5834	0.72732	0.65418	0.99523	0.90716	0.90785

4.3   Results of Prediction

The survival prediction in days has been evaluated for patients who underwent tumour resection. Tumour resection has been done for 101 out of 335 patients. The survival prediction result for the training dataset is 0.465 accuracy, 87198.948 mean square error (MSE), 26732.25 median square error (median SE), 170386.095 std SE, and 0.428 Spearman coefficient.

5   Conclusions and Future Work

In this paper, a voxel classification based segmentation algorithm using a neural network has been proposed. Preprocessing steps have been done to prepare the data for the neural network. The smaller size of the network allowed the use of large batch size for training and validation. The large batch size significantly reduced the training time to 12.5 min for 50 epochs on Google Colaboratory. However, the segmented results predicted the tumour and its sub-regions along with additional outliers. This is mostly due to intensity inhomogeneity, brain sub-structures with tumour-like hyperintensities, and a lack of neighbourhood information.

A linear regression model with volumetric features [7] and additional features for patient survival prediction in terms of days is presented and a training accuracy of 0.465 has been achieved. It is found that prediction models like linear regression are better than neural network based models for this task because of a small sample size.

In future work, bias correction [14] to improve the signal-to-noise ratio in the 3D MRI images needs to be studied. Also, experiments with different ways to select frames, voxels, and the need for additional preprocessing steps can be analysed. The neural network architecture is experimental and further study is needed to decide upon different architectures, the number of neurons, and hidden layers for optimum performance. Finally, there is a need for post-processing to remove segmented outliers which could give a better Dice score.

References

1. Bakas, S., et al.: Segmentation labels and radiomic features for the pre-operative scans of the TCGA-GBM collection. The Cancer Imaging Archive (2017)
2. Bakas, S., Akbari, H., Sotiras, A., et al.: Advancing the cancer genome atlas glioma MRI collections with expert segmentation labels and radiomic features. Sci. Data **4**, 170117 (2017)
3. Bakas, S., Akbari, H., Sotiras, et al.: Segmentation labels and radiomic features for the pre-operative scans of the TCGA-LGG collection. Cancer Imaging Arch. **286** (2017)
4. Bakas, S., Reyes, M., Jakab, A., et al.: Identifying the best machine learning algorithms for brain tumor segmentation, progression assessment, and overall survival prediction in the BRATS challenge (2018). http://arxiv.org/abs/1811.02629
5. Bauer, S., Wiest, R., et al.: A survey of MRI-based medical image analysis for brain tumor studies. Phys. Med. Biol. **58**(13), R97 (2013)
6. Chollet, F., et al.: Keras (2015). https://keras.io
7. Feng, X., Tustison, N., Meyer, C.: Brain tumor segmentation using an ensemble of 3D U-Nets and overall survival prediction using radiomic features. In: Crimi, A., Bakas, S., Kuijf, H., Keyvan, F., Reyes, M., van Walsum, T. (eds.) BrainLes 2018. LNCS, vol. 11384, pp. 279–288. Springer, Cham (2019). https://doi.org/10.1007/978-3-030-11726-9_25
8. Isensee, F., Kickingereder, P., Wick, W., Bendszus, M., Maier-Hein, K.H.: No new-net. In: Crimi, A., Bakas, S., Kuijf, H., Keyvan, F., Reyes, M., van Walsum, T. (eds.) BrainLes 2018. LNCS, vol. 11384, pp. 234–244. Springer, Cham (2019). https://doi.org/10.1007/978-3-030-11726-9_21
9. Kamnitsas, K., et al.: Ensembles of multiple models and architectures for robust brain tumour segmentation. In: Crimi, A., Bakas, S., Kuijf, H., Menze, B., Reyes, M. (eds.) BrainLes 2017. LNCS, vol. 10670, pp. 450–462. Springer, Cham (2018). https://doi.org/10.1007/978-3-319-75238-9_38

10. Kamnitsas, K., Ledig, C., Newcombe, V.F., et al.: Efficient multi-scale 3D CNN with fully connected CRF for accurate brain lesion segmentation. Med. Image Anal. **36**, 61–78 (2017)

11. Krex, D., Klink, B., Hartmann, C., Von Deimling, A., et al.: Long-term survival with glioblastoma multiforme. Brain **130**(10), 2596–2606 (2007)

12. Long, J., Shelhamer, E., Darrell, T., et al.: Fully convolutional networks for semantic segmentation. In: Proceedings of the IEEE Conference on Computer Vision and Pattern Recognition, pp. 3431–3440 (2015)

13. Menze, B.H., Jakab, A., Bauer, S., et al.: The multimodal brain tumor image segmentation benchmark (BRATS). IEEE Trans. Med. Imaging **34**(10), 1993–2014 (2015)

14. Myronenko, A.: 3D MRI brain tumor segmentation using autoencoder regularization. In: Crimi, A., Bakas, S., Kuijf, H., Keyvan, F., Reyes, M., van Walsum, T. (eds.) BrainLes 2018. LNCS, vol. 11384, pp. 311–320. Springer, Cham (2019). https://doi.org/10.1007/978-3-030-11726-9_28

15. Nyúl, L.G., Udupa, J.K., Zhang, X.: New variants of a method of MRI scale standardization. IEEE Trans. Med. Imaging **19**(2), 143–150 (2000)

16. Thaha, M.M., Kumar, K.P.M., Murugan, B.S., Dhanasekeran, S., Vijayakarthick, P., Selvi, A.S.: Brain tumor segmentation using convolutional neural networks in MRI images. J. Med. Syst. **43**(9), 1–10 (2019). https://doi.org/10.1007/s10916-019-1416-0

17. Ronneberger, O., Fischer, P., Brox, T.: U-Net: convolutional networks for biomedical image segmentation. In: Navab, N., Hornegger, J., Wells, W.M., Frangi, A.F. (eds.) MICCAI 2015. LNCS, vol. 9351, pp. 234–241. Springer, Cham (2015). https://doi.org/10.1007/978-3-319-24574-4_28

18. Van Meir, E.G., Hadjipanayis, C.G., Norden, A.D., et al.: Exciting new advances in neuro-oncology: the avenue to a cure for malignant glioma. CA: Cancer J. Clin. **60**(3), 166–193 (2010)

ONCOhabitats Glioma Segmentation Model

Javier Juan-Albarracín[1], Elies Fuster-Garcia[1,2](✉) (iD),
María del Mar Álvarez-Torres[1], Eduard Chelebian[1],
and Juan M. García-Gómez[1]

[1] Instituto Universitario de Tecnologías de la Información y Comunicaciones,
Universitat Politècnica de València, València, Spain
{jajuaall, juanmig}@itaca.upv.es,
{elfusgar, maaltor4}@upv.es, edchekoc@etsii.upv.es
[2] Department of Diagnostic Physics, Oslo University Hospital, Oslo, Norway

Abstract. ONCOhabitats is an open online service that provides a fully automatic analysis of tumor vascular heterogeneity in gliomas based on multiparametric MRI. Having a model capable of accurately segment pathological tissues is critical to generate a robust analysis of vascular heterogeneity. In this study we present the segmentation model embedded in ONCOhabitats and its performance obtained on the BRATS 2019 dataset. The model implements an residual-Inception U-Net convolutional neural network, incorporating several pre- and post- processing stages. A relabeling strategy has been applied to improve the segmentation of the necrosis of high-grade gliomas and the non-enhancing tumor of low-grade gliomas. The model was trained using 335 cases from the BraTS 2019 challenge training dataset and evaluated with 125 cases from the validation set and 166 cases from the test set. The results on the validation dataset in terms of the mean/median Dice coefficient are 0.73/0.85 in the enhancing tumor region, 0.90/0.92 in the whole tumor, and 0.78/0.89 in the tumor core. The Dice results obtained in the independent test are 0.78/0.84, 0.88/0.92 and 0.83/0.92 respectively for the same sub-compartments of the lesion.

Keywords: Glioma · Convolutional neural network · Segmentation

1 Introduction

Gliomas are one of the most common central nervous system (CNS) tumors. Gliomas comprise a very diverse group of CNS tumors that vary histologically from low grade (LGGs; grade II) to high grade (HGGs; Grades III, IV) [1]. Knowing the extent and the heterogeneity of the lesion is crucial to make a correct diagnosis, plan radiotherapy treatment, analyze the response to treatment, and monitor the progression of the disease.

Manual segmentation and volumetric studies of the different glioma tissues involves an arduous, time-consuming and often unaffordable task for humans, that is not often performed in clinical practice but only in some clinical studies.

In recent years, and with the emergence of new deep learning technologies, a substantial effort has been made to generate models capable of automatically delineate glioma pathologic tissues with high accurate confidence. An example of the effort

© Springer Nature Switzerland AG 2020
A. Crimi and S. Bakas (Eds.): BrainLes 2019, LNCS 11992, pp. 295–303, 2020.
https://doi.org/10.1007/978-3-030-46640-4_28

invested in this task is the creation of the multimodal Brain Tumour Segmentation (BRATS) challenge. Since 2012 and until now, numerous researchers have focused their efforts on generating more accurate brain tumor segmentation models, reaching computational models with a performance close to human expert labelling [2]. Nevertheless, there is still a need for more research to achieve completely reliable segmentation models that can handle the wide range of heterogeneous tumors that can arise in real clinical routine.

This work presents a segmentation model of gliomas that consists on a patch-based 3D U-net Convolutional Neural Network based on residual-Inception blocks. The preprocessing includes noise reduction, bias correction, and intensity normalization. A relabeling strategy was applied to differentiate HGG necrosis and LGG non-enhancing tumor in the training stage. Finally, a postprocessing stage was implemented to remove spurious or incoherent segmentation objects.

The proposed tumor segmentation model is included in the last version of ONCOhabitats [3] online platform (https://www.oncohabitats.upv.es), provided by the Polytechnic University of Valencia [4]. ONCOhabitats provides a fully automatic analysis of tumor vascular heterogeneity [5], based on four vascular habitats within the lesion from MRI images: the High Angiogenic Tumor (HAT), the Low Angiogenic Tumor (LAT), the Infiltrated Peripheral Edema (IPE) and the Vasogenic Peripheral Edema (VPE) [6, 7]. ONCOhabitats includes two main services: (1) glioma tissue segmentation based on CNN; and (2) vascular heterogeneity assessment. In addition, we provide to researchers and clinicians our computational resources, including a system able to process about 300 cases per day including image preprocessing and standardization, regions of interest (ROIs) segmentation, perfusion quantification and vascular heterogeneity assessment of the lesion.

2   Materials

To train the proposed model, only the images provided in the 2019 edition of the BRATS challenge were used [2, 8–11]. The training dataset includes 335 studies, each one composed by pre- and post-contrast T1-weighted MRI, as well as T2-weighted, T2-fluid attenuated inversion recovery (FLAIR) MRI. Additionally, the ground truth maps are provided, distinguishing between 3 labels: label 1, which encloses necrosis, non-enhancing tumor, cyst and hemorrhage tissues; label 2, which delineates the edema; and label 4 that represents the enhancing tumor. The validation dataset comprises 125 images while the test set is composed of 166 images, both including the same MRI sequences but without the ground truth maps. An online oracle is provided to evaluate the proposed models in a blind manner.

2.1   Preprocessing

BRATS2019 dataset preprocessing performed by the organizers includes: 1) voxel isotropic resampling to 1 mm^3, 2) intra-patient registration to the T1ce sequence and inter-patient registration to a common reference space, and 3) skull-stripping for cranium removal.

We have extended this preprocessing by including a denoising stage using the Adaptive Non-Local Means filter proposed in [12]. We employed search windows of $7 \times 7 \times 7$ and a patch window of $3 \times 3 \times 3$, with Rician noise model. Additionally, a bias field correction stage was performed using N4ITK software at different scale levels [13], with 150 B-splines. Finally, z-score normalization was performed for each image, only normalizing the voxels within the brain (i.e. excluding the background from the normalization).

3 Methods

We propose a patch-based 3D U-net Convolutional Neural Network based on residual-Inception blocks. The network takes as input 3D patches of $64 \times 64 \times 64$ of three channels being the T1 contrast enhanced, the T2 and the Flair sequences. T1 sequence was discarded due to a worsening of the results when including it in the learning process. Therefore, the network works with patches of $64 \times 64 \times 64 \times 3$. The architecture details are described below.

3.1 Architecture

A U-net with 4 levels of depth is designed. The encoding path includes 3 *downsampling blocks* consisting of Conv $3 \times 3 \times 3$ (stride $2 \times 2 \times 2$) + ReLU + Batch Normalization. Likewise, the decoding path incorporates 3 analogous *upsampling blocks* consisting of: TransposeConv $3 \times 3 \times 3$ (stride $2 \times 2 \times 2$) + ReLU + Batch Normalization. Hence, the downsampling and upsampling operations are learnt by the network instead of using Max Pooling or repeatable Upsampling operations.

The network is composed of 4 levels with 24, 48, 96, 192 filters at each level respectively. Each level contains a Residual-Inception module to capture features at different scales. The residual-Inception block has 4 parallel paths with the following structure:

- Conv $1 \times 1 \times 1$-NF + ReLU + Batch Normalization
- Conv $3 \times 3 \times 3$-NF + ReLU + Batch Normalization
- Conv $3 \times 3 \times 3$-NF + ReLU + Batch Normalization + Conv $3 \times 3 \times 3$-NF + ReLU + Batch Normalization
- Max Pooling $3 \times 3 \times 3$ (stride $1 \times 1 \times 1$) + Conv $1 \times 1 \times 1$-NF + ReLU + Batch Normalization,

where NF refers to the *Number of Filters* depending on the level of the U-net in which the Residual-Inception block is. The output of these 4 paths is then feed to a concatenation layer and the output is passed to a block of the form: *Conv $1 \times 1 \times 1$-NF + ReLU + Batch Normalization*, to compress the information extracted by the 4 paths. Finally, a residual connection is introduced by summing the input of the Residual-Inception block to the output. Figure 1 shows a diagram of the Residual-Inception block. Note that each *Simple block* (except the Max Pooling) includes a

Convolution + ReLU + Batch Normalization layers. Additionally, long-skip connection between symmetric levels are introduced to allow a better gradient flow during training process. Figure 2 shows a diagram of the network architecture used in the study.

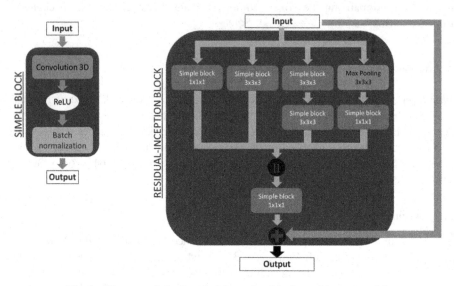

Fig. 1. Diagram of the Residual-Inception block used in our model.

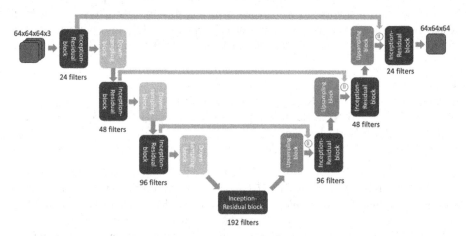

Fig. 2. Diagram of the 4-level residual network architecture. Residual-Inception blocks are used as feature extraction modules. Downsampling and upsampling operations are performed through strided conventional and transposed convolutions. Long concatenation-skip-connections are employed between symmetric levels.

3.2   Training Strategy

Label 1 in the BRATS 2019 dataset encloses a set of different glioma tissues, including necrosis, non-enhancing tumor, cyst, hemorrhage, etc. Such tissues largely differ in appearance in the MRI images, so in order to simplify the learning task for the network we decided to re-label the label 1 in all the LGG cases by label 3. Such re-labelling pursues the idea of associating the label 1 mostly to the necrosis tissue, typically present in HGG; and the label 3 to the non-enhancing tumor tissue, typically predominant in LGG.

We followed a balanced training strategy by creating batches containing a uniform proportion of patches containing predominantly edema (label 2), necrosis (label 1), enhancing tumor (label 4), non-enhancing tumor (label 3) and healthy tissues (label 0). Due to memory restrictions, batches of 4 samples was employed to train the network.

We also employed a combined loss consisting on the unweighted sum of cross-entropy and dice losses. Additionally, we trained the network with label smoothing with a factor 0.1, to relax the confidence in the labels. Adam optimizer was used with a starting learning rate of 1e−3. We trained the network 35k iterations.

3.3   Postprocessing

In order to remove spurious or incoherent segmentation components, we developed a simple postprocessing stage based on Connected Components (CC) analysis. As a *rule of thumb,* we always save the biggest CC as it is the most probable that contains the correct segmentation. The remaining CCs are analyzed and saved only if they met the following criteria:

1. The CC contains a number of voxels of class 4, class 3 or class 1 greater than the 5% of the size of the CC.
2. The CC has more than 1000 voxels.

Such simple post-processing mostly intends to discard erroneous CCs produced by magnetic bias field inhomogeneities in the images. Typically, these CCs are mainly labeled as class 2 (edema-like pattern), due to hyperintensities in Flair or T2 images. Thus, by the opposite, if the CC contains voxels segmented as enhancing tumor, non-enhancing tumor or necrosis, it can serve as an indicator of the confidence in the segmentation of the CC. Anyway, if the CC is big enough (more than 1000 voxels) we also assume that it is not an inhomogeneity artifact and the CC is saved for the final segmentation.

4   Results

The results obtained by ONCOhabitats glioma segmentation model on the independent validation dataset provided by BraTS 2019 challenge are summarized in Table 1. Dice, Sensitivity, Specificity and Hausdorff95 Distance metrics are reported.

Table 1. Summary of the results obtained by ONCOhabitats glioma segmentation model on the independent validation dataset for Enhancing Tumor (ET), Whole Tumor (WT) and Tumor Core (TC) regions.

	Dice			Sensitivity			Specificity			Hausdorff95		
	ET	WT	TC	ET	WT	TC	ET	WT	TC	ET	WT	TC
Mean	0.73	0.90	0.78	0.78	0.88	0.75	1.00	1.00	1.00	4.25	5.06	7.80
Std. Dev.	0.29	0.08	0.25	0.27	0.11	0.27	0.00	0.01	0.00	7.50	6.59	11.7
Median	0.85	0.92	0.89	0.87	0.91	0.87	1.00	1.00	1.00	2.24	3.16	3.39
25QT	0.73	0.89	0.73	0.77	0.85	0.64	1.00	0.99	1.00	1.41	2.24	2.00
75QT	0.90	0.94	0.94	0.95	0.95	0.94	1.00	1.00	1.00	3.32	5.10	9.26

Additionally, box plot of the distribution of the Dice, sensitivity and specificity metrics for the cases on the independent validation dataset evaluated on the Enhancing Tumor (ET), Whole Tumor (WT) and Tumor Core (TC) regions are presented in Fig. 3.

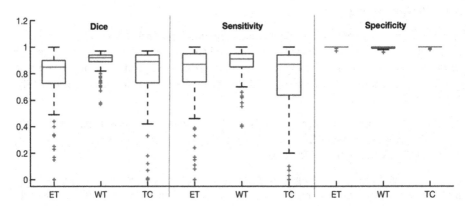

Fig. 3. Box plot showing the distribution of the Dice, sensitivity and specificity metrics for the cases on the independent validation dataset evaluated on the Enhancing Tumor (ET), Whole Tumor (WT) and Tumor Core (TC) regions.

Similarly, the results obtained by the ONCOhabitats model on the test dataset are presented in Table 2.

Table 2. Summary of the results obtained by ONCOhabitats glioma segmentation model on the test dataset for Enhancing Tumor (ET), Whole Tumor (WT) and Tumor Core (TC) regions.

	Dice		
	ET	WT	TC
Mean	0.78	0.88	0.83
Std. Dev.	0.22	0.11	0.25
Median	0.84	0.92	0.92
25QT	0.77	0.87	0.86
75QT	0.91	0.95	0.95

Figure 4 shows a comparison between the results of our model in the validation and the test set. An overall stable performance of the mean Dice is demonstrated, indicating that the model is robust against unseen samples and suggest no overfitting. Moreover, the ET and TC regions showed an improved performance in the test dataset with respect to the validation dataset. Finally, comparing our Dice results in the Whole Tumor sub-compartment with the Validation Leaderboard ranking, there is a small difference of 0.01687 Dice points with respect to the 1st place team, but using a small and therefore fast network.

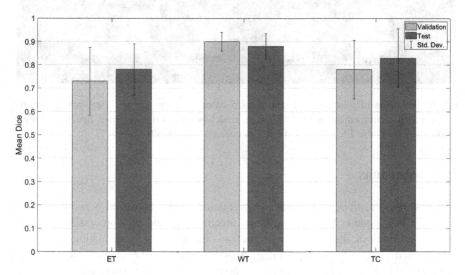

Fig. 4. Bar plot showing the mean Dice for the cases on the independent Validation and Test dataset evaluated on the Enhancing Tumor (ET), Whole Tumor (WT) and Tumor Core (TC) regions. The error bars represent the standard deviation.

Finally, Fig. 5 shows the segmentation results of several cases of the test dataset.

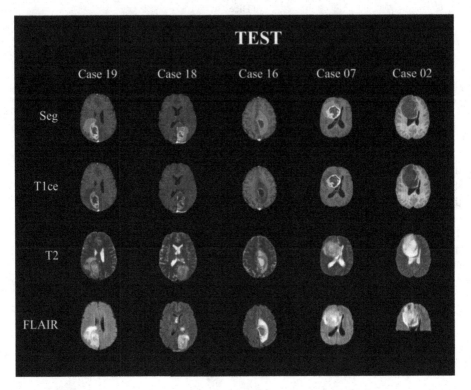

Fig. 5. Examples of glioma segmentation of 5 cases from the test set. First row shows the segmentation performed by the ONCOhabitats model over the T1ce sequence. Second, third and fourth rows show the T1ce, T2 and FLAIR sequences respectively.

5   Conclusions

In this work, we propose a glioma segmentation model based on a residual U-Net residual CNN together with an additional imaging pre- and post-processing stages to remove spurious or incoherent segmentation objects. This segmentation model has been trained using a relabeling strategy aimed to improve the segmentation of HGG necrosis and LGG non-enhancing tumor. The proposed model is included in the current version of ONCOhabitats open online service (https://www.oncohabitats.upv.es).

The results obtained show and improvement on the performance of the previous segmentation model included on ONCOhabitats reported in [3]. This allows to significantly improve the other services provided by ONCOhabitats, such as the vascular heterogeneity assessment service, since they use as basis the glioblastoma segmentation module.

Acknowledgements. This work was partially supported by: MTS4up project (National Plan for Scientific and Technical Research and Innovation 2013–2016, No. DPI2016-80054-R) (JMGG); H2020-SC1-2016-CNECT Project (No. 727560) (JMGG) and H2020-SC1-BHC-2018-2020 (No. 825750) (JMGG) and CaixaImpulse program from Fundació Bancaria "La Caixa" (LCF/TR/CI16/10010016). MMA-T was supported by DPI2016-80054-R (Programa Estatal de Promoción del Talento y su Empleabilidad en I+D+i). We gratefully acknowledge the support of NVIDIA Corporation with the donation of the Titan V GPU used for this research. EF-G was supported by the European Union's Horizon 2020 research and innovation programme under the Marie Skłodowska-Curie grant agreement No. 844646.

References

1. Louis, D.N., et al.: The 2016 world health organization classification of tumors of the central nervous system: a summary. Acta Neuropathol. **131**(6), 803–820 (2016)
2. Menze, B.H., et al.: The multimodal brain tumor image segmentation benchmark (BRATS). IEEE Trans. Med. Imaging **34**(10), 1993–2024 (2015)
3. Juan-Albarracín, J., Fuster-Garcia, E., García-Ferrando, G.A., García-Gómez, J.M.: ONCOhabitats: a system for glioblastoma heterogeneity assessment through MRI. Int. J. Med. Inform. **128**, 53–61 (2019)
4. ONCOHabitats - Glioblastoma segmentation - MRI: ONCOHabitats. https://www.oncohabitats.upv.es/. Accessed 09 Aug 2019
5. Juan-Albarracín, J., et al.: Glioblastoma: vascular habitats detected at preoperative dynamic susceptibility-weighted contrast-enhanced perfusion MR imaging predict survival. Radiology **287**(3), 944–954 (2018)
6. Fuster-Garcia, E., Juan-Albarracín, J., García-Ferrando, G.A., Martí-Bonmatí, L., Aparici-Robles, F., García-Gómez, J.M.: Improving the estimation of prognosis for glioblastoma patients by MR based hemodynamic tissue signatures. NMR Biomed. **31**(12), e4006 (2018)
7. Álvarez-Torres, M., Juan-Albarracín, J., Fuster-Garcia, E., et al.: Robust association between vascular habitats and patient prognosis in glioblastoma: an international multicenter study. J. Magn. Reson. Imaging (2019). https://doi.org/10.1002/jmri.26958
8. Bakas, S., et al.: Advancing the cancer genome atlas glioma MRI collections with expert segmentation labels and radiomic features. Sci. Data **4**, 170117 (2017)
9. Bakas, S., et al.: Identifying the best machine learning algorithms for brain tumor segmentation, progression assessment, and overall survival prediction in the BRATS challenge. arXiv:1811.02629. [cs, stat] (November 2018)
10. Bakas, S., et al.: Segmentation labels for the pre-operative scans of the TCGA-GBM collection. The Cancer Imaging Archive (2017)
11. Bakas, S., et al.: Segmentation labels for the pre-operative scans of the TCGA-LGG collection. The Cancer Imaging Archive (2017)
12. Coupé, D.L., Manjón, P., Robles, J.V., Collins, M.: Adaptive multiresolution non-local means filter for three-dimensional magnetic resonance image denoising. IET Image Process. **6**(5), 558–568 (2012)
13. Tustison, N.J., et al.: N4ITK: improved N3 bias correction. IEEE Trans. Med. Imaging **29**(6), 1310–1320 (2010)

Brain Tumor Segmentation with Uncertainty Estimation and Overall Survival Prediction

Xue Feng[1]([⊠]) [iD], Quan Dou[1], Nicholas Tustison[2], and Craig Meyer[1,2]

[1] Biomedical Engineering, University of Virginia,
Charlottesville, VA 22903, USA
xf4j@virginia.edu
[2] Radiology and Medical Imaging, University of Virginia,
Charlottesville, VA 22903, USA

Abstract. Accurate segmentation of different sub-regions of gliomas including peritumoral edema, necrotic core, enhancing and non-enhancing tumor core from multimodal MRI scans has important clinical relevance in diagnosis, prognosis and treatment of brain tumors. However, due to the highly heterogeneous appearance and shape, segmentation of the sub-regions is very challenging. Recent development using deep learning models has proved its effectiveness in the past several brain segmentation challenges as well as other semantic and medical image segmentation problems. Most models in brain tumor segmentation use a 2D/3D patch to predict the class label for the center voxel and variant patch sizes and scales are used to improve the model performance. However, it has low computation efficiency and also has limited receptive field. U-Net is a widely used network structure for end-to-end segmentation and can be used on the entire image or extracted patches to provide classification labels over the entire input voxels so that it is more efficient and expect to yield better performance with larger input size. In this paper we developed a deep-learning-based segmentation method using an ensemble of 3D U-Nets with different hyper-parameters. Furthermore, we estimated the uncertainty of the segmentation from the probabilistic outputs of each network and studied the correlation between the uncertainty and the performances. Preliminary results showed effectiveness of the segmentation model. Finally, we developed a linear model for survival prediction using extracted imaging and non-imaging features, which, despite the simplicity, can effectively reduce overfitting and regression errors.

Keywords: Brain tumor segmentation · Ensemble · Uncertainty estimation · Deep learning · Survival prediction · Linear regression

1 Introduction

Gliomas are the most common primary brain malignancies, with different degrees of aggressiveness, variable prognosis and various heterogeneous histological sub-regions, i.e. peritumoral edema, necrotic core, enhancing and non-enhancing tumor core. This intrinsic heterogeneity of gliomas is also portrayed in their radiographic phenotypes, as their sub-regions are depicted by different intensity profiles disseminated across

© Springer Nature Switzerland AG 2020
A. Crimi and S. Bakas (Eds.): BrainLes 2019, LNCS 11992, pp. 304–314, 2020.
https://doi.org/10.1007/978-3-030-46640-4_29

multimodal MRI (mMRI) scans, reflecting differences in tumor biology. Quantitative analysis of imaging features such as volumetric measures after manual/semi-automatic segmentation of the tumor region has shown advantages in image-based tumor phenotyping over traditionally used clinical measures such as largest anterior-posterior, transverse, and inferior-superior tumor dimensions on a subjectively-chosen slice [1, 2]. Such phenotyping may enable assessment of reflected biological processes and assist in surgical and treatment planning. To compare and evaluate different automatic segmentation algorithms, the Multimodal Brain Tumor Segmentation Challenge (BraTS) 2019 was organized using multi-institutional pre-operative MRI scans for the segmentation of intrinsically heterogeneous brain tumor sub-regions [3–5]. More specifically, the dataset used in this challenge includes multiple-institutional clinically-acquired pre-operative multimodal MRI scans of glioblastoma (GBM/HGG) and low-grade glioma (LGG) containing a) native (T1) and b) post-contrast T1-weighted (T1Gd), c) T2-weighted (T2), and d) Fluid Attenuated Inversion Recovery (FLAIR) volumes [6, 7]. 335 training volumes with annotated GD-enhancing tumor, peritumoral edema and necrotic and non-enhancing tumor. In addition, the segmentation uncertainty, which represents how confident the model is on the automatically segmented labels, is valuable in providing feedback to end users and in a more accurate evaluation for the segmentation quality. Furthermore, to pinpoint the clinical relevance of this segmentation task, BraTS'19 also included the task to predict patient overall survival from images together with the patient age and resection status. To tackle these two tasks, this study is performed with two goals: 1) provide pixel-by-pixel label maps for the three sub-regions and background and estimate the uncertainty of the model; 2) estimate the survival days.

Convolutional neural network (CNN) based models have proven their effectiveness and superiority over traditional medical image segmentation algorithms and are quickly becoming the mainstream in BraTS challenges. Due to the highly heterogeneous appearance and shape of brain tumors, small patches are usually extracted to predict the class for the center voxel. To improve model performance, multi-scale patches with different receptive field sizes are often used in the model [8]. In contrast, U-Net is a widely used convolutional network structure that consists of a contracting path to capture context and a symmetric expanding path that enables precise localization with 3D extension [8, 9]. It can be used on the entire image or extracted patches to provide class labels for all input voxels when padding is used. Furthermore, instead of picking the best network structure, an ensemble of multiple models, trained on different dataset or different hyper-parameters, can generally improve the segmentation performance over a single model due to the averaging effect. In this study we propose to use an ensemble of 3D U-Nets with different hyper-parameters trained on non-uniformly extracted patches for brain tumor segmentation. During testing, a sliding window approach is used to predict class labels with adjustable overlap to improve accuracy. The probabilistic outputs of each network will be used to estimate model uncertainty. With the segmentation labels, we will develop a linear model for survival prediction using extracted imaging features and additional non-imaging features since the linear models can effectively reduce overfitting and thus regression errors.

2 Methods

For the brain tumor segmentation task, the steps in our proposed method include pre-processing of the images, patch extraction, training multiple models using a generic 3D U-Net structure with different hyper-parameters, deployment of each model for full volume prediction and final ensemble modeling. The uncertainty is estimated from the probabilistic outputs of the networks. For the survival task, the steps include feature extraction, model fitting, and deployment. Details are described as follows.

2.1 Image Pre-processing

As MR images do not have standard pixel intensity values, to reduce the effects from different contrasts and different subjects, each 3D image was normalized to 0 to 1 separately by subtracting the min values and divided by the pixel intensity range. After normalization, for each subject, images of all contrast were fused to form the last dimension so that the whole input image size becomes $155 \times 240 \times 240 \times 4$.

2.2 Non-uniform Patch Extraction

For simplicity, we will use foreground to denote all tumor pixels and background to denote the rest. There are several challenges in directly using the whole images as the input to a 3D U-Net: 1) the memory of a moderate GPU is often 12 Gb so that in order to fit the model into the GPU, the network needs to greatly reduce the number of features and/or the layers, which often leads to a significant drop in performance as the expressiveness of the network is much reduced; 2) the training time will be greatly prolonged since more voxels contribute to calculation of the gradients at each step and the number of steps cannot be proportionally reduced during optimization; 3) as the background voxels dominate the whole image, the class imbalance will cause the model to focus on background if trained with uniform loss, or prone to false positives if trained with weighted loss that favors the foreground voxels. Therefore, to more effectively utilize the training data, smaller patches were extracted from each subject. As the foreground labels contain much more variability and are the main targets to segment, more patches from the foreground voxels should be extracted.

In implementation, during each epoch, a random patch was extracted from each subject using non-uniform probabilities. The valid patch centers were first calculated by removing edges to make sure each extracted patch was completely within the whole image. The probability of each valid patch center $p_{i,j,k}$ was calculated using the following equation:

$$p_{i,j,k} = \frac{s_{i,j,k}}{\sum_{i,j,k} s_{i,j,k}} \tag{1}$$

in which $s_{i,j,k} = 1$ for all voxels with maximal intensity lower than the 1st percentile, $s_{i,j,k} = 6$ for all foreground voxels and $s_{i,j,k} = 3$ for the rest. The patch center was then randomly selected based on the calculated probability and the corresponding patch was extracted. Since normal brain images are symmetric along the left-right direction, a random flip along this direction was made after patch extraction. No other augmentation was applied.

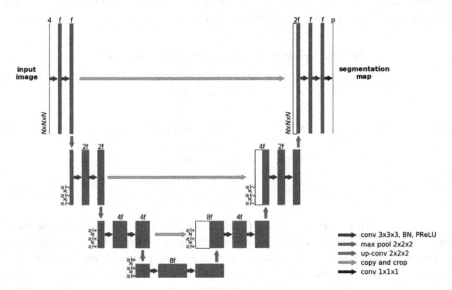

Fig. 1. 3D U-Net structure with 3 encoding and 3 decoding blocks.

2.3 Network Structure and Training

A 3D U-Net based network was used as the general structure, as shown in Fig. 1. Zero padding was used to make sure the spatial dimension of the output is the same with the input. For each encoding block, a VGG like network with two consecutive 3D convolutional layers with kernel size 3 followed by the activation function and batch norm layers were used. The parametric rectilinear function (PReLU), given as:

$$f(x) = \max(0, x) - \alpha\max(0, -x) \tag{2}$$

was used with trainable parameter α as the activation function. The number of features was doubled while the spatial dimension was halved with every encoding block, as in conventional U-Net structure. To improve the expressiveness of the network, a large number of features were used in the first encoding block. Dropout with ratio 0.5 was added after the last encoding block. Symmetric decoding blocks were used with skip-connections from corresponding encoding blocks. Features were concatenated to the de-convolution outputs. The extracted segmentation map of the input patch was expanded to the multi-class the ground truth labels (3 foreground classes and the background). Weighted/non-weighted cross entropy was used as the loss function.

The number of encoding/decoding blocks, the weights in the loss function and the patch size were chosen as the tunable hyper-parameters when constructing multiple models. Due to memory limitations, for a larger patch size, the number of features needs to be reduced. In current implementation, due to constraint in computational resources, six models were trained, with detailed parameters shown in Table 1. N denotes the input size, M denotes the number of encoding/decoding blocks and f denotes the input features at the first layer. For weighted loss, 1.0 was used for background and 2.0 was used for each class of foreground voxels.

Table 1. Detailed parameters for all 6 3D U-Net models.

Model #	M	N	f	Loss type
1	3	64	96	Uniform
2	3	64	96	Weighted
3	4	64	96	Uniform
4	4	96	96	Weighted
5	3	80	64	Uniform
6	3	80	64	Weighted

Training was performed on a Nvidia Titan Xp GPU with 12 Gb memory. 640 epochs were used. As mentioned earlier, during each epoch, only one patch was extracted every subject. Subject orders were randomly permuted every epoch. The Tensorflow framework was used with Adam optimizer. Batch size was set to 1 during training. During testing, as a smaller batch size was very sensitive to the running statistics, all batch norm layers did not use the running statistics but the statistics of the batch itself. This is usually called a layer normalization as it normalizes each feature map with its own mean and standard deviation. A learning rate of 0.0005 was used without further adjustments during training. The total training time was about 60 h.

2.4   Volume Prediction Using Each Model

Due to the fact that the entire image cannot fit into the memory during deployment, a sliding window approach needs to be used to get the output for each subject. However, as significant padding was made to generate the output label map at the same size as the input, boundary voxels of a patch were expected to yield unstable predictions when sliding the window across the whole image without overlaps. To alleviate this problem, a stride size at a fraction of the window size was used and the output probability was averaged. In implementation, the deployment window size was chosen to be the same as the training window size, and the stride was chosen as ½ of the window size. For each window, the original image and left-right flipped image were both predicted, and the average probability after flipping back the output of the flipped input was used as the output. Therefore, each voxel, except for a few on the edge, will be predicted 16 times when sliding across all directions. Although smaller stride sizes can be used to further improve the accuracy with more averages, the deployment time will be increased 8 times for every ½ reduction of the window size and thus will quickly become unmanageable. Using the parameters as mentioned on the same GPU, it took about 1 min to generate the output for the entire volume per subject. Instead of performing a thresholding on the probability output to get the final labels, the direct probability output was saved for each model to the disk.

2.5 Ensemble Modeling

The ensemble modeling process was rather straightforward. The probability output of all classes from each model was read from the disk and the final probability was calculated via simple averaging. The class with the highest probability was selected as the final segmentation label of each voxel.

2.6 Uncertainty Estimation

The research of Gal et al. [11] provides a Bayesian framework to estimate model uncertainty utilizing Monte Carlo dropout during testing. For the same testing sample, multiple outputs are generated, from which the mean probability or the variation of different probability maps can be calculated to obtain the uncertainty estimation. In this study, however, the ensemble of 3D U-Nets has the intrinsic advantage of producing several outputs for one testing sample. Two volume-based uncertainty calculation methods based on the given probability maps were implemented in this study. The first one uses the mean probability of all foreground voxels, which is given by the summation of all foreground voxels' mean probabilities divided by the total number of foreground voxels. The second one uses the mean probability variation of all foreground voxels. For each foreground voxel, the probability variation is defined as the standard deviation of the 6 generated voxel-wise probability maps. The uncertainty is then calculated by the summation of all foreground voxels' probability variations divided by the total number of foreground voxels. In order to compare the performances of two uncertainty measures, the Pearson's correlation coefficient between the uncertainty measure and the actual dice score is calculated. The one with higher correlation will be used to estimate voxel-wise uncertainty. The uncertainty is then normalized to be between 0 and 100 based on its original range. For the first measure (mean probability), the following equation describes the normalization step:

$$uncertainty = 200 \times (0.5 - |p - 0.5|) \tag{3}$$

in which p denotes the voxel's mean probability for one foreground class calculated from 6 probability maps. If the prediction is certain, the mean probability should be close to 0 or 1, and the uncertainty value should be close to 0.

2.7 Survival Prediction

To predict the post-surgery survival time measured in days, extracted images features and non-image features were used to construct a linear regression model. 6 image features were calculated from the ground truth label maps during training and the predicted label maps during validation. For each foreground class, the volume (V) by summing up the voxels and the surface area (S) by summing up the magnitude of the gradients along three directions were obtained, as described in the following equations

$$V_{ROI} = \sum_{i,j,k} s_{i,j,k} \qquad (4)$$

$$S_{ROI} = \sum_{i,j,k} s_{i,j,k} \sqrt{(\frac{\partial s}{\partial i})^2 + (\frac{\partial s}{\partial j})^2 + (\frac{\partial s}{\partial k})^2} \qquad (5)$$

in which ROI denotes a specific foreground class and $s_{i,j,k} = 1$ for voxels that are classified to belong to this ROI and $s_{i,j,k} = 0$ otherwise.

Age and resection status were used as non-imaging clinical features. As there were two classes of resection status and many missing values of this status, a two-dimensional feature vector was used to represent the status, given as GTR: (1, 0), STR: (0, 1) and NA: (0, 0). A linear regression model after normalizing the input features to zero mean and unit standard deviation was fit with the training data. As the input feature size is 9, the risk for overfitting is greatly reduced.

3   Results

3.1   Brain Tumor Segmentation

All 335 training subjects were used in the training process. 94 subjects were provided as validation and other 166 subjects were provided as testing. The dice indexes, sensitivities and specificities, 95 Hausdorff distances of the enhanced tumor (ET), whole tumor (WT) and tumor core (TC) were automatically calculated after submitting to the CBICA's Image Processing Portal. Table 2 shows the mean dice scores and 95 Hausdorff distances of ET, WT and TC for the training, validation and testing datasets. Sensitivity and specificity are highly correlated with the dice indexes so that they are not included.

Table 2. Performances of the ensemble on the training, validation and testing datasets.

Dataset	Dice_ET	Dice_WT	Dice_TC	Dist_ET	Dist_WT	Dist_TC
Training	0.7917	0.9094	0.8362	4.0186	3.8009	5.6451
Validation	0.7403	0.9061	0.8025	4.5864	4.2516	6.7645
Testing	0.7758	0.8810	0.8280	25.2965	9.0868	26.9982

3.2   Uncertainty Measure

Figure 2 shows the correlation coefficients between each volume-based uncertainty measure and the actual dice score for three foreground classes based on all validation subjects. The mean probability of all foreground voxels has higher correlation with dice score compared with the mean probability variation for all ROIs, indicating that the mean probability can serve as a good measurement of volume-based uncertainty, which conforms to the results in [12].

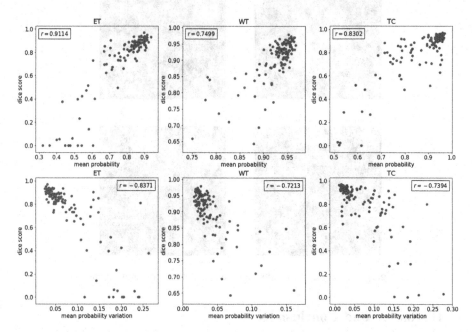

Fig. 2. Correlation between each uncertainty measure and the actual dice score.

Figure 3 shows one slice with segmentation results and the calculated uncertainty maps. At the boundary of each ROI, the corresponding uncertainty map shows higher values, indicating that the model is uncertain about the voxel label. After submitting the normalized uncertainty measures, the mean dice indexes of the validation dataset for WT, TC and ET were 0.897 ± 0.054, 0.790 ± 0.251, and 0.749 ± 0.272. Comparing with the original dice indexes, the performances did not change significantly, if not becoming slightly worse.

3.3 Survival Prediction

All 259 training subjects with survival data were used in the training process. 29 cases were evaluated after submitting to the CBICA's Image Processing Portal. The accuracy was 0.31, MSE was 107639.326, median SE was 77906.27, std SE was 109586.733 and Spearman Coefficient was 0.204. The performance on the validation dataset is not as accurate as other top teams in this task, however, our method achieved an accuracy of 0.55 in the testing dataset and was ranked 3rd (tie) overall. The results suggest that a linear model is robust against overfitting.

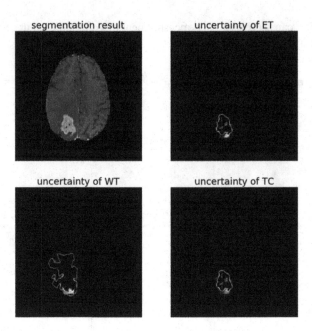

Fig. 3. Segmentation results and the calculated uncertainty maps.

4   Discussion and Conclusions

In this paper we developed a brain tumor segmentation method using an ensemble of 3D U-Nets. Intensity normalization was used as pre-processing. 6 networks were trained with different number of encoding/decoding blocks, input patch sizes and different weights for loss. The preliminary results showed an improvement with ensemble modeling. For survival prediction, we used a simple linear regression by combining radiomics features from images such as volumes and surface areas of each sub-region and non-imaging clinical features.

For segmentation, it is noted that the median metrics are significantly higher than the mean metrics. For example, the median dice indexes were 0.867, 0.923 and 0.904 for ET, WT and TC in the final ensembled model. It makes sense in that the theoretical maximum dice index is 1 and minimum dice index is 0. However, we noted that in several cases, the dice indexes are as low as 0 for ET and TC and 0.6 for WT. It is mostly due to the low sensitivity meaning that the model is not able to recognize the corresponding tumor regions. The possible reason for these failed regions is that their characteristics deviate a lot from the training dataset. This is also encouraging in that for majority of the cases, the segmentation quality is very high.

In the 3D U-Net model, we found that the batch norm layer was helpful in improving the model stability and performance. However, different with the canonical application of the batch norm layer, in which the batch statistics is used in training and the global statistics is used in deployment, it performed much better with batch statistics in deployment than global statistics. Since the batch size is 1, a per-channel normalization is actually performed by subtracting its own mean. One possible explanation could be that by doing such normalization, the model focuses on the differences of neighboring pixels in one channel and ignores the absolute values, which may help the segmentation process. However, further investigation is needed to figure out the exact reason.

Compared with the patch-based model that only predicts the center pixel, when predicting the segmentation label maps for the full patch, different pixels are very likely to have different effective receptive field sizes due to the zero padding in the edge. We argue that a pixel should still be able to be predicted even based on partial receptive field, which, for the very edge pixel, corresponds to only half of the maximal receptive field. Furthermore, the significant overlap in the sliding windows during deployment can improve the accuracy with more averages.

In the current implementation, 6 networks were trained due to limitations in computation time. It is expected with more networks, the results can be further improved, although the marginal improvement is expected to decrease.

To measure the model uncertainty, two volume-based uncertainty measurements based on the probability maps generated by 6 networks were proposed and compared. A simple calculation method based on the mean probability of all foreground voxels was adopted to evaluate voxel-wise uncertainty. This method takes advantage of the ensemble modeling, and does not require any modification to the network structure or training procedure. Comparison with other uncertain measure could be performed in future work.

For the survival prediction task, since it is very likely to overfit with such a small dataset and we argue that as many other features may play more important roles in overall survival such as histological and genetic features but unfortunately, they are not available in this challenge, a linear regression model was the safest option to minimize the test errors, although at the cost of its expressiveness. Further exploration of those additional features through clinical collaboration is expected to improve the accuracy of survival prediction.

In conclusion, we developed an ensemble of 3D U-Nets for brain tumor segmentation. The network hyper-parameters are varied to obtain multiple trained models. A linear regression model was also developed for the survival prediction task. The code is available at https://github.com/xf4j/brats18.

References

1. Kumar, V., Gu, Y., Basu, S., Berglund, A., Eschrich, S.A., Schabath, M.B., et al.: Radiomics: the process and the challenges. Magn. Reson. Imaging **30**(9), 1234–1248 (2012). https://doi.org/10.1016/j.mri.2012.06.010
2. Gillies, R.J., Kinahan, P.E., Hricak, H.: Radiomics: images are more than pictures, they are data. Radiology **278**(2), 563–577 (2016). https://doi.org/10.1148/radiol.2015151169
3. Menze, B.H., Jakab, A., Bauer, S., Kalpathy-Cramer, J., Farahani, K., Kirby, J., et al.: The multimodal brain tumor image segmentation benchmark (BRATS). IEEE Trans. Med. Imaging **34**(10), 1993–2024 (2015). https://doi.org/10.1109/TMI.2014.2377694

4. Bakas, S., Akbari, H., Sotiras, A., Bilello, M., Rozycki, M., Kirby, J.S., et al.: Advancing the cancer genome atlas glioma MRI collections with expert segmentation labels and radiomic features. Nat. Sci. Data **4**, 170117 (2017). https://doi.org/10.1038/sdata.2017.117

5. Bakas, S., Reyes, M., Jakab, A., Bauer, S., Rempfler, M., Crimi, A., et al.: Identifying the Best Machine Learning Algorithms for Brain Tumor Segmentation, Progression Assessment, and Overall Survival Prediction in the BRATS Challenge. arXiv preprint arXiv:1811.02629 (2018)

6. Bakas, S., Akbari, H., Sotiras, A., Bilello, M., Rozycki, M., Kirby, J., et al.: Segmentation labels and radiomic features for the pre-operative scans of the TCGA-GBM collection. The Cancer Imaging Archive (2017). https://doi.org/10.7937/K9/TCIA.2017.KLXWJJ1Q

7. Bakas, S., Akbari, H., Sotiras, A., Bilello, M., Rozycki, M., Kirby, J., et al.: Segmentation labels and radiomic features for the pre-operative scans of the TCGA-LGG collection. The Cancer Imaging Archive (2017). https://doi.org/10.7937/K9/TCIA.2017.GJQ7R0EF

8. Kamnitsas, K., Ledig, C., Newcombe, V.F.J., Simpson, J.P., Kane, A.D., Menon, D.K., et al.: Efficient multi-scale 3D CNN with fully connected CRF for accurate brain lesion segmentation. Med. Image Anal. **36**, 61–78 (2017). https://doi.org/10.1016/j.media.2016.10.004

9. Ronneberger, O., Fischer, P., Brox, T.: U-Net: Convolutional Networks for Biomedical Image Segmentation. arXiv preprint arXiv:1505.04597 (2015)

10. Cicek, O., Abdulkadir, A., Lienkamp, S.S., Brox, T., Ronneberger, O.: 3D U-Net: Learning Dense Volumetric Segmentation from Sparse Annotation. arXiv preprint arXiv:1606.06650 (2016)

11. Gal, Y., Ghahramani, Z.: Dropout as a Bayesian Approximation: Representing Model Uncertainty in Deep Learning. arXiv preprint arXiv:1506.02142 (2015)

12. Pan, H., Feng, Y., Chen, Q., Meyer, C., Feng, X.: Prostate Segmentation from 3D MRI Using a Two-Stage Model and Variable-Input Based Uncertainty Measure. arXiv preprint arXiv:1903.02500 (2019)

Cascaded Global Context Convolutional Neural Network for Brain Tumor Segmentation

Dong Guo[1], Lu Wang[1], Tao Song[2], and Guotai Wang[1(✉)]

[1] School of Mechanical and Electrical Engineering,
University of Electronic Science and Technology of China, Chengdu, China
guotai.wang@uestc.edu.cn
[2] SenseTime Research, Shanghai, China

Abstract. A cascade of global context convolutional neural networks is proposed to segment multi-modality MR images with brain tumor into three subregions: enhancing tumor, whole tumor and tumor core. Each network is a modification of the 3D U-Net consisting of residual connection, group normalization and deep supervision. In addition, we apply Global Context (GC) block to capture long-range dependency and inter-channel dependency. We use a combination of logarithmic Dice loss and weighted cross entropy loss to focus on less accurate voxels and improve the accuracy. Experiments with BraTS 2019 validation set show the proposed method achieved average Dice scores of 0.77338, 0.90712, 0.83911 for enhancing tumor, whole tumor and tumor core, respectively. The corresponding values for BraTS 2019 testing set were 0.79303, 0.87962, 0.82887 for enhancing tumor, whole tumor and tumor core, respectively.

Keywords: Brain tumor · Segmentation · Convolutional neural network

1 Introduction

Gliomas are the most common primary brain tumors, which arises from glial cells [17]. Gliomas can be categorized into two subtypes: low-grade gliomas (LGG) and high-grade gliomas (HGG). Low-grade gliomas indicate a relatively promising prognosis. On the contrary, high-grade gliomas have a worse prognosis [19]. In the diagnosis of brain tumors, Magnetic Resonance Imaging (MRI) is a powerful and useful methods for brain tumor analysis. The MRI sequences usually consist of several modalities, such as T1-weighted, contrast enhanced T1-weighted (T1c), T2-weighted and Fluid Attenuation Inversion Recovery (FLAIR). Different modalities provide complementary information to differentiate glioma subregions. For example, FLAIR has a good contrast for the whole peritumoral edema and T1c highlights the tumor without peritumoral edema.

© Springer Nature Switzerland AG 2020
A. Crimi and S. Bakas (Eds.): BrainLes 2019, LNCS 11992, pp. 315–326, 2020.
https://doi.org/10.1007/978-3-030-46640-4_30

Automatic segmentation of brain tumors makes a contribution to better diagnosis and treatment planning. However, this segmentation is challenging because (1) Tumor structures vary considerably across patients in terms of size, shape, and location. The shapes of brain tumor and its subregions are quite irregular, especially enhancing tumor core. Many small lumps, scattering in this region, bring difficulties to accurate segmentation of some small enhancing tumor cores. (2) The boundaries between adjacent structures are often ambiguous.

In recent years, methods based on convolutional neural networks (CNNs) dominate the state-of-the-art performance on brain tumor segmentation [12,18, 23]. DeepMedic [14], a 3D CNN model with fully connected Conditional Random Field (CRF), predicts a segmentation incorporating both local and non-local contextual information. However, DeepMedic works on local image patches and therefore is less efficient compared with more recent works [12,18,23]. With the success of encoder-decoder network architectures applied to semantic segmentation, especially FCN [16] and U-Net [21], all top performing methods [12,13,18,23] in BraTS challenge are based on encoder-decoder networks since 2017. Kamnitsas et al. [13] brought together a variety of CNN architectures and explored Ensembles of Multiple Models and Architectures (EMMA) to make their segmentation reliable. Wang et al. [23] used a cascade of FCN-based networks consisting of multiple layers of anisotropic and dilated convolution filters, which is complex mainly because of cascade but helps a lot to improve segmentation accuracy. Myronenko [18] used a encoder-decoder structure of CNN with a large patch size of $160 \times 192 \times 128$ and a variational auto-encoder branch to regularize the shared encoder. Isensee et al. [12] modified the U-Net [21] and used a combination of Dice loss and cross entropy loss, which demonstrated the effectiveness of a well trained U-Net. Besides, there are some novel ideas recently proposed for brain tumor segmentation. Wang et al. [24] provided voxel-wise and structure-wise uncertainty information of the segmentation result which helps to improve segmentation accuracy. Wang et al. [11] proposed global attention multiscale feature fusion module (GMF) and local dense multi-scale feature fusion module (LMF) to exploit both local dense features and global context information. Cheng et al. [6] decomposed the input modalities into the appearance code and content code to enhance the robustness of multimodal learning framework when some modalities are missed. Xu et al. [29] settled multi-tasks into corresponding branches with a shared feature extractor and implicitly involved subregions correlations as attention messages in a single model. However their attention mechanism only focused on the relationships of sub-regions.

In this work, we decompose the brain tumor segmentation task into three subtasks where the whole tumor, the tumor core and the enhancing tumor core are segmented respectively. For each subtask, we follow the success of the encoder-decoder structure of CNN and keep the skip connection of U-Net. We introduce the Global Context (GC) block [5] to the decoder, which can capture long-range dependency and inter-channel dependency. And we also make use of residual connection [9], group normalization [27] and deep supervision [28]. Moreover, we use a modified Dice loss function to focus on less accurate labels, and we add it to a weighted cross entropy loss to further improve the accuracy.

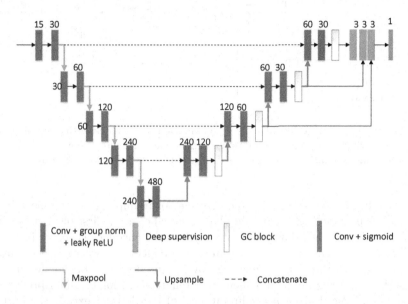

Fig. 1. The proposed network for brain tumor segmentation. Each block encompasses two convolutions with leaky ReLU and group normalization. The number besides each block means output channel number of convolution. A Global Context (GC) block is used at each resolution level of the decoder except the bottom one to capture the global context information. Deep supervision is used to employ multi-scale features for the final prediction.

2   Methods

2.1   Global Context Network Architecture

In our proposed method, we use a 3D U-Net [7] as our backbone due to its great performance in medical image segmentation, and separately trained three networks to hierarchically segment whole tumor, tumor core and enhancing tumor, as it is more challenging to segment these structures simultaneously with a single network due to their different sizes and appearances [23].

The proposed network for each subtask is shown in Fig. 1. Our network is a modification of the 3D U-Net [7]. Both encoder part and decoder part use residual blocks on account of their identity mapping to address the model degradation problem for very deep networks [9]. Each block encompasses two convolutions with leaky Rectified Linear Unit (ReLU) and group normalization. Group normalization performs better than its batch normalization counterpart when batch

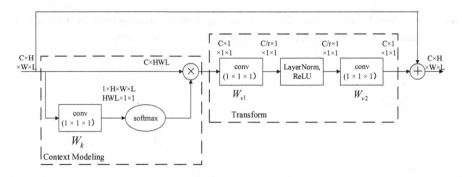

Fig. 2. Global Context (GC) block is a combination of SE block with simplified non-local block, which can capture long-range dependency and inter-channel dependency. $\otimes$ denotes matrix multiplication and $\oplus$ denotes broadcast elementwise addition.

size is small [27]. We utilize max-pooling for downsampling and trilinear interpolation for upsampling. We set the channel number in the first block as 30, and it is doubled after each down-sampling layer in the encoder, as shown in Fig. 1.

To learn representative high-level and low-level features, we apply deep supervision to the last three end of decoders by upsampling their output features to the resolution of the input and then concatenate them for the final prediction. A GC block is used at each resolution level of the decoder except the bottom one to capture the global context information, which will be introduced in the following.

2.2 Global Context Aggregating

The GC block [5] is a combination between simplified non-local block [25] and Squeeze-and-Excitation (SE) block [10]. These two blocks are shown as the context modeling part and transform part in Fig. 2, respectively. In this paper, we combine the GC block with the 3D U-Net by inserting it at multiple scales of the decoding path, as shown in Fig. 1. Inspired by attention mechanisms that learn to focus on the relevant image regions [8], non-local network [25] is a self-attention method which can capture long-range dependencies and thus helps to obtain better segmentation results. However, the original implementation [25] has a large consumption of memory and it is not practical to use the non-local block at each resolution level of the decoder. Therefore, we apply a simplified version [5], which is more memory efficient without reducing the performance. Moreover, another self-attention method SE [10] block can capture the inter-channel dependency and then adaptively recalibrate channel-wise feature responses.

The non-local network generally implies the importance of the corresponding positions to the query position through the query-specific attention weights. Instead of aggregating query-specific global context to each query position as the original version of non-local block [25] does, the simplified non-local block [5] explicitly uses a query-independent attention map for all query positions so that the attention coefficients for every query position are almost the same. The query-independent attention map here is a channel-wise global context. And then, we add the global context features: query-independent attention map to each query position. The simplified non-local is defined as:

$$y_i = x_i + W_v \sum_{j=1}^{N} \frac{\exp(W_k x_j)}{\sum_{l=1}^{N} \exp(W_k x_l)} x_j \tag{1}$$

where x and y denote the input and output of the simplified non-local block respectively, i and j denote two positions and N is the number of positions in the feature map. W_k and W_v are linear transformation matrices and implemented through $1 \times 1 \times 1$ convolution here.

Before the global context is added to each query position, the global context goes to SE block to recalibrate channel-wise feature responses, where layer normalization eases optimization in the bottleneck transform. By integrating GC blocks in all decoder blocks, we can capture the long-range dependency with a slight increase of computational cost.

2.3 Logarithmic Dice and Weighted Cross Entropy

Inspired by the previous work [26], we propose a loss function which helps to obtain accurate results:

$$Loss = L_{\text{Dice}} + L_{\text{Cross-Entropy}} \tag{2}$$

It is a combination of the logarithmic Dice loss [26] and weighted cross entropy. The logarithmic Dice loss gets a bigger decreasing gradient magnitude as dice increases, therefore we can converge to a lower loss and focus on less accurate voxels. It intrinsically focuses more on less accurately segmented structures. Based on the previous work [26] and our experience, the exponent parameter is set to 0.3 here. The weighted cross entropy is inspired by the heatmap in human pose estimation [22] and it can guide the network to focus more on the target region:

$$L_{\text{Dice}} = (-\log(\text{Dice}))^{0.3} \tag{3}$$

$$L_{\text{Cross-Entropy}} = -\frac{1}{N} \sum_{i=0}^{N} (y_i * \log(p_i) + (1 - y_i) * \log(1 - p_i)) * h_i \tag{4}$$

where i denotes the pixel position. y_i is the ground truth label at i, p_i is the probability of pixel i being the foreground, and h_i is the weight of pixel i:

$$h_i = \begin{cases} 1, & \text{if } i \in \mathscr{F} \\ \exp(-l_i^2/\sigma), & \text{otherwise} \end{cases} \tag{5}$$

$$\sigma = \frac{1}{2} r^{\frac{3}{2}} \tag{6}$$

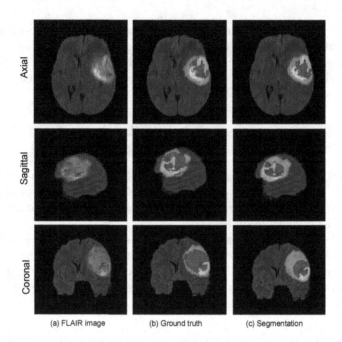

Fig. 3. A typical segmentation example of the brain tumor (HGG) from our local validation set, which is part of the official BraTS 2019 training set. The whole tumor includes all colours, the tumor core includes yellow and red, the enhancing tumor is shown in yellow. (Color figure online)

where $\mathscr{F}$ denotes the set of foreground voxels in the ground truth, and the weight for foreground voxels is 1. r denotes the average size of the foreground region along each axis. σ is a positive parameter that controls the weight for background voxels, which is inspired by Gaussian distribution. l_i is the distance to the center of foreground voxels and h_i gets lower when the voxels are further from the foreground voxels.

3   Experiments and Preliminary Results

Data and Implementation Details. We mainly used the Multimodal Brain Tumor Segmentation Challenge (BraTS) 2019[1] [1–4,17] training and validation set for experiments. BraTS focuses on the evaluation of state-of-the-art methods for the segmentation of brain tumors in 3D MRI scans [1–4,17]. The BraTS 2019 training set consists of 335 cases (259 HGG and 76 LGG) with four 3D MRI modalities (T1, T1c, T2 and FLAIR). Each case was annotated into 3 hetero- geneous histological sub-regions by expert raters: peritumoral edema, necrotic core and non-enhancing tumor core and enhancing tumor core. The evaluation

---

[1] http://www.med.upenn.edu/cbica/brats2019.html.

was based on the segmentation accuracy of three hierarchical regions: enhancing tumor (ET), tumor core (TC) which includes the ET, whole tumor (WT) which includes the TC. The BraTS 2019 validation set contains images from 125 patients with brain tumors of unknown grade. The segmentation masks are uploaded to the online evaluation platform and the segmentation performance is measured based on the Dice score, sensitivity, specificity and Hausdorff distance.

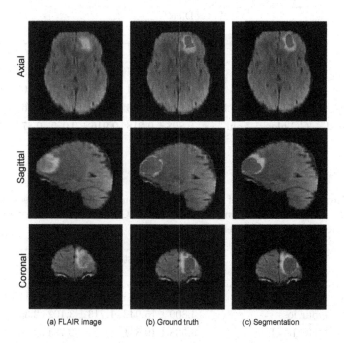

(a) FLAIR image   (b) Ground truth   (c) Segmentation

Fig. 4. A typical segmentation example of the brain tumor (LGG) from a our local validation set, which is part of the official BraTS 2019 training set. The whole tumor includes all colours, the tumor core includes yellow and red, the enhancing tumor is shown in yellow. (Color figure online)

We implemented our network in PyTorch [20]. We trained for 500 epochs and used Adaptive Moment Estimation (Adam) [15] for training, with initial learning rate 10^{-4}, which was reduced by half when validation performance has not improved for 30 epochs. We regularized with a $L2$ weight decay 10^{-5}, batch size 2. Training was implemented on two NVIDIA GeForce GTX 1080Ti GPUs. The training patch size was $128 \times 128 \times 128$ for each of the three binary segmentation tasks: whole tumor, tumor core and enhancing tumor.

For pre-processing, each image was normalized by its intensity mean value and standard deviation. Random crop, random elastic deformation, random rotation and random mirroring were used for data augmentation to alleviate the overfitting problem. At test time, we just segmented three subregions in sequence and test time augmentation by mirror flipping the input image. Finally, we used an ensemble of five models to improve the segmentation accuracy.

Table 1. Dice scores of our methods on our local validation set, which is a subset of the BraTS 2019 training set. EN, WT, TC denote enhancing tumor, whole tumor and tumor core respectively.

	Dice		
	ET	WT	TC
Baseline	0.757 ± 0.271	0.905 ± 0.064	0.862 ± 0.127
Baseline + our loss	0.781 ± 0.254	0.907 ± 0.062	0.858 ± 0.141
Baseline + GC block	0.792 ± 0.224	0.909 ± 0.062	0.865 ± 0.136
Baseline + GC block + our loss	0.798 ± 0.233	0.910 ± 0.066	0.869 ± 0.121
Cascaded + GC block + our loss	0.803 ± 0.218	0.916 ± 0.055	0.874 ± 0.099

Table 2. Hausdorff distances of our methods on our local validation set, which is a subset of the BraTS 2019 training set. EN, WT, TC denote enhancing tumor, whole tumor and tumor core respectively.

	Hausdorff distance (mm)		
	ET	WT	TC
Baseline	6.477 ± 23.483	1.727 ± 3.062	2.115 ± 3.027
Baseline + our loss	4.800 ± 15.942	1.713 ± 3.102	2.352 ± 3.229
Baseline + GC block	5.538 ± 19.341	1.584 ± 2.784	2.300 ± 3.693
Baseline + GC block + our loss	2.600 ± 6.770	1.483 ± 2.481	1.915 ± 2.430
Cascaded + GC block + our loss	5.633 ± 19.688	1.306 ± 2.020	1.948 ± 2.311

Segmentation Results. As a preliminary study, we first conducted our experiment with BraTS 2019 training images to validate the effectiveness of our GC block, from which we randomly selected 80% as the training set, and the remaining was used for validation. Our baseline is a multi-class 3D U-Net used to segment three subregions with a Dice loss function. We compared the baseline with or without GC block, the baseline with or without our loss function to investigate the effect of our GC block and our loss function individually. Table 1 and Table 2 show the results based on our local validation set, which is part of the BraTS 2019 training set. It can be observed that both GC block and our proposed loss function lead to improved segmentation results of enhancing tumor. These tables also show that our cascade of three binary segmentation models outperforms the baseline structure of a single multi-class segmentation model.

Figure 3 and Fig. 4 show examples of qualitative segmentation of our local validation images, which is from a subset of the BraTS 2019 training set. We visualize the results and labels in the FLAIR image. The green, red, yellow colors show the edema, non-enhancing and enhancing tumor cores, respectively. Figure 3 shows the segmentation of whole tumor and tumor core is close to the ground truth. However, the segmentation of enhancing tumor core is less accurate, which is potentially because of the small and irregular target shape. In Fig. 4, the LGG image does not contain enhancing tumors. The segmentation of whole tumor and tumor core is also close to the ground truth.

Table 3. Dice and Hausdorff measurements of our method on BraTS 2019 validation set. EN, WT, TC denote enhancing tumor, whole tumor and tumor core respectively.

	Dice			Hausdorff distance (mm)		
	ET	WT	TC	ET	WT	TC
Proposed (a single model)	0.77266	0.90304	0.83256	4.44100	7.09542	7.67823
Proposed (ensemble)	0.77338	0.90712	0.83911	4.30514	5.20383	7.15621
team SCAN	0.77656	0.90791	0.84640	3.37916	4.80370	6.43485
team Questionmarks	0.80211	0.90941	0.86473	3.14581	4.26398	5.43931

Table 4. Sensitivity and specificity measurements of our method on BraTS 2019 validation set. EN, WT, TC denote enhancing tumor, whole tumor and tumor core respectively.

	Sensitivity			Specificity		
	ET	WT	TC	ET	WT	TC
Proposed (a single model)	0.77204	0.91863	0.82286	0.99852	0.99409	0.99769
Proposed (ensemble)	0.77369	0.89893	0.82907	0.99851	0.99575	0.99767
team SCAN	0.77540	0.89128	0.84790	0.99858	0.99575	0.99717
team Questionmarks	0.80383	0.92371	0.86215	0.99843	0.99425	0.99739

We then trained our proposed network with the entire set of BraTS 2019 training images, and applied the trained model to the BraTS 2019 validation set. We also compared our method with the top-ranked methods. Team Questionmarks won the 1st place of the segmentation task of BraTS 2019. Table 3 presents Dice and Hausdorff measurements according to the online evaluation platform in BraTS 2019. It shows that our Dice accuracy performance was competitive according to the leaderboard[2]. Table 4 presents sensitivity and specificity measurements according to the online evaluation platform in BraTS 2019.

---

[2] https://www.cbica.upenn.edu/BraTS19/lboardValidation.html.

Table 5 presents quantitative results on the BraTS 2019 testing set. It includes the means, standard deviations, medians, 25 quantiles and 75 quantiles of Dice and Hausdorff measurements of enhancing tumor, whole tumor and tumor core. We achieved Dice scores of 0.79303, 0.87962 and 0.82887 for ET, WT and TC, respectively. The results of enhancing tumor and tumor core are close to our results on validation sets but scores for whole tumor are lower. It is of interest to reduce the overfitting problems of whole tumors' segmentation.

Table 5. Dice and Hausdorff measurements of our method on BraTS 2019 testing set. EN, WT, TC denote enhancing tumor, whole tumor and tumor core respectively.

	Dice			Hausdorff distance (mm)		
	ET	WT	TC	ET	WT	TC
Mean	0.79303	0.87962	0.82887	2.96669	6.35796	5.61115
StdDev	0.22185	0.14244	0.26101	6.08917	11.76922	13.13856
Median	0.84974	0.91866	0.92385	1.73205	3.00000	2.23607
25quantile	0.76702	0.87164	0.86590	1.00000	1.73205	1.41421
75quantile	0.91617	0.94839	0.95512	2.44949	5.65471	3.70763

4   Conclusion

We proposed a cascaded global context convolutional neural network to segment glioma subregions from multi-modality brain MR images. We train three subtasks separately so that each task is simpler and easier to train. Our GC block can capture long-range dependency and inter-channel dependency and helps to improve the segmentation accuracy. Experimental results show that our method achieved average Dice scores of 0.77338, 0.90712 and 0.83911 for enhancing tumor, whole tumor and tumor core, respectively on the BraTS 2019 validation set. The corresponding values for BraTS 2019 testing set were 0.79303, 0.87962 and 0.82887 for enhancing tumor, whole tumor and tumor core, respectively.

References

1. Bakas, S., et al.: Advancing the cancer genome atlas glioma MRI collections with expert segmentation labels and radiomic features. Nat. Sci. Data **4**, 170117 (2017)
2. Bakas, S., et al.: Segmentation labels and radiomic features for the pre-operative scans of the TCGA-LGG collection. The Cancer Imaging Archive (2017)
3. Bakas, S., et al.: Segmentation labels for the pre-operative scans of the TCGA-GBM collection. The Cancer Imaging Archive (2017)

4. Bakas, S., et al.: Identifying the best machine learning algorithms for brain tumor segmentation, progression assessment, and overall survival prediction in the brats challenge. arXiv preprint arXiv:1811.02629 (2018)

5. Cao, Y., Xu, J., Lin, S., Wei, F., Hu, H.: GCNet: Non-local networks meet squeeze-excitation networks and beyond. arXiv preprint arXiv:1904.11492 (2019)

6. Chen, C., Dou, Q., Jin, Y., Chen, H., Qin, J., Heng, P.-A.: Robust multimodal brain tumor segmentation via feature disentanglement and gated fusion. In: Shen, D., et al. (eds.) MICCAI 2019. LNCS, vol. 11766, pp. 447–456. Springer, Cham (2019). https://doi.org/10.1007/978-3-030-32248-9_50

7. Çiçek, Ö., Abdulkadir, A., Lienkamp, S.S., Brox, T., Ronneberger, O.: 3D U-Net: learning dense volumetric segmentation from sparse annotation. In: Ourselin, S., Joskowicz, L., Sabuncu, M.R., Unal, G., Wells, W. (eds.) MICCAI 2016. LNCS, vol. 9901, pp. 424–432. Springer, Cham (2016). https://doi.org/10.1007/978-3-319-46723-8_49

8. Fu, J., Zheng, H., Mei, T.: Look closer to see better: recurrent attention convolutional neural network for fine-grained image recognition. In: Proceedings of the IEEE Conference on Computer Vision and Pattern Recognition, pp. 4438–4446 (2017)

9. He, K., Zhang, X., Ren, S., Sun, J.: Deep residual learning for image recognition. In: Proceedings of the IEEE Conference on Computer Vision and Pattern Recognition, pp. 770–778 (2016)

10. Hu, J., Shen, L., Sun, G.: Squeeze-and-excitation networks. In: Proceedings of the IEEE Conference on Computer Vision and Pattern Recognition, pp. 7132–7141 (2018)

11. Wang, H., Wang, G., Liu, Z., Zhang, S.: Global and local multi-scale feature fusion enhancement for brain tumor segmentation and pancreas segmentation. In: Crimi, A., Bakas, S. (eds.) BrainLes 2019. LNCS, vol. 11992, pp. 80–88. Springer, Cham (2020). https://doi.org/10.1007/978-3-030-46640-4_8

12. Isensee, F., Kickingereder, P., Wick, W., Bendszus, M., Maier-Hein, K.H.: No New-Net. In: Crimi, A., Bakas, S., Kuijf, H., Keyvan, F., Reyes, M., van Walsum, T. (eds.) BrainLes 2018. LNCS, vol. 11384, pp. 234–244. Springer, Cham (2019). https://doi.org/10.1007/978-3-030-11726-9_21

13. Kamnitsas, K., et al.: Ensembles of multiple models and architectures for robust brain tumour segmentation. In: Crimi, A., Bakas, S., Kuijf, H., Menze, B., Reyes, M. (eds.) BrainLes 2017. LNCS, vol. 10670, pp. 450–462. Springer, Cham (2018). https://doi.org/10.1007/978-3-319-75238-9_38

14. Kamnitsas, K., et al.: Efficient multi-scale 3D CNN with fully connected CRF for accurate brain lesion segmentation. Med. Image Anal. 36, 61–78 (2017)

15. Kingma, D.P., Ba, J.L.: Adam: a method for stochastic optimization. In: 2015 International Conference on Learning Representations, pp. 1–15 (2015)

16. Long, J., Shelhamer, E., Darrell, T.: Fully convolutional networks for semantic segmentation. In: Proceedings of the IEEE Conference on Computer Vision and Pattern Recognition, pp. 3431–3440 (2015)

17. Menze, B.H., et al.: The multimodal brain tumor image segmentation benchmark (BRATS). TMI 34(10), 1993–2024 (2015)

18. Myronenko, A.: 3D MRI brain tumor segmentation using autoencoder regularization. In: Crimi, A., Bakas, S., Kuijf, H., Keyvan, F., Reyes, M., van Walsum, T. (eds.) BrainLes 2018. LNCS, vol. 11384, pp. 311–320. Springer, Cham (2019). https://doi.org/10.1007/978-3-030-11726-9_28

19. Ohgaki, H., Kleihues, P.: Population-based studies on incidence, survival rates, and genetic alterations in astrocytic and oligodendroglial gliomas. J. Neuropathol. Exp. Neurol. **64**(6), 479–489 (2005)

20. Paszke, A., et al.: Automatic differentiation in pytorch (2017)

21. Ronneberger, O., Fischer, P., Brox, T.: U-Net: convolutional networks for biomedical image segmentation. In: Navab, N., Hornegger, J., Wells, W.M., Frangi, A.F. (eds.) MICCAI 2015. LNCS, vol. 9351, pp. 234–241. Springer, Cham (2015). https://doi.org/10.1007/978-3-319-24574-4_28

22. Tompson, J.J., Jain, A., LeCun, Y., Bregler, C.: Joint training of a convolutional network and a graphical model for human pose estimation. In: Advances in Neural Information Processing Systems, pp. 1799–1807 (2014)

23. Wang, G., Li, W., Ourselin, S., Vercauteren, T.: Automatic brain tumor segmentation using cascaded anisotropic convolutional neural networks. In: Crimi, A., Bakas, S., Kuijf, H., Menze, B., Reyes, M. (eds.) BrainLes 2017. LNCS, vol. 10670, pp. 178–190. Springer, Cham (2018). https://doi.org/10.1007/978-3-319-75238-9_16

24. Wang, G., Li, W., Vercauteren, T., Ourselin, S.: Automatic brain tumor segmentation based on cascaded convolutional neural networks with uncertainty estimation. Front. Comput. Neurosci. **13**, 56 (2019)

25. Wang, X., Girshick, R., Gupta, A., He, K.: Non-local neural networks. In: Proceedings of the IEEE Conference on Computer Vision and Pattern Recognition, pp. 7794–7803 (2018)

26. Wong, K.C.L., Moradi, M., Tang, H., Syeda-Mahmood, T.: 3D segmentation with exponential logarithmic loss for highly unbalanced object sizes. In: Frangi, A.F., Schnabel, J.A., Davatzikos, C., Alberola-López, C., Fichtinger, G. (eds.) MICCAI 2018. LNCS, vol. 11072, pp. 612–619. Springer, Cham (2018). https://doi.org/10.1007/978-3-030-00931-1_70

27. Wu, Y., He, K.: Group normalization. In: Proceedings of the European Conference on Computer Vision (ECCV), pp. 3–19 (2018)

28. Xie, S., Tu, Z.: Holistically-nested edge detection. In: Proceedings of the IEEE International Conference on Computer Vision, pp. 1395–1403 (2015)

29. Xu, H., Xie, H., Liu, Y., Cheng, C., Niu, C., Zhang, Y.: Deep cascaded attention network for multi-task brain tumor segmentation. In: Shen, D., et al. (eds.) MICCAI 2019. LNCS, vol. 11766, pp. 420–428. Springer, Cham (2019). https://doi.org/10.1007/978-3-030-32248-9_47

Multi-task Learning for Brain Tumor Segmentation

Leon Weninger[✉], Qianyu Liu, and Dorit Merhof

Institute of Imaging and Computer Vision, RWTH Aachen University, Aachen,
Germany
leon.weninger@lfb.rwth-aachen.de

Abstract. Accurate and reproducible detection of a brain tumor and
segmentation of its sub-regions has high relevance in clinical trials and
practice. Numerous recent publications have shown that deep learning
algorithms are well suited for this application. However, fully supervised
methods require a large amount of annotated training data. To obtain
such data, time-consuming expert annotations are necessary. Furthermore, the enhancing core appears to be the most challenging to segment among the different sub-regions. Therefore, we propose a novel and
straightforward method to improve brain tumor segmentation by joint
learning of three related tasks with a partly shared architecture. Next to
the tumor segmentation, image reconstruction and detection of enhancing tumor are learned simultaneously using a shared encoder. Meanwhile,
different decoders are used for the different tasks, allowing for arbitrary
switching of the loss function. In effect, this means that the architecture
can partly learn on data without annotations by using only the autoencoder part. This makes it possible to train on bigger, but unannotated
datasets, as only the segmenting decoder needs to be fine-tuned solely
on annotated images. The second auxiliary task, detecting the presence
of enhancing tumor tissue, is intended to provide a focus of the network
on this area, and provides further information for postprocessing. The
final prediction on the BraTS validation data using our method gives
Dice scores of 0.89, 0.79 and 0.75 for the whole tumor, tumor core and
the enhancing tumor region, respectively.

Keywords: BraTS 2019 · Brain tumor segmentation · Multi-task
learning · U-Net

1 Introduction

Automatic brain tumor segmentation can provide massive support in clinical trials and practices, as manual annotation of 3D brain MRI images needs trained
personnel and is very labor intensive. For example, a recent study using a
deep learning tumor segmentation algorithm was carried out by Kickingereder
et al. [9], showcasing the clinical relevance of this task.

© Springer Nature Switzerland AG 2020
A. Crimi and S. Bakas (Eds.): BrainLes 2019, LNCS 11992, pp. 327–337, 2020.
https://doi.org/10.1007/978-3-030-46640-4_31

However, a high variety of segmentation algorithms exist, and different independent studies do not allow for direct comparisons of different algorithms. To address this need, the BraTS Challenge was launched [11]. It focuses on comparing the performance of automated segmentation algorithms on multi-institutional, pre-operative MRI scans. Next to this task, participants are asked to determine patient survival automatically.

In 2012, the BraTS Challenge appeared for the first time in conjunction with the International Conference Medical Image Computing and Computer Assisted Interventions (MICCAI). Since then, this challenge has yearly hold a worldwide competition of novel methods for automatic segmentation of brain tumors. Over 50 competitors participated in the BraTS Challenge in 2018.

A manually annotated training data set [3] is provided to the participants, while segmentations of a validation- and test data set remain unseen for the participants. Finally, the participants are ranked based on the quality of their predicted segmentations, which are compared to a manually annotated ground truth.

Motivated by the recent success of deep learning neural networks on different image segmentation tasks and the promising results from the BraTS challenge in the last few years, we present a novel method that seeks to improve the segmentation by multi-task learning and including unlabeled data.

2   Related Work

Extensive research has been presented in the field of brain tumor segmentation. In recent years, most publications rely on encoder-decoder type deep learning models. These types of models also performed best in the BraTS challenges of the last two years [4].

The best performing submissions of the BraTS challenge 2017 include the work of Kamnitsas et al. [8], which won the first place with an ensemble method, and the work of Wang et al. [17], who achieved the second place with a cascaded anisotropic architecture.

In 2018, Lachinov et al. [10] also employed a cascaded 3D U-Net architecture for the BraTS challenge. Within the cascaded structure, they implemented a multiple encoders U-Net architecture, which trained each encoder for one MRI modality. Comparing their results to the standard U-Net, they had significantly better predictions for the tumor core region, while the improvements for the whole tumor and enhancing tumor region are marginal. Finally, Sherman [16] demonstrated that training the V-Net architecture as proposed by Milletari et al. [12] with a multi-class Dice loss could achieve competitive Dice scores.

However, compared to the winners of the BraTS challenge 2017, the best performing teams of the BraTS challenge 2018 employed less complex architectures, which still outperformed previously proposed approaches. The first place went to Myronenko [14], who employed a customized encoder-decoder structure with an asymmetrically large encoder to extract deep image features. A variational autoencoder branch was also added to the overall architecture in order to regularize the shared encoder. The second place went to Isensee et al. [7] with a plain 3D U-Net.

A multi-task learning approach for weakly annotated data was presented Mlynarski et al. [13] in 2018. The network was jointly trained for segmentation and classification tasks with weakly-annotated 2-D images and voxelwise annotated 3D MRI scans. It was shown that such this multi-task approach could provide significant improvements over a segmentation-only network. In their system the segmentation network was trained on voxelwise annotations while the classification sub-network learned a simple tumor detection score. The two networks shared parameters. A combined loss function for training took both the perfectly annotated volumes as well as the weakly annotated images into account.

3   Data Set

Deep learning algorithms learn the features and the properties of the data set. Therefore, the quantity and quality of the training data set is very important for the performance of the algorithm.

The BraTS challenge 2019 training data consists of 335 annotated multimodal MRI scans of glioma patients, comprising 259 glioblastoma patients and 76 lower-grade glioma patients. As our method can leverage unlabeled data for training, we include the BraTS 2018 test data and the BraTS 2019 validation data as unlabeled data. The BraTS data originate from 19 different institutions worldwide that use various MRI scanners [1,2]. All the available multimodal MRI scans were pre-operative and uniformly pre-processed, for instance co-registered to the same anatomical template, interpolated to the same resolution ($1\,\text{mm}^3$), and skull-stripped. For each brain tumor patient, four different MRI sequences are available, from which precise information about the location, size, shape, and the sub-regions of the glioma can be extracted. T1, T2, FLAIR and a contrast-agent enhanced T1 image (T1GD) are available.

The ground truth labels have been annotated by more than one expert into three glioma sub-regions, and examined by experienced neuroradiologists. The three sub-regions are the enhancing tumor region (ET), visible in T1Gd, the necrotic and non-enhancing tumor core (NET) region with increased brightness as visualized by T2 weighted images, and the peritumoral edema (ED), which is typically hyperintense in the FLAIR image.

4   Methods

4.1   Preprocessing

Since the output of MRI scanners is not quantitative, and the BraTS data set originates from different institutions using various scanners, further preprocessing of the data is necessary for intensity normalization. Z-score normalization is chosen, which linearly scales the input images. Thus, it keeps the relative tissue contrast constant even for unusually high or low intensities. With z-score normalization, each MRI scan is subtracted by its mean μ and divided by its standard deviation σ individually.

Due to the increasing depth of the network, the image size for the input needs to be adjusted as a trade-off. This motivates a patch-based training with sequentially extracted patches of the dimension $128 \times 128 \times 128$. Before the patches are generated, each MRI scan is cropped to the brain mask, i.e. all voxels outside the mask are discarded such that only the brain region is left (Fig. 1).

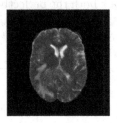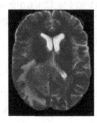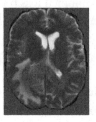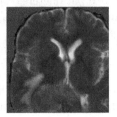

Fig. 1. Preprocessing of a T2 slice: Original data, cropped, z-score normalized, extracted patch (from left to right).

4.2 Multi-task Learning Based Segmentation

A multi-task learning (MTL) network needs to minimize a sum of several differently weighted loss functions. This leveraging process during the training prevents the trained network from overfitting on the one hand, and, on the other hand, encourages the network to focus on features that are important for all tasks. Incorporating similar but unlabeled data into the training data set, as well as determining the presence or absence of the enhancing tumor region, are to a certain extent related to the original segmentation task. Overall, the multi-task loss

$$L_{MTL} = w_1 \cdot L_{L2} + w_2 \cdot L_{KL} + w_3 \cdot L_{class} + w_4 \cdot L_{dice}$$

with fixed weights w_k, variational autoencoder losses L_{L2} and L_{KL}, classification loss L_{class}, and segmentation loss L_{dice} is employed. An overview over the network is given in Fig. 2, and the different branches are specified in the respective sections.

Segmentation. For the classical deep learning segmentation task, a 3D U-Net [6] with a depth of 4 is employed. The first layer consists of 28 feature maps. Compared to the original U-Net, normalization layers are replaced by group-norm layers and LeakyRelu is used for all activation functions. A multi-class Dice loss [12] was chosen as segmentation loss.

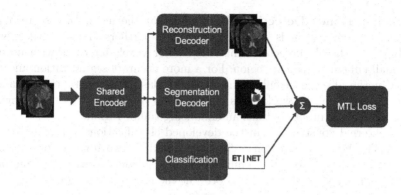

Fig. 2. Semi-supervised segmentation network architecture. ET: Enhancing, NET: Non-Enhancing

Autoencoder Reconstruction. Using an autoencoding branch as auxiliary task in brain tumor segmentation was introduced by Myronenko [14] in the BraTS challenge 2018, where it was used for regularizing effects during training. However, an autoencoding branch provides further possibilities: training the network jointly with labeled and unlabeled data to make the network more robust against scanner variations. Baur et al. [5] experimented with different network architectures for unsupervised brain tumor segmentation, and demonstrated that fully convolutional variational autoencoder produces better results than their densely connected counterpart. Thus, a FCVAE architecture was chosen for this auxiliary task. The internal construction of the utilized FCVAE is listed in Table 1.

Table 1. Architecture of the decoder part for the FCVAE: GN stands for group normalization with a group size of four; UpLin is 3D trilinear upsampling.

Name	Operations	Input	Output	Kernel	Padding	Repeat
DownStage3	Conv3D	244	100×16^3	1	No	0
Mean μ	Conv3D	100	50×16^3	1	No	0
Std σ	Conv3D	100	50×8^3	1	No	0
Sample	Conv3D	50	100×16^3	1	No	0
DC4	GN+ReLu+Conv3D	100	224×16^3	3	Yes	1
Up3	Conv3D+UpLin	224	224×32^3	1	No	0
DC3	GN+ReLu+Conv3D	224	112×32^3	3	Yes	1
Up2	Conv3D+UpLin	112	112×64^3	1	No	0
DC2	GN+ReLu+Conv3D	112	56×128^3	3	Yes	1
Up1	Conv3D+UpLin	56	56×128^3	1	No	0
DC1	GN+ReLu+Conv3D	56	28×128^3	3	Yes	1
VAEOut	Conv3D	28	4×128^3	1	No	0

Enhancing Tumor Detection. We found that the mean Dice score of the enhancing tumor region is strongly affected by outliers. These outliers were mainly due to falsely detected enhancing tumor voxels for cases with no or a very small enhancing tumor region. For a more accurate segmentation and correction in the absence of enhancing tissue, a classification branch was added to the segmentation network. This classification network learns to distinguish whether the given input contains an enhancing tumor region.

The internal construction of the developed classification branch is listed in Table 2. The bottleneck region is first connected to two further convolutional layers, followed by two fully connected layers. Furthermore, a dropout layer (p = 0.5) is incorporated in order to prevent the network from overfitting.

Table 2. Architecture of the classification network after the bottleneck region.

Name	Operations	Input	Output
Bottleneck	Conv3D	448×8^3	256×8^3
Reduced	Conv3D	256×8^3	128×8^3
Dense1	Linear	128×8^3	32
Dropout	-	-	-
Dense2	Linear	32	2
Softmax	-	-	-

Training Procedure. A stratified five-fold cross-validation is used on the training dataset, i.e., five different networks are trained on equally sampled subsets of the data. Supervised and unsupervised examples are selected in random order. Evaluating different training hyperparameters, the following were chosen: Batch size = 2, learning rate = 0.0001, weight decay = 0.00001. During training, no data augmentation was used. The different loss functions were weighted as follows:

$$L_{MTL} = 0.1 \cdot L_{L2} + 0.1 \cdot L_{KL} + 0.1 \cdot L_{class} + 1 \cdot L_{dice}$$

Our contribution to the BraTS challenge was implemented using pyTorch [15]. Training and prediction is carried out on two Nvidia 1080 Ti GPUs, each with a memory size of 11 Gb.

Inference. During inference, the reconstruction part of the segmenter is deactivated. Meanwhile, the output of the classification network is saved, and used for correction of the segmentation result. If the classification network has identified an absence of enhancing tumor in the image and in addition none or only a marginal amount of enhancing voxels are detected using a threshold, all enhancing areas in the segmentation map will be suppressed.

Test-time augmentation is employed for both, the segmentation and the classification task. For this, the input images are mirrored with respect to each axis and the resulting predictions are averaged in order to obtain the final segmentation.

4.3   Survival Prediction

In the BraTS challenge 2018, we participated with a linear regression on patient-age only [19] for the survival prediction task. This basic approach won the third place in the challenge. Subsequently, a study was published that shows that radiomic features do well in predicting survival of brain tumor patients if the tumor is not totally resected, but are unreliable for gross total resection (GTR) patients [18]. Thus, we did not change our approach and submitted a linear regression on patient-age, based on GTR patients only.

5   Results

5.1   Segmentation

For the BraTS training dataset, we provide the results of the five-fold cross validation, i.e. the five different networks predict the subset that was excluded during training. The final prediction on the BraTS validation dataset is obtained by a majority vote of all independent predictions. The scores can be seen in Table 3, and were taken from the online submission system. Our results varied strongly depending on the patient.

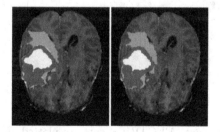 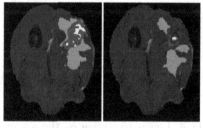

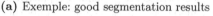

(a) Exemple: good segmentation results   (b) Exemple: bad segmentation results

Fig. 3. Qualitative segmentation result compared to the groundtruth using the multi-task network. Green: Edema, Yellow: Necrosis and non-enhancing tumor core, Red: Enhancing tumor. Our results are the right images of each image pair. (Color figure online)

An exemplary good as well as an exemplary bad result are presented in Fig. 3. It can be seen that the segmentation results are excellent in some cases, but can deviate strongly from the groundtruth in other cases.

Table 3. Mean results for the segmentation challenge. All measures are according to the online submission system.

Dataset	Dice			Hausdorff 95		
	ET	WT	TC	ET	WT	TC
Train set	0.65	0.82	0.76	9.20	11.4	10.5
Val set	0.75	0.89	0.79	6.14	5.76	9.11
Test set	0.75	0.85	0.78	5.76	7.98	8.25

5.2 Survival Prediction

In Fig. 4 the age-only linear regression approach is shown. The obtained model is plotted together with the 95% confidence interval on the BraTS 2019 training dataset. Only GTR patients were used for fitting, as the resection of tumor can strongly influence the survival time.

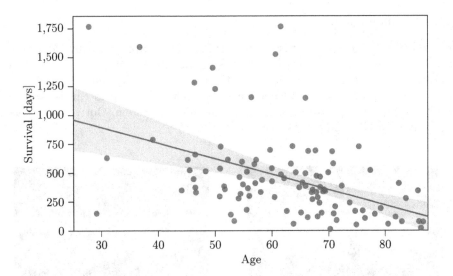

Fig. 4. Linear regression on the age of the patient. All GTR patients are plotted as single dots, and the obtained linear regression model as well as the 95% confidence interval is displayed.

Detailed scores for the survival task can be found in Table 4. These scores were extracted from the online submission system. While all scores on the train- and validation set are similar, the mean square error (MSE), median error (medianSE), and standard deviation (stdSE) are much higher for the test set than for the val and train set. However, the accuracy as well as the SpearmanR are similar for the three different data sets.

Table 4. Mean results for the survival prediction challenge. All measures are according to the online submission system.

Dataset	Accuracy	MSE	medianSE	stdSE	SpearmanR
Train set	0.485	88822	21135	181864	0.48
Val set	0.448	90109	36453	123542	0.27
Test set	0.533	395190	60264	1199488	0.36

6   Discussion and Conclusion

Reproducible and accurate segmentation of brain tumors for an appropriate treatment has high clinical relevance. Since manual annotation of the tumor area is tedious and prone to errors, an automatic or semi-automatic approach is desirable.

To address this need, a multi-task method is proposed, that trains the network for three related tasks: Tumor segmentation, image reconstruction and enhancing tumor detection. These three tasks share an encoder but have different decoder architectures. For the tumor segmentation task, an architecture similar to a 3D U-Net architecture was chosen, with small changes as suggested by recent advances in the field of deep learning. The brain reconstruction branch is based on a FCVAE, which has shown promising results on unlabeled data for brain tumor segmentation. The third part of the architecture, the enhancing tumor detection network, can reduce falsely detected enhancing tumor voxels during segmentation, especially for low-grade gliomas without enhancing tissue. In future, different variants of the existing architecture in terms of network depths, the amount of filter maps, and the spatial dimension of the latent space need to be evaluated in order to identify optimal settings. Further, an optimal weighting of the different loss functions still needs to be determined.

For the survival prediction task, we relied on a very simple model, a linear regression on the age of the patient. This model does not depend on the accuracy of the tumor segmentation step and is one of the most basic approaches possible. Thus, it should be robust against variations in the image data, and impossible to overfit on the training data. Still, the MSE, medianSE, and stdSE varied strongly between the validation- and test set. However, the obtained accuracy was even better on the test set than on the train- and validation sets.

References

1. Bakas, S., et al.: Segmentation labels and radiomic features for the pre-operative scans of the TCGA-LGG collection. The Cancer Imaging Archive (2017)
2. Bakas, S., et al.: Segmentation labels and radiomic features for the pre-operative scans of the TCGA-GBM collection. The Cancer Imaging Archive (2017)
3. Bakas, S., et al.: Advancing the cancer genome atlas glioma MRI collections with expert segmentation labels and radiomic features. Sci. Data **4**, 170117 (2017)

4. Bakas, S., et al.: Identifying the best machine learning algorithms for brain tumor segmentation, progression assessment, and overall survival prediction in the brats challenge. arXiv preprint arXiv:1811.02629 (2018)

5. Baur, C., Wiestler, B., Albarqouni, S., Navab, N.: Deep autoencoding models for unsupervised anomaly segmentation in brain MR images. In: Crimi, A., Bakas, S., Kuijf, H., Keyvan, F., Reyes, M., van Walsum, T. (eds.) BrainLes 2018. LNCS, vol. 11383, pp. 161–169. Springer, Cham (2019). https://doi.org/10.1007/978-3-030-11723-8_16

6. Çiçek, Ö., Abdulkadir, A., Lienkamp, S.S., Brox, T., Ronneberger, O.: 3D U-Net: learning dense volumetric segmentation from sparse annotation. In: Ourselin, S., Joskowicz, L., Sabuncu, M.R., Unal, G., Wells, W. (eds.) MICCAI 2016. LNCS, vol. 9901, pp. 424–432. Springer, Cham (2016). https://doi.org/10.1007/978-3-319-46723-8_49

7. Isensee, F., Kickingereder, P., Wick, W., Bendszus, M., Maier-Hein, K.H.: No new-net. In: Crimi, A., Bakas, S., Kuijf, H., Keyvan, F., Reyes, M., van Walsum, T. (eds.) BrainLes 2018. LNCS, vol. 11384, pp. 234–244. Springer, Cham (2019). https://doi.org/10.1007/978-3-030-11726-9_21

8. Kamnitsas, K., et al.: Ensembles of multiple models and architectures for robust brain tumour segmentation. In: Crimi, A., Bakas, S., Kuijf, H., Menze, B., Reyes, M. (eds.) BrainLes 2017. LNCS, vol. 10670, pp. 450–462. Springer, Cham (2018). https://doi.org/10.1007/978-3-319-75238-9_38

9. Kickingereder, P., et al.: Automated quantitative tumour response assessment of MRI in neuro-oncology with artificial neural networks: a multicentre, retrospective study. Lancet Oncol. **20**(5), 728–740 (2019)

10. Lachinov, D., Vasiliev, E., Turlapov, V.: Glioma segmentation with cascaded UNet. In: Crimi, A., Bakas, S., Kuijf, H., Keyvan, F., Reyes, M., van Walsum, T. (eds.) BrainLes 2018. LNCS, vol. 11384, pp. 189–198. Springer, Cham (2019). https://doi.org/10.1007/978-3-030-11726-9_17

11. Menze, B.H., Jakab, A., Bauer, S., et al.: The multimodal brain tumor image segmentation benchmark (BRATS). IEEE Trans. Med. Imaging **34**(10), 1993–2024 (2015)

12. Milletari, F., Navab, N., Ahmadi, S.A.: V-Net: fully convolutional neural networks for volumetric medical image segmentation. In: 2016 Fourth International Conference on 3D Vision (3DV), pp. 565–571. IEEE (2016)

13. Mlynarski, P., Delingette, H., Criminisi, A., Ayache, N.: Deep learning with mixed supervision for brain tumor segmentation. arXiv preprint arXiv:1812.04571 (2018)

14. Myronenko, A.: 3D MRI brain tumor segmentation using autoencoder regularization. In: Crimi, A., Bakas, S., Kuijf, H., Keyvan, F., Reyes, M., van Walsum, T. (eds.) BrainLes 2018. LNCS, vol. 11384, pp. 311–320. Springer, Cham (2019). https://doi.org/10.1007/978-3-030-11726-9_28

15. Paszke, A., et al.: Automatic differentiation in pytorch. In: NIPS-W (2017)

16. Sherman, R.: A volumetric convolutional neural network for brain tumor segmentation. arXiv preprint arXiv:1811.02654 (2018)

17. Wang, G., Li, W., Ourselin, S., Vercauteren, T.: Automatic brain tumor segmentation using cascaded anisotropic convolutional neural networks. In: Crimi, A., Bakas, S., Kuijf, H., Menze, B., Reyes, M. (eds.) BrainLes 2017. LNCS, vol. 10670, pp. 178–190. Springer, Cham (2018). https://doi.org/10.1007/978-3-319-75238-9_16

18. Weninger, L., Haarburger, C., Merhof, D.: Robustness of radiomics for survival prediction of brain tumor patients depending on resection status. Front. Comput. Neurosci. **13**, 73 (2019)
19. Weninger, L., Rippel, O., Koppers, S., Merhof, D.: Segmentation of brain tumors and patient survival prediction: methods for the BraTS 2018 challenge. In: Crimi, A., Bakas, S., Kuijf, H., Keyvan, F., Reyes, M., van Walsum, T. (eds.) BrainLes 2018. LNCS, vol. 11384, pp. 3–12. Springer, Cham (2019). https://doi.org/10.1007/978-3-030-11726-9_1

Brain Tumor Segmentation and Survival Prediction

Rupal R. Agravat[1]([✉])[iD] and Mehul S. Raval[2][iD]

[1] Ahmedabad University, Ahmedabad, Gujarat, India
rupal.agravat@iet.ahduni.edu.in
[2] Pandit Deendayal Petroleum University, Gandhinagar, Gujarat, India
mehul.raval@sot.pdpu.ac.in

Abstract. The paper demonstrates the use of the fully convolutional neural network for glioma segmentation on the BraTS 2019 dataset. Three-layers deep encoder-decoder architecture is used along with dense connection at the encoder part to propagate the information from the coarse layers to deep layers. This architecture is used to train three tumor sub-components separately. Sub-component training weights are initialized with whole tumor weights to get the localization of the tumor within the brain. In the end, three segmentation results were merged to get the entire tumor segmentation. Dice Similarity of training dataset with focal loss implementation for whole tumor, tumor core, and enhancing tumor is 0.92, 0.90, and 0.79, respectively. Radiomic features from the segmentation results predict survival. Along with these features, age and statistical features are used to predict the overall survival of patients using random forest regressors. The overall survival prediction method outperformed the other methods for the validation dataset on the leaderboard with 58.6% accuracy. This finding is consistent with the performance on the test set of BraTS 2019 with 57.9% accuracy.

Keywords: Brain tumor segmentation · Deep learning · Dense network · Overall survival · Radiomics features · U-net

1   Introduction

Early-stage brain tumor diagnosis can lead to proper treatment planning, which improves patient survival chances. Out of all types of brain tumors, Glioma is one of the most life-threatening brain tumors. It occurs in the glial cells of the brain. Depending on its severity and aggressiveness, glioma has grades ranging from grade I to grade IV. Grade I, II are Low-Grade Glioma (LGG), and grade III and IV are High-Grade Glioma (HGG). A Brain tumor can further be divided into constituent components like - Necrosis, Enhancing tumor, Non-enhancing tumor, and Edema. Tumor core consists of necrosis, enhancing tumor, non-enhancing tumor. In most cases, LGG does not contain enhancing tumor, whereas HGG contains necrosis, enhancing, and non-enhancing sub-components. Edema occurs from infiltrating tumor cells, as well as a biological response to the angiogenic

© Springer Nature Switzerland AG 2020
A. Crimi and S. Bakas (Eds.): BrainLes 2019, LNCS 11992, pp. 338–348, 2020.
https://doi.org/10.1007/978-3-030-46640-4_32

and vascular permeability factors released by the spatially adjacent tumor cells [3].

It is crucial to find tumor sub-components as it plays a vital role in treatment planning. Non-invasive Medical Resonance Imaging (MRI) is the most advisable imaging technique as it captures the functioning of soft tissue adequately compared to other imaging techniques. MR images are prone to inhomogeneity introduced by the surrounding magnetic field, which introduces the artifacts in the captured image. Besides, the appearance of various brain tissues is different in various modalities. Such issues increase the time in the study of the image.

Furthermore, the human interpretation of the image is non-reproducible as well as dependent on the expertise. It requires computer-aided MR image interpretation to locate the tumor. Also, even the initially detected tumor is completely resected, such patients have poor survival prognosis, as metastases may still redevelop, which leads to an open question to the accurate overall survival prediction.

Authors in [2] discussed the basic, generative, and discriminative techniques for brain tumor segmentation. Nowadays, Deep Neural Network (DNN) has gained more attention for the segmentation of biological images. In which, Convolution Neural Networks (CNN), like DeepMedic [16], U-net [25], V-Net [21], SegNet [4], ResNet [13], DenseNet [14] give state-of-the-art results for semantic segmentation. Out of all these methods, U-net is a widely accepted end-to-end segmentation architecture for brain tumors. In [17] the authors used an ensemble of various DNN architectures and supplied and utilized brain parcellation atlas for brain tumor segmentation. Connectomics data, parcellation information, and tumor mask were used to generate features for survival prediction. Authors of [5] supplied 3D patches to 3D U-net for tumor segmentation and used radiomics features for survival prediction. Biomedical image segmentation in [10] implemented using dense, residual, and inception modules. Authors in [22], used ResNet like blocks in encoder-decoder architecture with group normalization after the convolution layer and variational auto-encoder approach to cluster the features at the encoder part. Authors in [15] implemented a variation of 3D-U-net with leaky ReLU activation function, instance normalization with a multiclass dice loss function. In [19], the authors used densely connected dilated convolution stack for pooling free connections in U-net architecture. The ensemble of 6 3D U-Nets implemented in [12] with various input patch sizes and kernel sizes and an average of the segmentation output considered as the final output. Voxel volume, surface area, age, and resection status supplied to the linear regression model for OS prediction. Authors in [24] implemented FCN pre-trained on the VGG network for tumor segmentation. Three slices of the image volume act as three color channels in FCN and final segmentation considered the majority voting from the segmentation. Relative volumes of three sub-components, centroid coordinates of the tumor within the brain depending on the atlas created and centroid coordinates of the tumor core, were supplied to Random Forest (RF) classifier for OS prediction. The ensemble of Cascaded Anisotropic Convolutional Neural Network, DFKZ Net from German Cancer Research Center, and 3D U-Net with the majority voting for tumor segmentation and RF regressor on selected radiomic features for OS prediction implemented in [26]. [29] implemented cascaded 3D U-Nets for tumor segmentation and linear regressor for survival prediction on four features extracted from the network segmentation results.

All the approaches mentioned above use encoder-decoder deep learning architecture. Moreover, according to [1], inductive transfer learning improves network performance. In this paper, the U-net of [11,25] is implemented with reduced network depth. Reduction in network depth has reduced the number of network parameters. In addition to the depth reduction, the dense module at the encoder replaces the convolution module. The network training uses focal loss function. Initially, the network trains on the whole tumor, and then its weights are transfer to substructure network training. This transfer learning has improved the network training as well as the segmentation results.

The remaining paper is as follows: section two of the paper focuses on the BraTS 2019 dataset, section three demonstrates the proposed method section four provides implementation details, and section five shows the results. The last section covers the conclusion and future work.

2   Dataset

The dataset [8,9,20] contains 259 HGG and 76 LGG pre-operative scans. All the images have been segmented manually, by one to four raters, following the same annotation protocol to generate the ground truths. The annotations were approved by experienced neuro-radiologists [6,7]. Annotations have the enhancing tumor (ET label 4), the peritumoral edema (ED label 2), and the necrotic and non-enhancing tumor core (NCR/NET label 1). Images are co-registered to the same anatomical template, interpolated to the same resolution (1 mm x 1 mm x 1 mm), and skull-stripped. Features like age, survival days, and resection status for 213 HGG scans are provided separately for Overall Survival (OS). The validation dataset consists of 125 scans, with the same preprocessing as well as additional features, as mentioned for OS. The test dataset includes 166 scans.

3   Proposed Method

3.1   Task 1: Tumor Segmentation

A Fully Convolution Neural Network (FCNN) provides end-to-end semantic segmentation for the input of the arbitrary size and learns global information related to it. Our network is based on the network proposed by [11]. The network uses three-layer encoder-decoder architecture with the dense connections between the successive convolution layers and skip-connections across peer layers at the encoder side, as shown in Fig. 1. The network contains three dense modules and two convolution modules. Each convolution layer in the dense module is followed by ReLU activation function. Dense connections between the layers in the dense module allows to obtain additional inputs (collective knowledge) from all earlier layers and passes on its feature-maps to all subsequent layers. Dense connections allow the gradient to flow to the earlier layers directly, which provides in-depth supervision on preceding layers by the classification layer. Also, dense connections provide diversified features to the layers, which leads to having richer patterns identification capabilities. Each dense module generates 64, 128, and 256 feature maps, respectively. Each convolution module generates 128 and 64 feature maps applying 1×1 convolution at the end to generate a single probability map for binary classification of the sub-component.

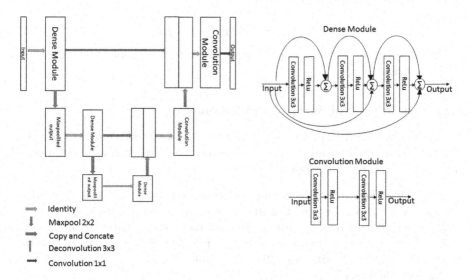

Fig. 1. Network architecture

Brain tumor segmentation task deals with highly imbalanced dataset where tumorous slices are less than non-tumorous slices; such an imbalance dataset reduces network accuracy. The approach of transfer learning mentioned in [23] deals with such an issue. Authors have shown the usefulness of the transfer learning for training a network with/without labels for similar or different tasks. Initially, we have trained the network for the whole tumor. The number of slices is more for the whole tumor compared to sub-components. This step provides tumor localization in the brain. The sub-component (i.e., edema, enhancing tumor and necrotic core) training uses whole tumor parameters for faster convergence and better localization.

We have trained the network separately with two types of loss functions: soft dice loss function and focal loss function.

– Soft Dice Loss: is a measure to find overlap between two regions.

$$SoftDiceLoss = 1 - \frac{2 \sum_{voxels} y_{true} y_{pred}}{\sum_{voxels} y_{pred}^2 + \sum_{voxels} y_{true}^2} \qquad (1)$$

y_{true} represents ground truth and y_{pred} represents network output probability. The dice loss function directly considers the predicted probabilities without converting into binary output. The numerator provides standard correct predictions between input and target, whereas the denominator provides individual separate correct predictions. This ratio normalizes the loss according to the target mask and allows learning even from the minimal spatial representation of the target mask.

- Focal Loss [18]: It is dependent on the network probability p_t. It balances negative and positive samples by tuning α. It also deals with easy and hard examples by focusing on parameter γ.

$$FL(p_t) = -\alpha_t(1 - p_t)^\gamma log(p_t) \qquad (2)$$

The modulating factor $(1 - p_t)^\gamma$ adjusts the rate at which easy examples are down-weighted.

3.2 Task 2: Overall Survival Prediction

OS prediction deals with predicting the number of days for which patients survive after providing appropriate treatment. We have used the following features to train Random Forest Regressor (RFR):

- **Statistical Features**: the amount of edema, amount of necrosis, amount of enhancing tumor, the extent of tumor and proportion of tumor
- **Radiomic Features**[28] **for necrosis**: Elongation, flatness, minor axis length, primary axis length, 2D diameter row, 2D diameter column, sphericity, surface area, 2D diameter slice, 3D diameter, and
- **Age** (available with BraTS dataset)

Necrosis plays a significant role in the treatment of tumors. Gross Total Resection (GTR) of necrosis is comparatively easy concerning enhancing tumor. Considering this, shape features of necrosis are extracted using a radiomics package [28]. In addition to those features, whole tumor statistical features from the segmentation results and age are considered to train RFR.

4 Implementation Details

4.1 Pre-processing

Pre-processing boosts network training and improves the performance. The Z-score normalization on individual MR sequence is applied where each sequence was subtracted by its mean from the data and divided by its standard deviation. Also, data is augmented by rotation, flip, elastic transformation, shear, shift, and zoom on MRI sequences.

4.2   Training

Input to the network is 2D slices from four modalities (T1, T2, T1c, FLAIR). The network is trained on 85% dataset as training images and 15% dataset as validation (part of training dataset). The first five and last ten slices of all the volumes do not contain any useful information. That is why the preparation of the training and validation datasets do not use such slices. The network is trained for two different loss functions separately; namely, 1) dice loss function and 2) focal loss function with $\alpha = 0.25$ and $\gamma = 0.5$. In both cases, initially, the network is trained for the whole tumor, and afterward, sub-component network training initializes with these weights. The network is trained for 30 epochs with batch size 10. Segmentation results for each sub-component (i.e., necrosis, enhancing, and edema) are combined based on the higher values of probabilities.

5   Results

Segmentation results are generated for dice loss function as well as the focal loss function. Evaluation metrics covers both the loss functions, i.e., in Table 1 and Table 2 for the training dataset and in Table 3 and Table 4 for the validation dataset. The results show that the implementation with focal loss improves the segmentation results; for the test dataset, results use that implementation only, and results are in Table 5. Table 6 shows a comparison of training dataset results of the proposed method with an average of the top ten methods according to the leader board. This comparison is irrespective of multiple submissions, as well as without the knowledge of the segmentation method used.

Table 1. DSC, Sensitivity and Hausdorff95 for BraTS 2019 training dataset with dice loss.

	DSC			Sensitivity			Hausdorff95		
	ET	WT	TC	ET	WT	TC	ET	WT	TC
Mean	0.74	0.89	0.85	0.73	0.83	0.80	5.42	6.41	5.82
StdDev	0.25	0.10	0.17	0.22	0.13	0.19	13.13	6.25	7.73
Median	0.83	0.92	0.90	0.78	0.87	0.86	2	4.90	4
25quantile	0.72	0.88	0.85	0.66	0.81	0.77	1.41	3.46	2.83
75quantile	0.89	0.94	0.93	0.87	0.91	0.90	3.16	7.31	6

Figure 2 and Fig. 3 show the segmentation of a tumorous slice with different loss functions.

RFR trains on features extracted from the 213 ground truth images. In the trained RFR, features of network segmented images are supplied, to predict OS days. If the network fails to identify/segment necrosis from the image, then the feature extractor considers the absence of the necrosis and marks all the features

Table 2. DSC, Sensitivity and Hausdorff95 for BraTS 2019 validation dataset with dice loss.

	DSC			Sensitivity			Hausdorff95		
	ET	WT	TC	ET	WT	TC	ET	WT	TC
Mean	0.60	0.70	0.63	0.59	0.63	0.61	11.69	14.33	17.10
StdDev	0.33	0.23	0.30	0.31	0.25	0.30	20.31	18.24	22.33
Median	0.75	0.80	0.75	0.70	0.73	0.73	3.61	7.81	8.25
25quantile	0.33	0.51	0.45	0.33	0.44	0.38	2	5.20	4.58
75quantile	0.85	0.88	0.88	0.84	0.83	0.87	10.18	13.45	16.28

Table 3. DSC, Sensitivity and Hausdorff95 for BraTS 2019 training dataset with focal loss.

	DSC			Sensitivity			Hausdorff95		
	ET	WT	TC	ET	WT	TC	ET	WT	TC
Mean	0.79	0.92	0.90	0.79	0.90	0.88	4.07	4.23	3.75
StdDev	0.25	0.09	0.12	0.21	0.12	0.14	11.66	6.39	7.79
Median	0.87	0.95	0.93	0.85	0.94	0.92	1.41	2.24	2
25quantile	0.81	0.91	0.89	0.77	0.89	0.88	1	1.41	1.41
75quantile	0.92	0.96	0.96	0.91	0.96	0.95	1.73	4.24	3

Table 4. DSC, Sensitivity and Hausdorff95 for BraTS 2019 validation dataset with focal loss.

	DSC			Sensitivity			Hausdorff95		
	ET	WT	TC	ET	WT	TC	ET	WT	TC
Mean	0.59	0.73	0.65	0.59	0.67	0.64	9.62	12.80	15.37
StdDev	0.34	0.24	0.30	0.33	0.25	0.31	15.83	16.86	19.90
Median	0.76	0.84	0.78	0.71	0.75	0.76	3.60	7.48	7.81
25quantile	0.29	0.65	0.51	0.33	0.54	0.41	1.93	4.58	4
75quantile	0.85	0.89	0.88	0.86	0.88	0.88	7.98	12.80	16.15

Table 5. DSC, Sensitivity and Hausdorff95 for BraTS 2019 test dataset with focal loss.

	DSC			Hausdorff95		
	ET	WT	TC	ET	WT	TC
Mean	0.64	0.72	0.66	55.11	41.30	57.12
StdDev	0.33	0.29	0.36	125.68	99.06	122.20
Median	0.78	0.84	0.86	2.24	6.40	5.51
25quantile	0.55	0.67	0.51	1.41	3.81	2.45
75quantile	0.86	0.91	0.92	10.47	13.24	18.45

Table 6. Comparison of DSC, Sensitivity and Hausdorff95 for BraTS 2019 training dataset with average of top 10 teams.

	DSC			Sensitivity			Hausdorff95		
	ET	WT	TC	ET	WT	TC	ET	WT	TC
Average of top 10 teams	0.80	0.91	0.87	0.83	0.91	0.88	3.96	7.54	7.21
Proposed	0.79	0.92	0.90	0.79	0.90	0.88	4.07	4.23	3.75

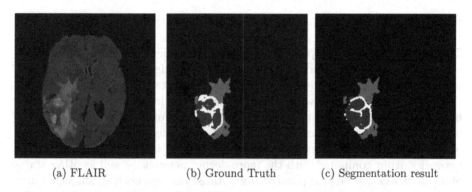

(a) FLAIR   (b) Ground Truth   (c) Segmentation result

Fig. 2. Segmentation result with dice loss

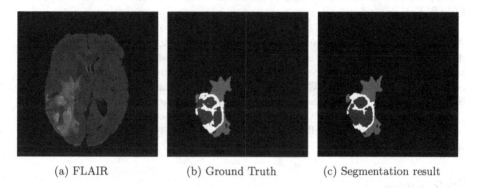

(a) FLAIR   (b) Ground Truth   (c) Segmentation result

Fig. 3. Segmentation result with focal loss

except age as zero. OS accuracy for training, validation, and test datasets of the images whose resection status is GTR are in Table 7. This method secured the first rank for the survival prediction task.

According to the study [27], gender plays a vital role in response to tumor treatment. The females respond to the post-operative treatment better compared to males, which improve their life expectancy. The inclusion of the 'gender' feature into the existing feature list can significantly improve OS accuracy.

Table 7. OS accuracy for training, validation and test dataset.

Dataset	Accuracy	MSE	MedianSE	StdSE	SpearmanR
Training	0.564	73144.54	22891.69	136542.535	0.604
Validation	0.586	105061.874	16460.89	188752.439	0.404
Test	0.579	374998.775	46483.36	1160428.922	0.434

6 Conclusion

The proposal uses three-layer deep U-net based encoder-decoder architecture for semantic segmentation. Each layer of the encoding side incorporates dense modules and decoding side convolution modules. The network achieves comparable DSC for training datasets with other methods of the leader board but generates little poor results for the validation dataset. In the future, pre-processing techniques, in addition to Z-score normalization and augmentation, better design of the decoding module, as well as post-processing, will be incorporated. The network output compared with the more deeper network as well as other state-of-art networks will be incorporated. Age, statistical, and necrosis shape features of the ground truth are provided to train RFR with five-fold cross-validation for OS prediction. Later, network segmentation for cases with GTR tests RFR for OS prediction.

Acknowledgement. The authors would like to thank NVIDIA Corporation for donating the Quadro K5200 and Quadro P5000 GPU used for this research, Dr. Krutarth Agravat (Medical Officer, Essar Ltd) for clearing our doubts related to medical concepts, Po-yu Kao, Ph.D. Candidate, Vision Research Lab, University of California, Santa Barbara for his continuous guidance during implementation difficulties, Ujjawal Baid for his help during BraTS-2019. The authors acknowledge continuous support from Professor Sanjay Chaudhary, Professor N. Padmanabhan, and Professor Manjunath Joshi for this work.

References

1. Agravat, R.R., Raval, M.S.: Prediction of overall survival of brain tumor patients. In: TENCON 2019–2019 IEEE Region 10 Conference (TENCON), pp 31–35. IEEE (2019)
2. Agravat, R.R., Raval, M.S.: Deep learning for automated brain tumor segmentation in MRI images. In: Soft Computing Based Medical Image Analysis, pp. 183–201. Elsevier (2018)
3. Akbari, H., et al.: Pattern analysis of dynamic susceptibility contrast-enhanced MR imaging demonstrates peritumoral tissue heterogeneity. Radiology **273**(2), 502–510 (2014)
4. Badrinarayanan, V., Kendall, A., Cipolla, R.: SegNet: a deep convolutional encoder-decoder architecture for image segmentation. IEEE Trans. Pattern Anal. Mach. Intell. **39**(12), 2481–2495 (2017)

5. Baid, U., et al.: Deep learning radiomics algorithm for gliomas (DRAG) model: a novel approach using 3D UNET based deep convolutional neural network for predicting survival in gliomas. In: Crimi, A., Bakas, S., Kuijf, H., Keyvan, F., Reyes, M., van Walsum, T. (eds.) BrainLes 2018. LNCS, vol. 11384, pp. 369–379. Springer, Cham (2019). https://doi.org/10.1007/978-3-030-11726-9_33

6. Bakas, S., et al.: Segmentation labels and radiomic features for the pre-operative scans of the TCGA-GBM collection. The Cancer Imaging Archive (2017)

7. Bakas, S., et al.: Segmentation labels and radiomic features for the pre-operative scans of the TCGA-LGG collection. The Cancer Imaging Archive, 286 (2017)

8. Bakas, S., et al.: Advancing the cancer genome atlas glioma MRI collections with expert segmentation labels and radiomic features. Sci. Data 4, 170117 (2017)

9. Bakas, S., et al.: Identifying the best machine learning algorithms for brain tumor segmentation, progression assessment, and overall survival prediction in the brats challenge. arXiv preprint arXiv:1811.02629 (2018)

10. Chen, L., Bentley, P., Mori, K., Misawa, K., Fujiwara, M., Rueckert, D.: DRINet for medical image segmentation. IEEE Trans. Med. Imaging 37(11), 2453–2462 (2018)

11. Dong, H., Yang, G., Liu, F., Mo, Y., Guo, Y.: Automatic brain tumor detection and segmentation using U-Net based fully convolutional networks. In: Valdés Hernández, M., González-Castro, V. (eds.) MIUA 2017. CCIS, vol. 723, pp. 506–517. Springer, Cham (2017). https://doi.org/10.1007/978-3-319-60964-5_44

12. Feng, X., Tustison, N., Meyer, C.: Brain tumor segmentation using an ensemble of 3D U-Nets and overall survival prediction using radiomic features. In: Crimi, A., Bakas, S., Kuijf, H., Keyvan, F., Reyes, M., van Walsum, T. (eds.) BrainLes 2018. LNCS, vol. 11384, pp. 279–288. Springer, Cham (2019). https://doi.org/10.1007/978-3-030-11726-9_25

13. He, K., Zhang, X., Ren, S., Sun, J.: Deep residual learning for image recognition. In: Proceedings of the IEEE conference on computer vision and pattern recognition, pp. 770–778 (2016)

14. Iandola, F., Moskewicz, M., Karayev, S., Girshick, R., Darrell, T., Keutzer, K.: DenseNet: implementing efficient convnet descriptor pyramids. arXiv preprint arXiv:1404.1869 (2014)

15. Isensee, F., Kickingereder, P., Wick, W., Bendszus, M., Maier-Hein, K.H.: No new-net. In: Crimi, A., Bakas, S., Kuijf, H., Keyvan, F., Reyes, M., van Walsum, T. (eds.) BrainLes 2018. LNCS, vol. 11384, pp. 234–244. Springer, Cham (2019). https://doi.org/10.1007/978-3-030-11726-9_21

16. Kamnitsas, K., et al.: Efficient multi-scale 3D CNN with fully connected CRF for accurate brain lesion segmentation. Med. Image Anal. 36, 61–78 (2017)

17. Kao, P.-Y., Ngo, T., Zhang, A., Chen, J.W., Manjunath, B.S.: Brain tumor segmentation and tractographic feature extraction from structural MR images for overall survival prediction. In: Crimi, A., Bakas, S., Kuijf, H., Keyvan, F., Reyes, M., van Walsum, T. (eds.) BrainLes 2018. LNCS, vol. 11384, pp. 128–141. Springer, Cham (2019). https://doi.org/10.1007/978-3-030-11726-9_12

18. Lin, T.Y., Goyal, P., Girshick, R., He, K., Dollár, P.: Focal loss for dense object detection. In: Proceedings of the IEEE International Conference on Computer Vision, pp. 2980–2988 (2017)

19. McKinley, R., Meier, R., Wiest, R.: Ensembles of densely-connected CNNs with label-uncertainty for brain tumor segmentation. In: Crimi, A., Bakas, S., Kuijf, H., Keyvan, F., Reyes, M., van Walsum, T. (eds.) BrainLes 2018. LNCS, vol. 11384, pp. 456–465. Springer, Cham (2019). https://doi.org/10.1007/978-3-030-11726-9_40

20. Menze, B.H., et al.: The multimodal brain tumor image segmentation benchmark (BRATS). IEEE Trans. Med. Imaging **34**(10), 1993–2024 (2014)
21. Milletari, F., Navab, N., Ahmadi, S.A.: V-Net: fully convolutional neural networks for volumetric medical image segmentation. In: 2016 Fourth International Conference on 3D Vision (3DV), pp. 565–571. IEEE (2016)
22. Myronenko, A.: 3D MRI brain tumor segmentation using autoencoder regularization. In: Crimi, A., Bakas, S., Kuijf, H., Keyvan, F., Reyes, M., van Walsum, T. (eds.) BrainLes 2018. LNCS, vol. 11384, pp. 311–320. Springer, Cham (2019). https://doi.org/10.1007/978-3-030-11726-9_28
23. Pan, S.J., Yang, Q.: A survey on transfer learning. IEEE Trans. Knowl. Data Eng. **22**(10), 1345–1359 (2009)
24. Puybareau, E., Tochon, G., Chazalon, J., Fabrizio, J.: Segmentation of gliomas and prediction of patient overall survival: a simple and fast procedure. In: Crimi, A., Bakas, S., Kuijf, H., Keyvan, F., Reyes, M., van Walsum, T. (eds.) BrainLes 2018. LNCS, vol. 11384, pp. 199–209. Springer, Cham (2019). https://doi.org/10.1007/978-3-030-11726-9_18
25. Ronneberger, O., Fischer, P., Brox, T.: U-Net: convolutional networks for biomedical image segmentation. In: Navab, N., Hornegger, J., Wells, W.M., Frangi, A.F. (eds.) MICCAI 2015. LNCS, vol. 9351, pp. 234–241. Springer, Cham (2015). https://doi.org/10.1007/978-3-319-24574-4_28
26. Sun, L., Zhang, S., Luo, L.: Tumor segmentation and survival prediction in glioma with deep learning. In: Crimi, A., Bakas, S., Kuijf, H., Keyvan, F., Reyes, M., van Walsum, T. (eds.) BrainLes 2018. LNCS, vol. 11384, pp. 83–93. Springer, Cham (2019). https://doi.org/10.1007/978-3-030-11726-9_8
27. Sun, T., Plutynski, A., Ward, S., Rubin, J.B.: An integrative view on sex differences in brain tumors. Cell. Mol. Life Sci. **72**(17), 3323–3342 (2015). https://doi.org/10.1007/s00018-015-1930-2
28. Van Griethuysen, J.J., et al.: Computational radiomics system to decode the radiographic phenotype. Cancer Res. **77**(21), e104–e107 (2017)
29. Weninger, L., Rippel, O., Koppers, S., Merhof, D.: Segmentation of brain tumors and patient survival prediction: methods for the BraTS 2018 challenge. In: Crimi, A., Bakas, S., Kuijf, H., Keyvan, F., Reyes, M., van Walsum, T. (eds.) BrainLes 2018. LNCS, vol. 11384, pp. 3–12. Springer, Cham (2019). https://doi.org/10.1007/978-3-030-11726-9_1

Improving Brain Tumor Segmentation with Multi-direction Fusion and Fine Class Prediction

Sun'ao Liu$^{(\boxtimes)}$ and Xiaonan Guo

University of Science and Technology of China, Hefei, China
lsa1997@mail.ustc.edu.cn

Abstract. Convolutional neural networks have been broadly used for medical image analysis. Due to its characteristics, segmentation of glioma is considered to be one of the most challenging tasks. In this paper, we propose a novel Multi-direction Fusion Network (MFNet) for brain tumor segmentation with 3D multimodal MRI data. Unlike conventional 3D networks, the feature-extracting process is decomposed and fused in the proposed network. Furthermore, we design an additional task called Fine Class Prediction to reinforce the encoder and prevent over-segmentation. The proposed methods finally obtain dice scores of 0.81796, 0.8227, 0.88459 for enhancing tumor, tumor core and whole tumor respectively on BraTS 2019 test set.

Keywords: Brain tumor segmentation · 3D convolution · Multi-direction fusion · Fine class prediction

1 Introduction

Convolutional Neural Networks (CNN) have made significant progress in several computer vision tasks. As a result, more and more CNN-based algorithms are proposed for medical image analysis, of which medical image segmentation is an important part. The Brain Tumor Segmentation Challenge (BraTS) [1–4,14] focuses on the segmentation of brain tumors, gliomas specifically. BraTS provides magnetic resonance imaging (MRI) scans with four modalities: T1, T1Gd, T2 and FLAIR. The participants are required to produce segmentation of three glioma sub-regions, which contains enhancing tumor (ET), tumor core (TC) and whole tumor (WT).

In recent years, fully convolutional networks like U-Net [20] have been widely used for brain tumor segmentation. The winner of BraTS 2017 [9] used Ensembles of Multiple Models and Architectures (EMMA) containing DeepMedic [10], FCN [13] and U-Net. The second place [21] used three cascaded networks to predict three tumor sub-regions separately. In BraTS 2018, large 3D networks like 3D U-Net [5] became popular and performed well. The winner [16] used a large amount of 3D convolution and auto-encoder as regularization. The second place [8] used 3D U-Net architecture and extra data for co-training.

© Springer Nature Switzerland AG 2020
A. Crimi and S. Bakas (Eds.): BrainLes 2019, LNCS 11992, pp. 349–358, 2020.
https://doi.org/10.1007/978-3-030-46640-4_33

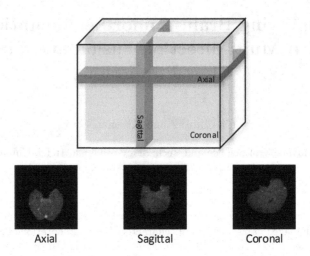

Fig. 1. Three planes of medical image.

In this paper, we combine the theories of deep learning and medical analysis and propose a new Multi-direction Fusion Network (MFNet) for glioma segmentation. Keeping the basic encoder-decoder structure, we replace 3D convolutional blocks with multi-direction fusion modules. The proposed module uses dilated pseudo-3D convolution rather than standard 3D convolution to decompose spatial information, reduce parameters and improve performance. Meanwhile, different information on three medically defined planes (i.e., sagittal, coronal and axial plane in Fig. 1) is extracted in parallel and integrated as final features. Furthermore, an additional Fine Class Prediction task is introduced for better local segmentation.

2   Methods

2.1   Network Structure

The overall structure of the Multi-direction Fusion Network (MFNet) is shown in Fig. 2. We keep the basic fully convolutional encoder-decoder architecture and skip-connection for semantic segmentation, while each level of the network contains a multi-direction fusion module rather than conventional convolutional block. The multimodal inputs are downsampled and encoded through three stride convolution layers. The decoder is divided into two parts. One part uses transpose convolution to upsample and finally recovers resolution for pixel-wise semantic segmentation. Another part only contains convolution layers for fine class prediction. Details will be shown in the following subsections.

2.2   Multi-direction Fusion Module

There are several ways to deal with the three dimensions of medical volumetric data. We notice that though experienced human experts label pixels on each

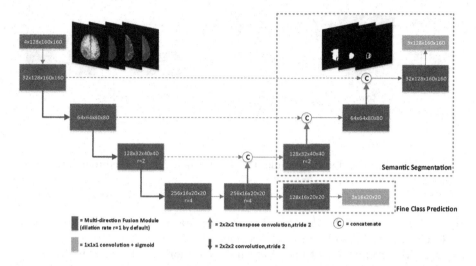

Fig. 2. Network with Multi-direction Fusion and Dilated Convolution (MFNet).

image, they can actually estimate the position of each slice in brain and thus know what kind of organs or tissues might appear. In other words, information from three dimensions is decomposed. Inspired by this fact, we introduce the pseudo-3D-A [18] module to decompose 3D convolution. It uses a $1 \times 3 \times 3$ convolution to extract 2D features first and then a $3 \times 1 \times 1$ convolution for features from the third dimension. In this way the network can decouple information as human experts do.

Note that there are three different planes in a volume, defined as axial, sagittal and coronal planes for medical data. These planes view brain from three directions and contain different spatial information. Meanwhile, the direction of convolutions in P3D can also be changed. Therefore we propose the Multi-direction Fusion (MF) module, which contains three different-direction P3D convolutions in parallel. Each of them focuses on features from one direction and information from the third dimension in addition. The complete Multi-direction Fusion module (see Fig. 3) contains residual connection [6] for better performance and quicker convergence. We also replace Batch Normalization (BN) [7] with Group Normalization (GN) [22] due to limited batch size.

2.3   Replace Pooling with Dilated Convolution

For semantic segmentation, local information is as important as global information to generate pixel-wise prediction. However, deep convolutional networks often use downsampling operations like max pooling or stride convolution to enlarge the receptive field and capture global information. Though we can use upsampling to recover resolution, it is inevitable to lose local spatial information during resolution changes. For example, the encoder of V-Net uses four stride convolution layers and can finally get a $194 \times 194 \times 194$ receptive field, while the

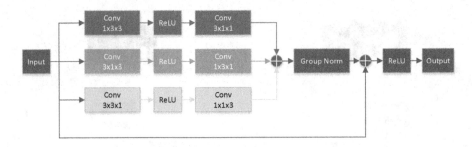

Fig. 3. Multi-direction Fusion Module. $1 \times 1 \times 1$ convolution is used to match dimensions for shortcut.

size of feature maps shrinks from $128 \times 160 \times 160$ to $8 \times 10 \times 10$. We believe that this is harmful to brain tumor segmentation, which has very small tumor sub-regions like enhancing tumor and thus needs more reliable local information.

To avoid this problem, dilated convolution [23] is applied in deep layers and the last stride convolution layer in V-Net is removed when we stack the proposed modules to form the MFNet (see Fig. 3). Note that dilated convolution is only applied to the 2D convolutions in our module, i.e., $1 \times 3 \times 3$, $3 \times 1 \times 3$ and $3 \times 3 \times 1$ convolutions. With dilated convolution, we extend the final receptive field from 78×78 to 158×158 and meanwhile keep the size of feature maps. Since our module uses branches in parallel, the receptive field is smaller than that of V-Net. But it can still cover the whole brain and is sufficient for brain tumor segmentation.

2.4   Fine Class Prediction

The segmentation of enhancing tumor is the most difficult part of brain tumor segmentation. The size of ET changes dramatically depending on the patient's condition. There exists many hard samples that can be over- or under-predicted and the dice scores fall to zero. To handle this problem, we propose the Fine Class Prediction (FCP).

Unlike traditional class prediction [19] which simply distinguishes the existence of tumor in the whole volume, we utilize the downsampling architecture of network and add a class prediction branch in the deepest layer. The size of feature maps is $16 \times 20 \times 20$, and each voxel represents an $8 \times 8 \times 8$ cube in the original volume. The network thus needs to make class predictions for each small cube rather than the whole input. We believe the encoder can learn more reliable representations to generate accurate local segmentation. Therefore, the decoder part is designed to be light and only contains one MF module for better back-propagation in the encoder part.

2.5 Hybrid Loss

For semantic segmentation task, we design a hybrid loss combining dice loss [15] and focal loss [12]:

$$L_{seg} = L_{dice} + \omega_1 L_{focal}, \tag{1}$$

$$L_{dice} = 1 - \frac{2 * \sum_i p_i g_i + \epsilon}{\sum_i p_i + \sum_i g_i + \epsilon}, \tag{2}$$

$$L_{focal} = - \sum_{x \in \Omega} (1 - p_{l(x)}(x))^\gamma \log(p_{l(x)}(x)), \tag{3}$$

where p_i, g_i are binary predicted label and ground truth respectively and ϵ is a smoothing constant. $l(\cdot)$ is the true class and $p_{l(\cdot)}(\cdot)$ is the predicted probability of the true class for each pixel.

For fine class prediction, we choose binary focal loss:

$$L_{class} = - \sum_c [I(c)(1 - p(c))^\gamma \log(p(c)) + (1 - I(c))p(c)^\gamma \log(1 - p(c))], \tag{4}$$

where $I(c)$ is ground truth and $p(c)$ is predicted probability of the existence of tumor sub-regions. The total loss is:

$$L_{total} = L_{seg} + \omega_2 L_{class} = L_{dice} + \omega_1 L_{focal} + \omega_2 L_{class}. \tag{5}$$

In our experiments, ω_1 and ω_2 are set to 1 for equal weights. The factor γ in focal loss is set to 2.

We further try to build tight connection between segmentation and class branches. Therefore, we explore the effect of a synchronous loss to constrain the predictions of two branches:

$$L_{sync} = - \sum_c [S(c) \log(p(c)) + (1 - S(c)) \log(1 - p(c))]. \tag{6}$$

$p(c)$ is defined in L_{class}, and $S(c)$ is calculated from segmentation prediction:

$$S(c) = I(\sum_{x \in C(c)} I(p(x) - \tau)), \tag{7}$$

where $I(\cdot)$ is indicative function, $\tau (= 0.5)$ is the binarization threshold. Here $x \in C(c)$ indicates that we use the summation of $8 \times 8 \times 8$ segmentation predictions in the cube which is represented by the voxel in fine class prediction branch as reference result to synchronize these two branches.

3 Experiments

3.1 Dataset

The training set of BraTS 2019 contains 259 high grade gliomas (HGG) and 76 low grade gliomas (LGG). In each sample there are four modalities and a segmentation map annotated by experts. All of them are co-registered to a common template, resampled to $1\,mm^3$ and skull-stripped, with the final size of $240 \times 240 \times 155$. 125 samples without grades and annotations are provided for validation. The results of validation set are evaluated by the online evaluation platform[1].

3.2 Implementation Details

All of our experiments are based on PyTorch [17] and trained on a single NVIDIA TITAN Xp for 200 epochs. We use Adam [11] optimizer with the initial learning rate 0.0003. The learning rate decreases 5% every 5 epochs and batch size is set to 1 due to limited memory.

During training, the grade of gliomas, i.e., HGG or LGG, is not distinguished. For data augmentation, we choose random rotation, random flip, random scaling and random noise. Input data is normalized based on non-zero voxels and randomly cropped to $160 \times 160 \times 128$. No additional data or post-processing method is used.

3.3 Results

Table 1 shows our single-model results on BraTS 2019 validation set (team name LSA). The performance of MFNet validates the effectiveness of the proposed Multi-direction Fusion Module. Furthermore, we report the results of MFNet with fine class prediction (MFNet+FCP). As shown in the table, introducing fine class prediction improves segmentation of enhancing tumor significantly.

Table 1. Results on BraTS 2019 validation set (Dice and HD95).

Model	Dice			HD95		
	WT	TC	ET	WT	TC	ET
MFNet	**0.90642**	**0.83626**	0.7596	**4.37827**	6.71071	3.72688
MFNet+FCP	0.90187	0.83333	0.77534	5.86396	6.61492	3.70443
MFNet+FCP+post	0.90352	0.83375	**0.77994**	5.5749	**6.49542**	**3.47492**
MFNet+FCP+sync+post	0.90603	0.83077	0.77959	4.97514	6.72179	3.57613

To make full use of the fine class prediction, we design a simple post-process method: set regions where the tumor sub-region doesn't exist according to fine

---

[1] https://ipp.cbica.upenn.edu/.

Table 2. Results on BraTS 2019 validation set (Spec. and Sens.).

Model	Sensitivity			Specificity		
	WT	TC	ET	WT	TC	ET
MFNet	**0.90793**	0.82875	0.76523	0.99512	**0.99728**	0.99849
MFNet+FCP	0.90544	**0.83686**	**0.78475**	0.995	0.99686	0.99834
MFNet+FCP+post	0.90415	0.82952	0.77996	0.99516	0.99708	0.99837
MFNet+FCP+sync+post	0.9053	0.82613	0.77194	**0.99521**	0.9972	**0.99852**

class prediction as background. This method (MFNet+FCP+post) mainly handles the over-segmentation problem. It further improves the results for enhancing tumor and whole tumor, but usually doesn't work well for tumor core. Therefore, we consider it proves that the main difficulty of tumor core segmentation is the under-segmentation of the necrotic and non-enhancing regions. The additional synchronous loss also improves the dice score of whole tumor. We believe this result benefits from the connection between two branches.

Results of sensitivity and specificity are shown in Table 2. The FCP module mainly improves sensitivity on enhancing tumor, which decreases the number of false negative samples. Results of specificity, however, are extremely high on all models. But we find this is because the great imbalance between the number of true negative and false positive samples. The former is much more and makes specificity abnormally high. Over-segmentation is still a tough problem for some samples.

Results on BraTS 2019 test set are shown in Table 3. We use an ensemble of three models: MFNet, MFNet+FCP+post and MFNet+FCP+sync+post.

Table 3. Results on BraTS 2019 test set.

	Dice			HD95		
	WT	TC	ET	WT	TC	ET
Mean	0.88459	0.8227	0.81796	5.57825	4.58734	2.39516
StdDev	0.12415	0.25656	0.18411	7.83032	7.15944	2.75358
Median	0.92041	0.91642	0.86307	3	2.23607	1.41421

Some segmentation samples are shown in Fig. 4. On the top row, MFNet over-predicts the region of enhancing tumor, while model with fine class prediction predicts correctly. The bottom row is a under-segmentation sample of MFNet. Adding FCP also makes network distinguish sub-regions better.

T1Gd	FLAIR	MFNet	MFNet+FCP	MFNet+FCP+post

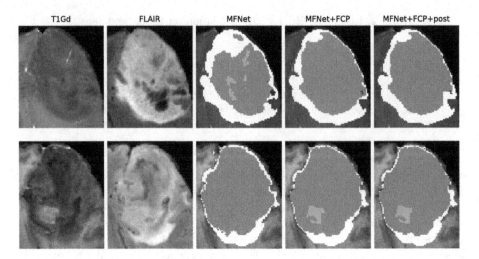

Fig. 4. Segmentation results on BraTS 2019 validation set. Blue: enhancing tumor (ET). Red: necrotic and non-enhancing tumor core (NCR/NET). Yellow: peritumoral edema (ED). TC contains ET and NCR/NET, while WT covers TC and ED. (Color figure online)

4 Conclusion

In this paper, we propose a novel network for brain tumor segmentation. The most important part, Multi-direction Fusion module, uses three branches of dilated pseudo-3D convolution in parallel to extract and assemble features from three direction of a MRI volume data simultaneously as a replacement of standard 3D convolution block. We further design the Fine Class Prediction branch which significantly improves the performance. Our MFNet shows strong competitiveness on the BraTS 2019 online validation set and obtains dice scores of 0.77994, 0.90352, 0.83375 for enhancing tumor, whole tumor and tumor core respectively. The final results on BraTS 2019 test set are 0.81796, 0.88459 and 0.8227.

Another advantage of proposed model that we don't emphasize in this paper is the reduction of parameters compared with conventional 3D networks, thanks to the multi-direction design. Since the concept of axial, sagittal and coronal planes is common for all medical data, we believe this method deserves further researches for not only brain tumor but also other medical image segmentation tasks, to achieve a better balance of effectiveness and efficiency.

References

1. Bakas, S., et al.: Segmentation labels and radiomic features for the pre-operative scans of the TCGA-GBM collection. Cancer Imaging Arch. (2017)
2. Bakas, S., et al.: Segmentation labels and radiomic features for the pre-operative scans of the TCGA-LGG collection. Cancer Imaging Arch. **286** (2017)

3. Bakas, S., et al.: Advancing the cancer genome atlas glioma MRI collections with expert segmentation labels and radiomic features. Sci. Data **4**, 170117 (2017)

4. Bakas, S., et al.: Identifying the best machine learning algorithms for brain tumor segmentation, progression assessment, and overall survival prediction in the brats challenge. arXiv preprint arXiv:1811.02629 (2018)

5. Çiçek, Ö., Abdulkadir, A., Lienkamp, S.S., Brox, T., Ronneberger, O.: 3D U-Net: learning dense volumetric segmentation from sparse annotation. In: Ourselin, S., Joskowicz, L., Sabuncu, M.R., Unal, G., Wells, W. (eds.) MICCAI 2016. LNCS, vol. 9901, pp. 424–432. Springer, Cham (2016). https://doi.org/10.1007/978-3-319-46723-8_49

6. He, K., Zhang, X., Ren, S., Sun, J.: Deep residual learning for image recognition. In: Proceedings of the IEEE Conference on Computer Vision and Pattern Recognition, pp. 770–778 (2016)

7. Ioffe, S., Szegedy, C.: Batch normalization: accelerating deep network training by reducing internal covariate shift. arXiv preprint arXiv:1502.03167 (2015)

8. Isensee, F., Kickingereder, P., Wick, W., Bendszus, M., Maier-Hein, K.H.: No new-net. In: Crimi, A., Bakas, S., Kuijf, H., Keyvan, F., Reyes, M., van Walsum, T. (eds.) BrainLes 2018. LNCS, vol. 11384, pp. 234–244. Springer, Cham (2019). https://doi.org/10.1007/978-3-030-11726-9_21

9. Kamnitsas, K., et al.: Ensembles of multiple models and architectures for robust brain tumour segmentation. In: Crimi, A., Bakas, S., Kuijf, H., Menze, B., Reyes, M. (eds.) BrainLes 2017. LNCS, vol. 10670, pp. 450–462. Springer, Cham (2018). https://doi.org/10.1007/978-3-319-75238-9_38

10. Kamnitsas, K., et al.: DeepMedic for brain tumor segmentation. In: Crimi, A., Menze, B., Maier, O., Reyes, M., Winzeck, S., Handels, H. (eds.) BrainLes 2016. LNCS, vol. 10154, pp. 138–149. Springer, Cham (2016). https://doi.org/10.1007/978-3-319-55524-9_14

11. Kingma, D.P., Ba, J.: Adam: a method for stochastic optimization. arXiv preprint arXiv:1412.6980 (2014)

12. Lin, T.Y., Goyal, P., Girshick, R., He, K., Dollár, P.: Focal loss for dense object detection. In: Proceedings of the IEEE international Conference on Computer Vision, pp. 2980–2988 (2017)

13. Long, J., Shelhamer, E., Darrell, T.: Fully convolutional networks for semantic segmentation. In: Proceedings of the IEEE Conference on Computer Vision and Pattern Recognition, pp. 3431–3440 (2015)

14. Menze, B.H., Jakab, A., Bauer, S., Kalpathy-Cramer, J., Farahani, K., Kirby, J., Burren, Y., Porz, N., Slotboom, J., Wiest, R., et al.: The multimodal brain tumor image segmentation benchmark (BRATS). IEEE Trans. Med. Imaging **34**(10), 1993–2024 (2014)

15. Milletari, F., Navab, N., Ahmadi, S.A.: V-net: fully convolutional neural networks for volumetric medical image segmentation. In: 2016 Fourth International Conference on 3D Vision (3DV), pp. 565–571. IEEE (2016)

16. Myronenko, A.: 3D MRI brain tumor segmentation using autoencoder regularization. In: Crimi, A., Bakas, S., Kuijf, H., Keyvan, F., Reyes, M., van Walsum, T. (eds.) BrainLes 2018. LNCS, vol. 11384, pp. 311–320. Springer, Cham (2019). https://doi.org/10.1007/978-3-030-11726-9_28

17. Paszke, A., et al.: Automatic differentiation in pytorch (2017)

18. Qiu, Z., Yao, T., Mei, T.: Learning spatio-temporal representation with pseudo-3D residual networks. In: Proceedings of the IEEE International Conference on Computer Vision, pp. 5533–5541 (2017)

19. Ren, X., et al.: Task decomposition and synchronization for semantic biomedical image segmentation. arXiv preprint arXiv:1905.08720 (2019)
20. Ronneberger, O., Fischer, P., Brox, T.: U-Net: convolutional networks for biomedical image segmentation. In: Navab, N., Hornegger, J., Wells, W.M., Frangi, A.F. (eds.) MICCAI 2015. LNCS, vol. 9351, pp. 234–241. Springer, Cham (2015). https://doi.org/10.1007/978-3-319-24574-4_28
21. Wang, G., Li, W., Ourselin, S., Vercauteren, T.: Automatic brain tumor segmentation using cascaded anisotropic convolutional neural networks. In: Crimi, A., Bakas, S., Kuijf, H., Menze, B., Reyes, M. (eds.) BrainLes 2017. LNCS, vol. 10670, pp. 178–190. Springer, Cham (2018). https://doi.org/10.1007/978-3-319-75238-9_16
22. Wu, Y., He, K.: Group normalization. In: Proceedings of the European Conference on Computer Vision (ECCV). pp. 3–19 (2018)
23. Yu, F., Koltun, V.: Multi-scale context aggregation by dilated convolutions. arXiv preprint arXiv:1511.07122 (2015)

An Ensemble of 2D Convolutional Neural Network for 3D Brain Tumor Segmentation

Kamlesh Pawar[1,2(✉)], Zhaolin Chen[1], N. Jon Shah[1,3],
and Gary F. Egan[1,2]

[1] Monash Biomedical Imaging, Monash University, Melbourne, Australia
kamlesh.pawar@monash.edu
[2] School of Psychological Sciences, Monash University, Melbourne, Australia
[3] Institute of Medicine, Research Centre Juelich, Juelich, Germany

Abstract. We propose an ensemble of 2D convolutional neural networks to predict the 3D brain tumor segmentation mask using the multi-contrast brain images. A pretrained Resnet50 and Nasnet-mobile architecture were used as an encoder, which was appended with a decoder network to create an encoder-decoder neural network architecture. The encoder-decoder network was trained end to end using T1, T1 contrast-enhanced, T2 and T2-Flair images to classify each pixel in the 2D input image to either no tumor, necrosis/non-enhancing tumor (NCR/NET), enhancing tumor (ET) or edema (ED). Separate Resent50 and Nasnet-mobile architectures were trained for axial, sagittal and coronal slices. Predictions from 5 inferences including Resnet at all three orientations and Nasnet-mobile at two orientations were averaged to predict the final probabilities and subsequently the tumor mask. The mean dice scores calculated from 166 were 0.8865, 0.7372 and 0.7743 for whole tumor, tumor core and enhancing tumor respectively.

Keywords: Convolutional neural network · Ensemble networks · Residual learning · Brain tumor segmentation

1 Introduction

Automated brain tumor segmentation [1–6] from magnetic resonance images is a challenging task due to variations in the acquisition protocol at different imaging sites. Different imaging parameters including field strength, acceleration factors, resolution, etc. causes variance in the MR images which makes it difficult for automated algorithms to accurately segment the brain tumor regions. Accurate segmentation of brain tumor or gliomas is an important task in grading and monitoring of the disease progression.

Brain tumor segmentation challenge (Brats) is an annual competition which provides manually segmented brain tumor dataset [7–11] to assess the performance of brain tumor segmentation algorithms. Brats challenge started with brain tumor segmentation and have been extended to the task of survival prediction and quantification of uncertainty in segmentation. In recent times, with the availability of large annotated dataset and compute power, most of the best performing algorithms in the challenge are

© Springer Nature Switzerland AG 2020
A. Crimi and S. Bakas (Eds.): BrainLes 2019, LNCS 11992, pp. 359–367, 2020.
https://doi.org/10.1007/978-3-030-46640-4_34

based on deep learning [12]. The best performing algorithms proposed different realizations of the encoder-decoder neural network architectures. Algorithms based on the variations of 3D Unet [13] have been used in the Brats challenge. Since the implementation of 3D Unet requires a large amount of memory, a patch-based approach is often used. The patch-based approach involves training the 3D Unet on a 3D patch of the image often a cube of 64 or 128 depending on the available memory and width/depth of the network. The large memory requirement of the 3D Unet restricts the width and depth of the network. Contrary a 2D Unet [14] requires comparatively less memory than the 3D counterpart at an expense of loss of information from the third spatial dimension. In 2D Unet each image slice is processed independently without considering any information from the rest of the slices within the volume.

In this work, we improve our previous method [15] and propose to use an ensemble of 2D encoder-decoder networks with each network predicting segmentation probabilities for a different orientation (axial, sagittal and coronal). The predicted probabilities from the different orientation are averaged to predict the final probability maps and the segmentation mask for the whole 3D volume. Since probability maps from each individual encoder-decoder are from a different orientation the final averaged probability may contain the 3D information. We hypotheses that false positives from one orientation will be suppressed, when its predicted probability is averaged with the probability map from another orientation.

2   Methods

2.1   Dataset

Manually segmented dataset of brain tumor MR images was provided by the organizers of BRATS challenge. The dataset consisted of two types of brain tumor images namely high-grade tumor (HGG) and low-grade tumor (LGG). Four different contrast T1, T2, T1 contrast-enhanced and T2 Flair image were provided with the manually segmented masks. The mask consisted of three different labels, the necrotic and non-enhancing tumor core (NCR/NET - label 1), the peritumoral edema (ED - label 2), GD-enhancing tumor (ET - label 4) and everything else is classified as label 0. A total of 335 subjects were present consisting of 259 HGG and 76 LGG. The whole dataset was divided into two, one for training and another for validation. Following was the composition of the training and local validation dataset.

- Training dataset: consisted of 288 HGG and 61 LGG subjects.
- Local validation dataset: consisted of 51 HGG and 61 LGG subjects.

Apart from the local validation dataset, another 125 cases were provided to validate the generalization of the model. These 125 cases were provided without the ground truth and segmentation performance was evaluated online using the CBICA Image Processing Portal (https://ipp.cbica.upenn.edu).

2.2 Network Architecture

Our approach consisted of using a 2D convolutional neural network on the individual slices of the whole 3D brain image. We performed end to end training with input being multi-contrast brain images (T1, T2, Tl-CE, T3-Flair) and output being the segmentation mask. The overview of the segmentation process is depicted in Fig. 1, we used an ensemble of 2D networks to predict the segmentation of the whole 3D volume.

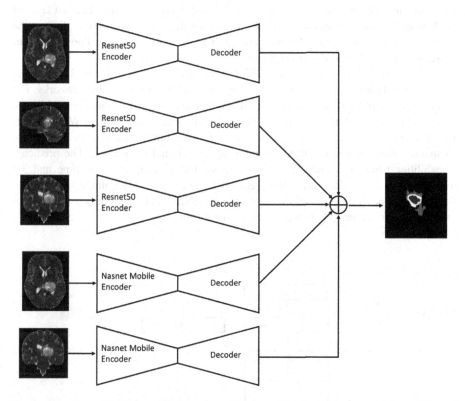

Fig. 1. Overview of the segmentation process. Six separate networks were trained consisting of three Resne50 encoder-decoder architecture for axial, sagittal and coronal orientations and three Nasnet mobile encoder-decoder architecture for axial, sagittal and coronal orientations. The probabilities from the individual predictions were averaged and the segmentation mask was generated from the averaged probabilities.

The encoder-decoder architecture similar to the Unet was the building blocks of the ensemble network. Specifically, we used five separately trained encoder-decoder networks in the ensemble. The encoder-decoder architectures consisted of three networks with Resnet50 [16] as encoder and two networks with Nasnet-mobile [17] as an encoder. A spate encoder-decoder network was trained for each orientation (axial, sagittal and coronal).

The decoder part in each of the network was the same and consisted of a series of convolution and upsampling operations. One block of a decoder is depicted in Fig. 2, which consists of a 2D upsampling operation by a factor of 2 using bilinear interpolation. The upsampling layer increases the spatial dimension of the features using bilinear interpolation and increased size features are then concatenated with the features from the encoder part having the same spatial dimension. The concatenated features are then passed through the two blocks of convolution, batch normalization, spatial dropout and rectilinear activation (ReLU). The number of features for each convolutional layer in the decoder was 256 at each scale except the last scale where it was 128. The convolution kernel size was always 3 × 3 in the decoder network. The last layer of the decoder network consisted of four features, a softmax activation was applied on the last layer, which converts the features into probability maps corresponding to the four classes (NCR/NET, ED, ET or no tumor).

As depicted in Fig. 1, at the time of inference individual slices of the 3D image at different orientations, were processed through the 2D encoder-decoder networks. For Resnet50 predictions were made for axial, sagittal and coronal orientations while for Nasnet-mobile predictions were made for axial and coronal orientations. With Nasnet-mobile encoder-decoder we did not find performance improvement with the sagittal orientation hence it was not used for inference on sagittal orientation. The predicted probabilities from individual 2D slices were stacked to form a 3D volume and 3D volumes from all orientations were averaged for the whole 3D volume. An argmax along the channel dimension on the averaged probability map classified each pixel into one of the four classes (NCR/NET, ED, ET or no tumor).

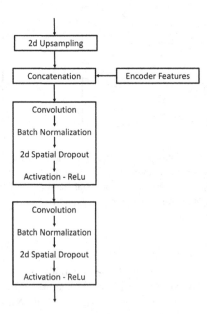

Fig. 2. One block of the decoder network, which first upsamples the input features by a factor of 2 using bilinear interpolation and concatenate the upsampled features with the same scale features form the encoder network. The concatenated features are passed through two blocks of convolution with batch normalization and ReLU activation.

2.3 Pre-processing

The spatial dimension of the input image was 240 × 240 × 155. However, all the 2D networks were trained on 256 × 256 images. The 2D input images were first zero-padded symmetrically to make the 2D input to be 256 × 256.

Since the data was sourced from multiple sites, a preprocessing is required to normalize the images. We used a simple pre-processing of normalizing the mean and standard deviation of the whole 3D volume to zero mean and unity standard deviation using Eq. 1.

$$x_{pp} = \frac{(x - \bar{x})}{std(x)} \tag{1}$$

where x_{pp} is the preprocessed 3D volume, x is the input 3D volume, $\bar{x}$ is the mean of the input volume and $std(x)$ is the standard deviation of input.

2.4 Training

The training of the network was performed on the Keras [18] deep learning library with Tensorflow backend. The adaptive stochastic gradient descent Adam optimizer was used for training the network with a batch size of 4 and initial learning rate of 0.0001. We considered the training of 2000 batches as one epoch. The learning rate was decreased with a step decay of 0.96 per epoch. All the networks were trained for 100 epochs and the network for which the average dice score was maximum on the local validation dataset was chosen as the best model and used for inference on the no ground truth validation dataset.

The loss function used to train the Resnet encoder-decoder architecture consisted of a weighted sum of categorical cross-entropy and soft dice loss. The soft dice loss is defined as:

$$dice\_loss = \frac{2 * \sum p_p * p_t}{\sqrt{\sum p_p^2 * p_t^2}} \tag{2}$$

where p_p is the predicted probability map and p_t is the true probability map.

We trained the Resnet50 encoder-decoder with the weighted sum of categorical cross-entropy loss and dice loss with a weight of 1.0 for cross-entropy and weight of 0.1 for dice loss. For the Nasnet-mobile encoder-decoder, only categorical cross-entropy was used as a loss function.

The results of segmentation were evaluated using the dice score, sensitivity (true positive rate) and specificity (true negative rate) and Hausdorff distance (95%). The evaluation on the validation dataset was calculated using online web-portal provided by the BRATS organizers.

3 Results and Discussion

The individual predictions of the trained networks were ensemble as depicted in Fig. 1 and the segmentation masks were uploaded to the online validation portal. Figure 3 shows the dice scores for the 125 validation subjects and 166 test subjects calculated by the online portal. The median dice scores were higher than the mean dice scores for both the test and validation dataset, suggesting that few difficult cases were segmented by the network with lower accuracy. The dice scores for the test dataset were higher than the validation dataset and also the sample size for the test dataset was larger 166 compared to 125 for the validation dataset. Higher dice score for test dataset (sample size 166) suggests that the algorithm works well for most of the cases but does require further improvements to accurately predict the segmentation for a few subjects that were segmented with lower accuracy.

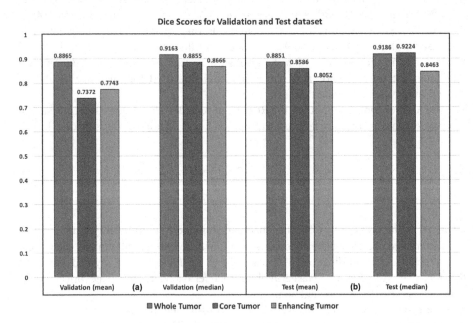

Fig. 3. Bar plot showing the mean dice score using the proposed method; **(a):** bar plot for validation dataset from 125 subjects **(b):** bar plot for test dataset from 166 subjects

Table 1 shows the mean and median dice score, sensitivity and Hausdorff distance on the 125 validation subjects and Table 2 shows the mean and median dice score and Hausdorff distance for 166 test subjects.

Table 1. Quantitative score on validation dataset of 125 subjects calculated using the online IPP portal

	Whole tumor	Core tumor	Enhancing tumor
Dice score (mean)	0.8865	0.7372	0.7743
Dice score (median)	0.9163	0.8855	0.8666
Hausdorff distance	4.2348	5.7720	8.1844
Sensitivity (mean)	0.8602	0.6996	0.7786
Sensitivity (median)	0.9054	0.8510	0.8510

Table 2. Quantitative score on test dataset of 166 subjects calculated using the online IPP portal

	Whole tumor	Core tumor	Enhancing tumor
Dice score (mean)	0.8851	0.8586	0.8052
Dice score (median)	0.9186	0.9224	0.8463
Hausdorff distance	6.4109	4.6700	3.4515

The medians of the dice scores are higher than that of the mean for all three categories of the tumor. The higher median indicates that there are a few harder cases where the algorithm fails to perform well. Usually, the performance on the enhancing tumor class is more challenging compared to the other two classes. However, it is worthwhile to note that the performance of our algorithm on the enhancing tumor class is comparatively higher compared to the tumor core class. This suggests that there is a scope of improvement for the core tumor class, which may require further training and fine-tuning of the network. Two representative segmentations are shown in Fig. 4, one for a highly accurate prediction with average dice score of 0.9508 (Fig. 3 (a)) and another for less accurate segmentation with average dice score of 0.6301 (Fig. 3 (b)).

This work aimed to reduce the memory footprints of the 3D networks by transforming it into multiple 2D networks. This transformation constitutes a trade-off between the computational complexity and memory requirements, the proposed approach reduces the memory footprints but increases the computational complexity. For instance, a 3D network of similar architecture would require 3 times more computation compared to 2D counterpart. However, an ensemble of five 2D networks makes the computational complexity to be 5 times than the single 2D network.

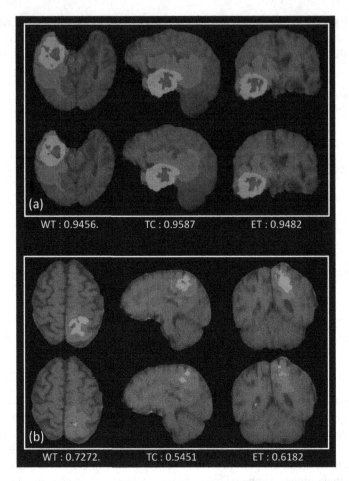

Fig. 4. Segmentation results for two representative images with red color: NCR/NET, orange color: edema and white color: ET. The bottom of the figure shows the dice score for whole tumor (WT), tumor core (TC) and enhancing tumor (ET) respectively. **(a):** first row shows ground truth segmentation; second row shows predicted segmentation results for one of the highly accurate prediction; **(b):** first row shows ground truth segmentation; second row shows predicted segmentation results for one of the less accurate prediction. (Color figure online)

4   Conclusion

In this work, we have presented an approach to predict the brain tumor segmentation for the whole 3D volume using an ensemble to the 2D CNN. Specifically, we used resnet50 and nasnet-mobile architectures for the predictions. The results are promising with the average dice score of 0.8851, 0.8586 and 0.8052 for whole tumor, core tumor and enhancing tumor respectively.

References

1. Sharma, N., Aggarwal, L.M.: Automated medical image segmentation techniques. J. Med. Phys./Assoc. Med. Phys. India **35**, 3 (2010)
2. Pham, D.L., Xu, C., Prince, J.L.: Current methods in medical image segmentation. Annu. Rev. Biomed. Eng. **2**, 315–337 (2000)
3. Corso, J.J., Sharon, E., Dube, S., El-Saden, S., Sinha, U., Yuille, A.: Efficient multilevel brain tumor segmentation with integrated bayesian model classification. IEEE Trans. Med. Imaging **27**, 629–640 (2008)
4. Angelini, E.D., Clatz, O., Mandonnet, E., Konukoglu, E., Capelle, L., Duffau, H.: Glioma dynamics and computational models: a review of segmentation, registration, and in silico growth algorithms and their clinical applications. Curr. Med. Imaging Rev. **3**, 262–276 (2007)
5. Gupta, M.P., Shringirishi, M.M.: Implementation of brain tumor segmentation in brain mr images using k-means clustering and fuzzy c-means algorithm. Int. J. Comput. Technol. **5**, 54–59 (2013)
6. Liu, J., Li, M., Wang, J., Wu, F., Liu, T., Pan, Y.: A survey of MRI-based brain tumor segmentation methods. Tsinghua Sci. Technol. **19**, 578–595 (2014)
7. Bakas, S., et al.: Advancing the cancer genome atlas glioma MRI collections with expert segmentation labels and radiomic features. Sci. Data **4**, 170117 (2017)
8. Bakas, S., et al.: Identifying the best machine learning algorithms for brain tumor segmentation, progression assessment, and overall survival prediction in the BRATS challenge. arXiv preprint arXiv:1811.02629 (2018)
9. Bakas, S., et al.: Segmentation labels and radiomic features for the pre-operative scans of the TCGA-LGG collection. The Cancer Imaging Archive 286 (2017)
10. Menze, B.H., et al.: The multimodal brain tumor image segmentation benchmark (BRATS). IEEE Trans. Med. Imaging **34**, 1993–2024 (2015)
11. Bakas, S., et al.: Segmentation labels and radiomic features for the pre-operative scans of the TCGA-GBM collection. The Cancer Imaging Archive (2017)
12. LeCun, Y.A., Bengio, Y., Hinton, G.E.: Deep learning. Nature **521**, 436–444 (2015)
13. Kamnitsas, K., Ledig, C., Newcombe, V.F., Simpson, J.P., Kane, A.D., Menon, D.K., Rueckert, D., Glocker, B.: Efficient multi-scale 3D CNN with fully connected CRF for accurate brain lesion segmentation. Med. Image Anal. **36**, 61–78 (2017)
14. Ronneberger, O., Fischer, P., Brox, T.: U-Net: convolutional networks for biomedical image segmentation. In: Navab, N., Hornegger, J., Wells, William M., Frangi, Alejandro F. (eds.) MICCAI 2015. LNCS, vol. 9351, pp. 234–241. Springer, Cham (2015). https://doi.org/10.1007/978-3-319-24574-4_28
15. Pawar, K., Chen, Z., Shah, N.J., Egan, G.: Residual encoder and convolutional decoder neural network for glioma segmentation. In: Crimi, A., Bakas, S., Kuijf, H., Menze, B., Reyes, M. (eds.) BrainLes 2017. LNCS, vol. 10670, pp. 263–273. Springer, Cham (2018). https://doi.org/10.1007/978-3-319-75238-9_23
16. He, K.M., Zhang, X.Y., Ren, S.Q., Sun, J.: Deep residual learning for image recognition. In: Proceedings of the CVPR IEEE, pp. 770–778 (2016)
17. Zoph, B., Vasudevan, V., Shlens, J., Le, Q.V.: Learning transferable architectures for scalable image recognition. In: Proceedings of the IEEE Conference on Computer Vision and Pattern Recognition, pp. 8697–8710 (Year)
18. Chollet, F.: Keras (2015)

An Integrative Analysis of Image Segmentation and Survival of Brain Tumour Patients

Sebastian Starke[1,2]([✉]), Carlchristian Eckert[2], Alex Zwanenburg[2,3,4,5,6], Stefanie Speidel[3,4,5,6], Steffen Löck[2], and Stefan Leger[2,3,4,5,6]

[1] Computational Science Group, Helmholtz-Zentrum Dresden - Rossendorf, Dresden, Germany
s.starke@hzdr.de
[2] OncoRay - National Center for Radiation Research in Oncology, Faculty of Medicine and University Hospital Carl Gustav Carus, Helmholtz-Zentrum Dresden - Rossendorf, Technische Universität Dresden, Dresden, Germany
[3] National Center for Tumor Diseases (NCT), Partner Site Dresden, Dresden, Germany
[4] German Cancer Research Center (DKFZ), Heidelberg, Germany
[5] Faculty of Medicine and University Hospital Carl Gustav Carus, Technische Universität Dresden, Dresden, Germany
[6] Helmholtz Association/Helmholtz-Zentrum Dresden - Rossendorf (HZDR), Dresden, Germany

Abstract. Our contribution to the BraTS 2019 challenge consisted of a deep learning based approach for segmentation of brain tumours from MR images using cross validation ensembles of 2D-UNet models. Furthermore, different approaches for the prediction of patient survival time using clinical as well as imaging features were investigated. A simple linear regression model using patient age and tumour volumes outperformed more elaborate approaches like convolutional neural networks or radiomics-based analysis with an accuracy of 0.55 on the validation cohort and 0.51 on the test cohort.

Keywords: UNet · Segmentation · Radiomics · Linear regression · Deep-learning · Ensemble · Survival analysis

1 Methods

1.1 Data Set

The data set used in the BraTS 2019 challenge [1] includes multi-institutional, clinically acquired, pre-operative multi-modal magnetic resonance (MR) images of glioblastoma and lower grade glioma patients. The training and validation

---

S. Starke and C. Eckert—Shared first authorship.
S. Löck and S. Leger—Shared senior authorship.

© Springer Nature Switzerland AG 2020
A. Crimi and S. Bakas (Eds.): BrainLes 2019, LNCS 11992, pp. 368–378, 2020.
https://doi.org/10.1007/978-3-030-46640-4_35

cohorts consisted of 332 and 125 patients, respectively. Imaging for each patient comprises a native (T1), a post-contrast T1-weighted (T1c), a T2-weighted (T2), and a T2 fluid attenuated inversion recovery (FLAIR) MR sequence [2,3]. All imaging data sets were manually segmented to define the contrast-enhancing tumour, the peritumoral edema and the necrotic and non-enhancing tumour core regions [4,5]. In addition, patient age and resection status were collected as clinical variables to predict overall survival of the patients.

All data were part of the BraTS 2019 challenge. No external data were used.

1.2   Image Pre-processing

The N4ITK bias correction algorithm was applied to the T1, T1c, T2 and FLAIR images to reduce MR intensity non-uniformity [6]. Subsequently, a non-local means denoising approach was used to reduce image noise [7].

Prior to segmentation, each MR sequence in the imaging data set was normalised using z-score normalisation. For this purpose, the surrounding air was excluded. Furthermore, to reduce the computation time, all images were cropped to the smallest possible volume that included the brain tissue of all patients. Subsequently, to be divisible by 2^5, the images were padded with zeros, leading to a final input shape of $160 \times 192 \times 160$ for each of the four sequences.

1.3   Brain Tumour Segmentation

Network Structure and Training. A 2D-UNet architecture [8] was applied for the segmentation task and co-registered axial slices of each sequence were used as input (Fig. 1). Each of the five encoding blocks contained two consecutive 2D convolutional layers, each followed by a batch normalisation and a Leaky-ReLU activation function ($\alpha = 10^{-2}$), as well as by a max-pooling layer. Furthermore, symmetric decoding blocks were used with skip-connections from corresponding encoding blocks, resulting in a model with 70 750 468 trainable parameters. For the output, we use the one-hot encoded segmentation and optimise on the individual layers with a sigmoid activation.

The Sørensen-Dice coefficient loss was combined with an L2-norm regularisation term of 10^{-7} and used as loss function to reduce model overfitting. Network parameters were optimised using the Adam optimiser. The optimiser had an initial learning rate of 10^{-3}. The learning rate was reduced through multiplication by 0.25 if loss did not improve for three consecutive epochs.

We used a 5-fold cross validation scheme to train the model for a maximum number of 100 epochs. In case the loss did not improve for six consecutive epochs, model training was stopped early. Due to GPU memory limitations, models were trained using a batch size of 64 image slices. Data augmentation was performed during training to improve generisability of the model. The image was randomly rotated, scaled and cropped using the framework by F. Isensee et al. (www.github.com/MIC-DKFZ/batchgenerators). Training resulted in five models which were subsequently combined into an ensemble for inference on the validation data. To that end, the sigmoid predictions produced by each model were averaged and the label with the highest probability was used as the final label for each voxel.

Above, we used 2D axial slices for training. This potentially limits the information contained in the sagittal and coronal directions. We therefore created a second ensemble of 15 models using the same approach as described above, but using sagittal and coronal slices in addition to axial slices. During inference, each model is fed slices of the appropriate direction.

All segmentation models were trained using the Keras framework with the Tensorflow backend.

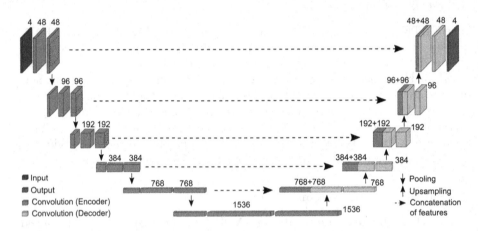

Fig. 1. Structure of the 2D-UNet used for prediction of segmentation masks from four MR sequences. A block represents a combination of convolutional layer, batch normalisation and Leaky-ReLU activation function. Numbers of filters are provided above each block.

1.4   Survival Prediction

We investigated different approaches for the prediction of overall survival after surgery. We created:

- two linear regression models based on clinical information and tumour volumes.
- three different deep-learning based models.
- a linear regression model based on clinical information, tumour volumes and survival time predicted by a deep-learning model.
- a conventional radiomic model based on handcrafted image features computed from the segmentations of each sequence.

The classification performance of the models into the three survivor classes ('short' for patients with survival <10 months, 'mid' for patients with survival between 10 and 15 months, 'long' for patients surviving more than 15 months) were measured by the accuracy (ACC). In addition, mean square error (MSE), median square error (medianSE), standard deviation of the square error (stdSE) and the Spearman correlation coefficient (SpearmanR) were computed. The model that achieved the highest ACC in the validation cohort will be applied to the final test cohort.

Linear Regression Model Based on Clinical Features. To predict the post-surgery survival time, two linear regression models were constructed.

Patient age and survival time were moderately correlated in the training cohort (Pearson's $\rho = -0.486$). Correlations between survival time and tumour volumes were weak, from -0.06 for enhancing tumour to 0.11 for edema, despite higher tumour volume often being related to worse outcome prognosis. Therefore we evaluated two regression models. The first model included patient age as the single predictor (LM-Age). The second model also included patient age and, in order to capture basic imaging information, additionally the three volume parameters for the enhancing, edema and necrotic areas as predictors (LM-Age+Vol). Volume was measured by summation of the segmentation masks. No feature normalisation was performed.

Deep-Learning Based Multi-task Regression Model (DL-MTR). A segmentation UNet (as presented in Fig. 1) has likely learned relevant features of the tumours during training. Transfer-learning may exploit these features in a multi-task learning problem to simultaneously predict log-transformed survival time and survival class labels (short-, mid- and long-term survivors).

For this purpose, the pre-trained encoder branch of the segmentation UNet of the first of the five cross validation folds was extended by a global-max-pooling layer followed by two dense layers (size: 64 and 16, respectively) with Leaky-ReLU activation ($\alpha = 0.2$). Finally, the model outputs were computed based on two dense layers (sizes 1 and 3 for regression and classification, respectively) with a linear output for the log-transformed survival time and a softmax function for the class label prediction. As with segmentation, a batch size of 64 image slices was used together with the Adam optimiser. Mean-absolute error and categorical cross-entropy were used as loss functions for the regression and classification tasks, respectively. Imaging from 32 axial slices per patient, consisting of the tumour slice with the largest total tumour region and the 15 slices below and 16 slices above this slice, were used as input. Survival time for each patient was predicted by averaging the predicted log-survival times for each slice after exponential transform. The predicted class label was not considered for the final patient prediction since classification loss served only as an auxiliary optimisation function during training.

During the training phase, the encoder branch of the pre-trained UNet was fine-tuned based on a 10-fold cross validation that was repeated three times. Fine-tuning was performed in two steps. First, to adjust the network to the specific prediction task, only the extended part of the UNet was trained for 10 epochs with a learning rate of 1e−3. Subsequently, all layers of the model were fine-tuned using a lower learning rate of 1e−5 for 20 epochs. Finally, for each patient, an ensemble prediction was constructed using the median predicted survival time of all 30 models (DL-MTR-Ensemble-30).

In addition, the survival time predicted by the model ensemble was included as a fifth predictor in the linear regression model for age and volume (LM-Age+Vol+DL-MTR-Ensemble-30).

A further ensemble, containing seven models, was created by using only those models that achieved an accuracy score of at least 0.4 on the validation folds (DL-MTR-Ensemble-7).

Deep-Learning Based Cox Proportional Hazards Model and Elastic-Net (DL-Cox-ElasticNet). The second deep-learning model was also based on transfer learning. In this case, the encoder branch of the developed UNet was re-trained to predict hazard values instead of the survival times directly. Therefore, an optimisation of the partial log-likelihood of the Cox-proportional hazard model was performed using the 32 selected slices for each patient in the training cohort. All 30 models were used this time to compute an ensemble hazard score for each slice of a patient by averaging over model outputs. To obtain survival times, an elastic net regression model was trained on the ensemble average of predicted hazards for each patient.

Radiomic-Based Prediction Model. Radiomic imaging features were computed for each of the segmentations in every MR sequence. Nine additional images were created for each sequence by spatial filtering of the base image to emphasise image characteristics such as edges and blobs. Eight of nine additional images were created using a stationary coiflet-1 wavelet high-/low-pass filter along each of the three spatial dimensions [9,10]. The remaining image was created by averaging five images that were individually filtered using a Laplacian of Gaussian filter with a kernel width of 1.0, 2.0, 3.0, 5.0, and 6.0 mm, respectively [11]. Subsequently, 18 statistical, 38 histogram-based and 95 texture features were extracted from each tumour segmentation within each MR image sequence (base image and 9 transformed images). 28 morphological features were computed within the T1 MR base image only. The extracted texture features were based on the following texture matrices: grey-level co-occurrence matrix (GLCM) [12], grey-level run length matrix (GLRLM) [13,14], neighbourhood grey tone difference matrix [15], grey-level size zone matrix [16], grey-level distance zone matrix [17] and neighbourhood grey level dependence matrix [18]. For this purpose the tumour segmentations were discretised using 32 quantisation levels before calculation of texture matrices and the intensity histogram [19,20]. GLCM and GLRLM-based features were first calculated for each of the thirteen different spatial directions and subsequently averaged. All features were calculated using a volumetric approach, and not slice by slice. Thus, 24927 features were computed in total per patient. Image pre-processing and feature extraction were performed according to the guidelines of the image biomarker standardisation initiative [21].

Four feature selection methods and three learning algorithms found as most reliable in a previous systematic evaluation were used for predictive modelling [22,23]. The following feature selection methods were applied: Spearman correlation (Spearman) [24], mutual information maximisation (MIM), mutual information feature selection (MIFS) [25] and minimum redundancy maximum relevance (MRMR) [26]. For model building we used random forest (RF) [27], boosting trees gaussian linear model (BT-Gaussian) and boosted generalised gaussian linear model (BGLM-Gaussian) [28].

The subset of patients ($n = 101$) with gross total resection status was used as the training cohort for creating radiomic models using our in-house modelling framework [23]. Models were developed and selected in two stages: (I) an internal cross-validation for model selection and (II) the final model development.

(I) The model selection step was based on a 5-times repeated 5-fold cross-validation with four major processing steps: feature pre-processing, feature selection, hyper-parameter optimisation, model building. All features in the training folds were standardised by z-normalisation. Subsequently, hierarchical clustering was performed to identify mutually redundant features [29]. Features with an average intra-cluster Spearman correlation of >0.90 were replaced by a new meta-feature. This meta-feature was created by sample-wise averaging of the values of all features in the same cluster. During the feature selection step the most relevant features are identified in the training fold. Prior to feature selection the training fold was bootstrapped 50 times using the .632 bootstrap method with replacement to be able to select stable, relevant features. Subsequently, the top 20 best features were aggregated according their rank and their frequency of occurrence over all bootstrap samples. Model hyper-parameters, such as the number of input features, were subsequently optimised using a grid-search in pre-defined parameter space. Finally, the learning algorithms were trained on 50 bootstrap samples for the training fold using the top rank features as well as the optimised hyper-parameter set.

(II) The average internal validation accuracy for the predicted survival class was used to select the final combination of feature selection method and model from 12 potential combinations. To determine the predicted survival class, we first averaged the predicted survival time for the validation fold over the 50 models [30]. Patients in the validation fold were then assigned to one of three survival classes, and the accuracy computed. This was repeated until an accuracy score was obtained for each cross-validation step.

The combination of feature selection method and model with the highest average accuracy was then used to develop a final model. For this purpose only a subset of features was used, namely those that were included at least once into a model for the selected combination during the cross-validation steps in stage I. Then, the steps described below were repeated using the full training cohort of 101 patients, with the exception that feature selection and model development were conducted using 1000 bootstraps.

S. Starke et al.

2 Results and Discussion

2.1 Brain Tumour Segmentation

Table 1 presents the performance of the ensemble models for segmentation on the validation cohort, as computed by the competition website. The larger ensemble comprised of axial, coronal and sagittal directions showed slightly worse results for the mean Dice score, but a considerable improvement for the Hausdorff distance (95th percentile).

This is probably due to the fact that the larger ensemble is prone to eliminate small structures, which leads to less false positive regions outside the actual tumour, but at the same time creates false negatives for small foreground objects (see Fig. 2 and Table 2).

Table 3 provides segmentation performance of the axial ensemble on the test cohort. Compared to the validation cohort, we observed similar dice coefficients for all tumour sub-regions but increased Hausdorff distances for enhancing tumour and tumour core.

Table 1. Performances of the different ensemble models for the segmentation on the validation cohort.

Model	Dice			Hausdorff95		
	enh.	whole	core	enh.	whole	core
Axial slices						
Mean	71.04	85.11	71.03	6.57	8.85	10.28
StdDev	28.94	12.94	29.62	10.07	13.16	13.18
Median	82.01	89.48	84.65	2.24	4.47	5.48
Axial+coronal+sagittal slices						
Mean	70.51	85.17	70.68	4.73	7.29	7.71
StdDev	30.95	15.19	31.39	7.83	10.78	10.81
Median	83.89	90.05	85.48	2.24	4.12	4.36

Table 2. Quantitative results of the comparison of two different ensemble models applied to patient BraTS19_CBICA_AMU_1 of the validation cohort.

Model	Dice			Hausdorff95		
	enh.	whole	core	enh.	whole	core
Axial slices	83.48	92.76	91.45	2.83	5.39	21.40
Axial+coronal+sagittal slices	84.27	93.13	94.83	2.00	4.90	2.83

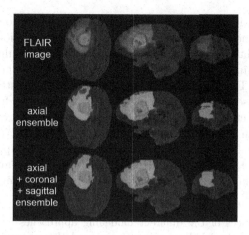

Fig. 2. Qualitative results of the comparison of two different ensemble models. The ensemble model based on axial slices segmented areas in the frontal part of the brain falsely as necrotic tumour, while the ensemble model using all three directions correctly segments the area as edema. The image refers to patient BraTS19_CBICA_AMU_1 of the validation cohort.

Table 3. Performance of the axial ensemble on the test cohort.

Model	Dice			Hausdorff95		
	enh.	whole	core	enh.	whole	core
Axial slices						
Mean	78.33	87.19	82.03	20.36	7.50	22.07
StdDev	22.10	11.29	24.58	79.86	29.32	79.61
Median	84.20	90.45	91.13	2.00	3.61	2.83

2.2 Survival Prediction

Table 4 shows the performance measures for all considered modelling approaches. The linear regression model using patient age and the tumour volumes (LM-Age+Vol) achieved the highest performance (ACC = 0.55) on the external validation cohort.

The different deep-learning based models obtained higher ACC values on the training cohort compared to the LM-Age+Vol model. This is likely due to overfitting, as the accuracy of the predictions on the validation data cohort was considerably lower (ACC $\leq$ 0.41). Interestingly, direct use of deep-neural networks as predictors worked slightly worse than using their outputs in an additional regression model (here: ElasticNet). Compared to training, the DL-MTR models also showed a large MSE for the validation cohort, which may be indicative of instability of the trained models. Also, incorporation of the ensemble prediction into the linear regression model as an additional predictor affected validation performance negatively.

For the conventional radiomics approach, the combination of MIM feature selection and RF model algorithm achieved the best validation accuracy in the internal cross-validation (ACC = 0.38). The accuracy of this model combination on the entire training cohort and the external validation cohort were 0.58 and 0.38, respectively. The developed radiomics signature consisted of a first-order statistic feature based on the enhancing tumour region and a first-order statistic meta-feature extracted from the edema region. Both features were computed and extracted from wavelet-transformed images. Based on these findings, the LM-Age+Vol model was selected as a final model to be applied to the final test cohort (Table 4). This model achieved an ACC of 0.51 and SpearmanR of 0.416, similar to the training and validation cohort. However, we note a large

Table 4. Performances of the different modelling approaches for the prediction of the survival time on the whole training (train) and the external validation (valid) cohort. Best validation results for each metric are marked in bold. Furthermore, the performance of the final model (LM-Age+Vol) on the test cohort is shown.

Model	Performance metrics				
	ACC	MSE	medianSE	stdSE	SpearmanR
LM-Age					
Train	0.48	88822.0	21135.3	181864.0	0.479
Valid	0.45	90109.0	36453.5	**123542.2**	0.265
LM-Age+Vol					
Train	0.48	85456.6	28490.4	164291.8	0.465
Valid	**0.55**	**88463.9**	31817.6	142106.5	0.277
Test	0.51	424151.8	50412.7	1256964.5	0.416
DL-MTR-Ensemble-7					
Train	0.70	58090.0	6509.8	159519.0	0.826
Valid	0.38	2.47e+25	43215.0	1.3e+26	0.256
DL-MTR-Ensemble-30					
Train	0.70	58160.5	6192.7	164442.8	0.849
Valid	0.34	1.13e+24	42090.1	5.97e+24	0.248
LM-Age+Vol+DL-MTR-Ensemble-30					
Train	0.75	37638.2	5830.8	92566.3	0.852
Valid	0.38	447543.8	67653.0	829805.3	0.298
DL-Cox-ElasticNet					
Train	0.69	50341.1	11756.0	108240.3	0.782
Valid	0.41	126882.9	81595.1	145380.7	**0.354**
RF-MIM					
Train	0.58	79782.3	16522.9	199069.9	0.692
Valid	0.38	130423.3	**23532.1**	280367.6	0.116

increase for MSE on the test cohort, which may be due to the imperfect volume estimations coming from our segmentation model.

Acknowledgment. The author SLe is supported by the Federal Ministry of Education and Research (BMBF-13GW0211D).

References

1. Menze, B.H., et al.: The multimodal brain tumor image segmentation benchmark (BRATS). IEEE Trans. Med. Imaging **34**(10), 1993–2024 (2014)
2. Bakas, S., et al.: Advancing the cancer genome atlas glioma MRI collections with expert segmentation labels and radiomic features. Sci. Data **4**, 170117 (2017)
3. Bakas, S., Reyes, M., Jakab, A., Bauer, S., Rempfler, M., Crimi, et al.: Identifying the best machine learning algorithms for brain tumor segmentation, progression assessment, and overall survival prediction in the BRATS challenge. arXiv e-prints arXiv:1811.02629, November 2018
4. Bakas, S., et al.: Segmentation labels and radiomic features for the pre-operative scans of the TCGA-LGG collection (2017). https://doi.org/10.7937/K9/TCIA. 2017.GJQ7R0EF
5. Bakas, S., et al.: Segmentation labels and radiomic features for the pre-operative scans of the TCGA-GBM collection (2017). https://doi.org/10.7937/K9/TCIA. 2017.KLXWJJ1Q
6. Tustison, N.J., et al.: N4ITK: improved N3 bias correction. IEEE Trans. Med. Imaging **29**(6), 1310 (2010)
7. Manjón, J.V., Coupé, P., Martí-Bonmatí, L., Collins, D.L., Robles, M.: Adaptive non-local means denoising of MR images with spatially varying noise levels. J. Magn. Reson. Imaging **31**(1), 192–203 (2010)
8. Ronneberger, O., Fischer, P., Brox, T.: U-Net: convolutional networks for biomedical image segmentation. In: Navab, N., Hornegger, J., Wells, W.M., Frangi, A.F. (eds.) MICCAI 2015. LNCS, vol. 9351, pp. 234–241. Springer, Cham (2015). https://doi.org/10.1007/978-3-319-24574-4_28
9. Aerts, H.J., et al.: Decoding tumour phenotype by noninvasive imaging using a quantitative radiomics approach. Nat. Commun. **5**, 4006 (2014)
10. Vallières, M., Freeman, C.R., Skamene, S.R., El Naqa, I.: A radiomics model from joint FDG-PET and MRI texture features for the prediction of lung metastases in soft-tissue sarcomas of the extremities. Phys. Med. Biol. **60**(14), 5471 (2015)
11. Coroller, T.P., et al.: CT-based radiomic signature predicts distant metastasis in lung adenocarcinoma. Radiother. Oncol. **114**(3), 345–350 (2015)
12. Haralick, R.M., Shanmugam, K., Dinstein, I., et al.: Textural features for image classification. IEEE Trans. Syst. Man Cybern. **3**(6), 610–621 (1973)
13. Galloway, M.M.: Texture analysis using grey level run lengths. NASA STI/Recon Technical report N 75 (1974)
14. Dasarathy, B.V., Holder, E.B.: Image characterizations based on joint gray level-run length distributions. Pattern Recognit. Lett. **12**(8), 497–502 (1991)
15. Amadasun, M., King, R.: Textural features corresponding to textural properties. IEEE Trans. Syst. Man Cybern. **19**(5), 1264–1274 (1989)
16. Thibault, G., et al.: Texture indexes and gray level size zone matrix application to cell nuclei classification. Pattern Recognit. Inf. Process., 140–145 (2009)

17. Thibault, G., Angulo, J., Meyer, F.: Advanced statistical matrices for texture characterization: application to cell classification. IEEE Trans. Biomed. Eng. **61**(3), 630–637 (2014)

18. Sun, C., Wee, W.G.: Neighboring gray level dependence matrix for texture classification. Comput. Vis. Graph. Image Process. **23**(3), 341–352 (1983)

19. Gómez, W., Pereira, W.C.A., Infantosi, A.F.C.: Analysis of co-occurrence texture statistics as a function of gray-level quantization for classifying breast ultrasound. IEEE Trans. Med. Imaging **31**(10), 1889–1899 (2012)

20. Clausi, D.A.: An analysis of co-occurrence texture statistics as a function of grey level quantization. Can. J. Remote Sens. **28**(1), 45–62 (2002)

21. Zwanenburg, A., Leger, S., Vallières, M., Löck, S.: Image biomarker standardisation initiative. arXiv preprint arXiv:1612.07003 (2016)

22. Parmar, C., Grossmann, P., Bussink, J., Lambin, P., Aerts, H.J.: Machine learning methods for quantitative radiomic biomarkers. Sci. Rep. **5**, 13087 (2015)

23. Leger, S., et al.: A comparative study of machine learning methods for time-to-event survival data for radiomics risk modelling. Sci. Rep. **7**(1), 13206 (2017)

24. Spearman, C.: Correlation calculated from faulty data. Br. J. Psychol. 1904-1920 **3**(3), 271–295 (1910)

25. Battiti, R.: Using mutual information for selecting features in supervised neural net learning. IEEE Trans. Neural Netw. **5**(4), 537–550 (1994)

26. Peng, H., Long, F., Ding, C.: Feature selection based on mutual information criteria of max-dependency, max-relevance, and min-redundancy. IEEE Trans. Pattern Anal. Mach. Intell. **27**(8), 1226–1238 (2005)

27. Ishwaran, H., Kogalur, U.B., Blackstone, E.H., Lauer, M.S.: Random survival forests. Ann. Appl. Stat. **2**(3), 841–860 (2008)

28. Hothorn, T., Bühlmann, P., Kneib, T., Schmid, M., Hofner, B.: Model-based boosting 2.0. J. Mach. Learn. Res. **11**(Aug), 2109–2113 (2010)

29. Parmar, C., et al.: Radiomic feature clusters and prognostic signatures specific for lung and head & neck cancer. Sci. Rep. **5**, 11044 (2015)

30. Dieterich, T.G.: Ensemble methods in machine learning. In: Kittler, J., Roli, F. (eds.) MCS 2000. LNCS, vol. 1857, pp. 1–15. Springer, Heidelberg (2000). https://doi.org/10.1007/3-540-45014-9_1

Triplanar Ensemble of 3D-to-2D CNNs with Label-Uncertainty for Brain Tumor Segmentation

Richard McKinley[✉], Michael Rebsamen, Raphael Meier, and Roland Wiest

Support Centre for Advanced Neuroimaging, University Institute of Diagnostic and Interventional Neuroradiology, Inselspital, Bern University Hospital, Bern, Switzerland
richard.mckinley@insel.ch

Abstract. We introduce a modification of our previous 3D-to-2D fully convolutional architecture, DeepSCAN, replacing batch normalization with instance normalization, and adding a lightweight local attention mechanism. These networks are trained using a previously described loss function which mo els label noise and uncertainty. We present results on the validation dataset of the Multimodal Brain Tumor Segmentation Challenge 2019.

1 Introduction

Brain Tumor segmentation has become a benchmark problem in medical image segmentation, due to the existence since 2012 of a long-running competition, BRATS [4,19], together with a large curated dataset [1–3] of annotated images. Both fully-automated and semi-automatic approaches to brain-tumor segmentation are accepted to the challenge, with supervised learning approaches dominating the fully-automated part of the challenge. A good survey of approaches which dominated BRATS up to 2013 can be found here [5]. More recently, CNN-based approaches have dominated the fully-automated approaches to the problem [7,11,22].

We present a network architecture for semantic segmentation, incorporating dense blocks, [8] and dilated convolutions [26]: it is based on our 3rd-place entry to the BRaTS 2018 segmentation challenge [16], which has also been applied to the segmentation of MS lesions [18], and brain anatomy [17]. In this paper we describe a variant of the architecture with the addition of a lightweight attention mechanism, and in which batch normalization is replaced by instance normalization. The bulk of the network is composed of 2D convolutions, but to provide 3D context the initial layers of the network are 3D convolutions. As a result, the network has an anisotropic receptive field, which is intended to take advantage of the symmetries of the human brain. The network is trained on sagittal, coronal, and axial views of the brain: the final result of the network is derived by ensembling over those three views at test time. The network is trained using a

© Springer Nature Switzerland AG 2020
A. Crimi and S. Bakas (Eds.): BrainLes 2019, LNCS 11992, pp. 379–387, 2020.
https://doi.org/10.1007/978-3-030-46640-4_36

heteroscedastic loss function, previously described in [16,17]. We report preliminary results on the validation portion of the BRATS 2018 dataset.

2 Heteroscedastic Classification Models

Heteroscedastic classification networks (those which predict the variance of their preactivation outputs) were introduced in [12]. In that paper heteroscedastic classification was shown to improve street-scene segmentation: this increase in performance can be attributed to learned loss attenuation, in which gradients from examples with possibly erroneous labels are attenuated. Aside from our previous work on brain segmentation [17], use of heteroscedastic classification networks in medical image segmentation has been so far limited, with authors focusing on uncertainty derived from dropout [9,10,15,23] or test-time augmentation [25]. Predictive variance was explored, together with other measures of uncertainty, as a method of filtering MS lesion segmentations by Nair et al. [20]. A multi-task network using a homoscedastic (per task rather than per example) measure of task uncertainty was presented by Bentaib et al. [6].

The term "heteroscedastic regression" refers to regression models which do not assume constant variance of residuals, but rather predict both the mean and the variance of the predicted quantity [21]. This notion of uncertainty is distinct from Bayesian Uncertainty (for example as approximated using Bayesian Dropout techniques); the two were contrasted and presented in a combined form by Kendall and Gal [12]. In regression modelling, heteroscedasticity can be modelled by assuming that model outputs form a Gaussian distribution, and predicting the mean and variance of that distribution. The form of the loss function is such that gradients from training examples with high predicted variance contribute less than those with low predicted variance: heteroscedastic regression therefore performs a sort of learned loss annealing.

For classification problems, the correct notion of heteroscedasticity is not immediately clear. To make the subject easier to grasp, we restrict to the topic of binary classification. Given a network output $p \in [0,1]$, and a ground truth value $x \in \{0,1\}$, we want to model some quantity that represents the uncertainty of p as a model output predicting x. This is somewhat confusing at first sight, as the value of p already contains, in some sense, a measure of certainty. Some insight may be gained, however, from viewing a classifier as a latent variable model in *logit space*. In the predictive variance method of Kendall and Gal [12], the logit output logit(p) of the network (i.e. the output of the network before application of a sigmoid nonlinearity) is assumed to follow a Gaussian distribution with nonconstant variance. For each example the network outputs a probability, p, and a log variance, $\log(\sigma^2)$. Unlike for heteroscedastic regression the loss function cannot be computed as an analytic function of p, σ^2 and x, the true label. Instead the loss is approximated by averaging a loss not involving σ^2 over T Monte-Carlo samples, in each of which the logit is perturbed by a normally distributed noise term with mean zero and s.d. σ.

In this paper we make use of an alternative heteroscedastic network, in which the uncertainty in the logit is directly modelled by the probability q of disagreement with the ground truth, or 'label-flip'. This has the advantage that the loss function is analytic in p, q and x. For each example of a binary classification problem, the network outputs a $p \in (0,1)$ denoting class membership, and an output $q \in (0, 0.5)$ predicting the probability that the ground truth and classifier disagree. If $x \in \{0,1\}$ is the label of the voxel, according to the ground truth, the *label-flip loss* at to that voxel is

$$\text{BCE}(p, (1-x)*q + x*(1-q)) + \text{BCE}(q, z) \tag{1}$$

where z is the indicator function for disagreement between the classifier (thresholded at the $p = 0.5$ level) and the ground truth, and BCE is binary cross-entropy. Unlike for predictive variance this loss can be formulated in closed form and is differentiable, and so can be used directly in backpropogation. Label-flip loss can be seen as a form of loss attenuation: the loss at voxels with low label noise is dominated by the first loss term, and the loss at voxels with substantial label noise is dominated by the second term. It can also be seen as learned *label smoothing*: a hard labels are replaced by a soft labels according to the uncertainty in the data [24].

Both of the above notions of heteroscedastic network may be easily formulated in either a binary or multi-class setting. In this paper, we focus on the binary case for two reasons: ease of presentation, and because we may easily translate between the two notions in the binary setting. Specifically, if a model predicts a logit $\text{logit}(p)$, and a variance σ^2, the associated flip-probability is $\Phi(|\text{logit}(p)|/\sigma)$: the probability that a draw from a normal distribution with mean $\text{logit}(p)$ and variance σ^2 has a different sign to $\text{logit}(p)$. Conversely, a label-flip probability q can be viewed as a variance. The expression of this variance is simplified if we assume that the logit of p follows not a Gaussian but a logistic distribution (as is the standard assumption in classical statistical learning theory). The probability that a sample drawn from a logistic distribution with mean $\text{logit}(p)$ and scale s has a different sign to $\text{logit}(p)$ (and therefore produces a label flip) is

$$\frac{1}{1 + e^{-|\text{logit}(p)|/s}} = \text{logit}^{-1}(-\text{logit}(p)/s) \tag{2}$$

From the moments of the logistic distribution, we can derive that associated variance is $(\text{logit}(p)^2 \pi^2)/3\text{logit}(q)^2$.

3 Application to Brain Tumor Segmentation

3.1 Data Preparation and Homogenization

The raw values of MRI sequences cannot be compared across scanners and sequences, and therefore a homogenization is necessary across the training examples. In addition, learning in CNNs proceeds best when the inputs are standardized (i.e. mean zero, and unit variance). To this end, the nonzero intensities in the training, validation and testing sets were standardized, this being done across individual volumes rather than across the training set. This achieves both standardization and homogenization.

3.2   The DeepSCAN Architecture with Attention

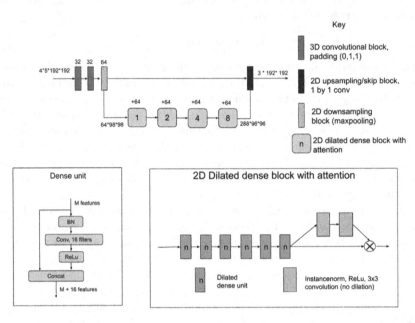

Fig. 1. The DeepSCAN classifier, as applied in this paper to Brain Tumor Segmentation

Our model architecture (shown in Fig. 1) was implemented in Pytorch: it consists of an initial phase of 3D convolutions to reduce a non-isotropic 3D patch to 2D, followed by a swallow encoder/decoder network using densely connected dilated convolutions in the bottleneck. This architecture is very similar to that used in our BraTS 2018 submission: principal differences are that we use Instance normalization rather than Batch normalization, and that we add a simple local attention mechanism between dilated dense blocks.

We use multi-task rather than multi-class classification: each tumor region (Whole tumor, tumor core, enhancing tumor) is treated as a separate binary classification problem. To combat data imbalance between foreground and background problem, we use focal loss [13] with parameter $\gamma = 2$. The loss functions for our heteroscedastic networks use focal loss as a base loss function as a base loss function. Inputs to the network (5*196*196 patches) were sampled randomly from either axial, sagittal or coronal direction. We perform simple data augmentation: reflection about the (approximate) midline, rotation around a random principal axis through a random angle, and global shifting/rescaling of voxel intensities. The network was trained with RMSprop, using a batch size of 2 and a cosine annealing learning rate schedule with restarts [14], where the learning rate was varied from 10^{-4} to 10^{-7} every 20000 learning steps:(we refer to this rather loosely as an epoch).

Models were trained using five-fold cross-validation. We trained our model for 20 epochs without the uncertainty loss and then for a further 80 epochs with a sum of focal loss and label-flip uncertainty loss. We also trained the same model without uncertainty loss (i.e. just focal loss) for 100 epochs, and an example of our previous model (with batch normalization, and without attention) for 100 epochs. Smoothed curves showing the cross-validated dice coefficients for the three experiments, stratified by LGG and HGG, are shown in Fig. 2.

Final segmentations were derived by ensembling axial, sagittal and coronal views by averaging logits.

4   Variance and Model Ensembling

After cross-validation, we have five trained classifiers. We applied these classifiers to the validation data in the saggital, axial and coronal directions, and averaged their logit outputs. Given the 30 different model outputs, we want to ensemble them in a way which respects uncertainty, meaning that

1. Confident predictions count more than unconfident predictions
2. We can provide a confidence/flip-probability for the ensembled output.

Our chosen ensembling method is *weighted ensembling of logits*: if m_i denotes the ith logit, then the ensembled logit is

$$m_{ens} = \Sigma_{i=1}^{30} w_i m_i$$

where

$$\Sigma_{i=1}^{30} w_i = 1$$

The *variance* of this ensemble is then given by

$$\sigma_{ens}^2 = \Sigma_{i=1}^{30} w_i \sigma_i^2$$

(if we make the simplifying assumption that all 30 models make independent errors).

To convert the variance of the ensemble to a flip-probability, we assume that the weighted ensemble follows a Gaussian distribution, with mean m_{ens} and variance σ_{ens}^2: the flip probability q_{ens} is then $\Phi(|m_{ens}|/\sigma_{ens})$.

We compared weighting with a fixed weight of 1/30, weighting by variance, and weighting by a weight defined as follows:

$$w_i = |n_i/m_i|$$

where m_i is the logit output, and n_i the logit of the flip probability. The latter gave best performance over our cross-validation.

Subsequently, we filtered the results of our classifier according to principles derived from tumor biology and our uncertainty measure. Specifically

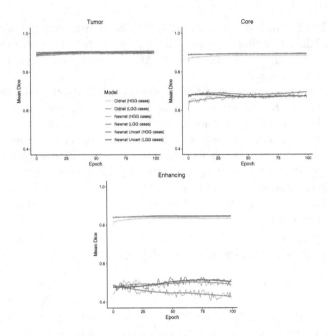

Fig. 2. Learning curves of the three model variants trained: Oldnet (Our entry to BRATS 2018), Newnet (The network with instance normalization and attention) and Newnet Uncert (The network with instance normalization, attention, and label flip uncertainty from epoch 20). Results are split between LGG and HGG cases.

- Connected components of whole tumor or tumor core or that were primarily uncertain (median label flip probability > 0.3) were deleted.
- Connected components of enhancing tumor that were primarily very uncertain (median label flip probability > 0.4) were deleted.
- Small components (< 10 voxels) of any tissue class were deleted
- If no tumor core was detected (primarily a problem in LGGs), the whole tumor was assumed to consist of tumor core.
- If a tumor core was detected, then tumor components containing no tumor core were deleted.

These heuristics were hand-crafted, by examining mistake made on the worst-performing cases in the validation set.

5   Results

Results of our classifier, as applied to the official BraTS validation data from the 2019 challenge, as generated by the official BraTS validation tool, before and after filtering for uncertainty and plausible tumor morphology, are shown in Fig. 1.

Table 1. Results on the BRATS 2019 validation set using the online validation tool. Raw output denotes the ensembled results of the five classifiers derived from cross-validation.

	Dice ET	Dice WT	Dice TC	Hausdorff95 ET	Hausdorff95 WT	Hausdorff95 TC
Raw output	0.75	0.91	0.81	4.35	5.97	6.21
Filtered Output	0.77	0.91	0.83	3.92	4.52	6.27

This method was used as our entry into the BraTS 2019 challenge: 166 additional cases were supplied, to which we applied our method. Results, as provided by the challenge organizers, are shown in Table 2. The method was ranked third in the 2019 challenge.

An uncertainty component to the challenge was also run in 2019: this method was mentioned by the organizers as being among the four best-performing methods in this part of the challenge, but no further details were released.

Table 2. Results on the BRATS 2019 testing set.

Label	Dice ET	Dice WT	Dice TC	Hausdorff95 ET	Hausdorff95 WT	Hausdorff95 TC
Mean	0.81	0.89	0.83	2.74	4.85	3.99
StdDev	0.20	0.11	0.25	5.21	6.90	6.01
Median	0.86	0.92	0.92	1.41	2.91	2.24
25quantile	0.78	0.88	0.87	1.00	1.73	1.41
75quantile	0.91	0.95	0.96	2.24	4.56	3.74

Acknowledgements. This work was supported by the Swiss Personalized Health Network (SPHN, project number 2018DRI10). Calculations were performed on UBELIX (http://www.id.unibe.ch/hpc), the HPC cluster at the University of Bern.

References

1. Bakas, S., et al.: Advancing the cancer genome atlas glioma MRI collections with expert segmentation labels and radiomic features. Nat. Sci. Data **4**, 170117 (2017)
2. Bakas, S., et al.: Segmentation labels and radiomic features for the pre-operative scans of the TCGA-GBM collection. Cancer Imaging Arch. (2017)
3. Bakas, S., et al.: Segmentation labels and radiomic features for the pre-operative scans of the TCGA-LGG collection. Cancer Imaging Arch. (2017)
4. Bakas, S., et al.: Identifying the best machine learning algorithms for brain tumor segmentation, progression assessment, and overall survival prediction in the brats challenge (2018). ArXiv abs/1811.02629
5. Bauer, S., Wiest, R., Nolte, L.L., Reyes, M.: A survey of mri-based medical image analysis for brain tumor studies. Phys. Med. Biol. **58**(13), R97–129 (2013)

6. BenTaieb, A., Hamarneh, G.: Uncertainty driven multi-loss fully convolutional networks for histopathology. In: Cardoso, M.J., et al. (eds.) LABELS/CVII/STENT -2017. LNCS, vol. 10552, pp. 155–163. Springer, Cham (2017). https://doi.org/10.1007/978-3-319-67534-3_17

7. Havaei, M., et al.: Brain tumor segmentation with deep neural networks. Med. Image Anal. **35**, 18–31 (2017)

8. Huang, G., Liu, Z., van der Maaten, L., Weinberger, K.Q.: Densely connected convolutional networks. In: Proceedings of the IEEE Conference on Computer Vision and Pattern Recognition (2017)

9. Jungo, A., et al.: On the Effect of Inter-observer Variability for a Reliable Estimation of Uncertainty of Medical Image Segmentation. In: Frangi, A.F., Schnabel, J.A., Davatzikos, C., Alberola-López, C., Fichtinger, G. (eds.) MICCAI 2018. LNCS, vol. 11070, pp. 682–690. Springer, Cham (2018). https://doi.org/10.1007/978-3-030-00928-1_77

10. Jungo, A., Meier, R., Ermis, E., Herrmann, E., Reyes, M.: Uncertainty-driven sanity check: application to postoperative brain tumor cavity segmentation. In: Proceedings of MIDL (2018)

11. Kamnitsas, K., et al.: Efficient multi-scale 3d cnn with fully connected crf for accurate brain lesion segmentation. Med. Image Anal. **36**, 61–78 (2017)

12. Kendall, A., Gal, Y.: What uncertainties do we need in bayesian deep learning for computer vision? In: NIPS (2017)

13. Lin, T.Y., Goyal, P., Girshick, R.B., He, K., Dollár, P.: Focal loss for dense object detection. In: 2017 IEEE International Conference on Computer Vision (ICCV), pp. 2999–3007 (2017)

14. Loshchilov, I., Hutter, F.: SGDR: stochastic gradient descent with warm restarts. In: ICLR (2017)

15. McClure, P., et al.: Knowing what you know in brain segmentation using deep neural networks (2018). http://arxiv.org/abs/1812.01719

16. McKinley, R., Meier, R., Wiest, R.: Ensembles of densely-connected CNNs with label-uncertainty for brain tumor segmentation. In: Crimi, A., Bakas, S., Kuijf, H., Keyvan, F., Reyes, M., van Walsum, T. (eds.) BrainLes 2018. LNCS, vol. 11384, pp. 456–465. Springer, Cham (2019). https://doi.org/10.1007/978-3-030-11726-9_40

17. McKinley, R., Rebsamen, M., Meier, R., Reyes, M., Rummel, C., Wiest, R.: Few-shot brain segmentation from weakly labeled data with deep heteroscedastic multi-task networks (2019). arXiv e-print, https://arxiv.org/abs/1904.02436

18. McKinley, R., et al.: Simultaneous lesion and neuroanatomy segmentation in multiple sclerosis using deep neural networks (2019). ArXiv abs/1901.07419

19. Menze, B.H., Menze, B.H., et al.: The multimodal brain tumor image segmentation benchmark (BRATS). IEEE Trans. Med. Imaging **34**(10), 1993–2024 (2015)

20. Nair, T., et al.: Exploring uncertainty measures in deep networks for multiple sclerosis lesion detection and segmentation. In: Proceedings of MICCAI (2018)

21. Nix, D.A., Weigend, A.S.: Estimating the mean and variance of the target probability distribution. In: IEEE ICNN 1994, vol. 1 (1994)

22. Pereira, S., Pinto, A., Alves, V., Silva, C.A.: Brain tumor segmentation using convolutional neural networks in MRI images. IEEE Trans. Med. Imaging **35**(5), 1240–1251 (2016)

23. Roy, A.G., Conjeti, S., Navab, N., Wachinger, C.: Inherent brain segmentation quality control from fully convnet monte carlo sampling. In: Frangi, A.F., Schnabel, J.A., Davatzikos, C., Alberola-López, C., Fichtinger, G. (eds.) MICCAI 2018. LNCS, vol. 11070, pp. 664–672. Springer, Cham (2018). https://doi.org/10.1007/978-3-030-00928-1_75

24. Szegedy, C., et al.: Rethinking the Inception architecture for computer vision. In: CVPR (2016)
25. Wang, G., et al.: Aleatoric uncertainty estimation with test-time augmentation for medical image segmentation with convolutional neural networks. Neurocomputing **338**, 34–45 (2019)
26. Yu, F., Koltun, V.: Multi-scale context aggregation by dilated convolutions. In: Proceedings of International Conference on Learning Representations (ICLR 2017) (2017)

Memory Efficient Brain Tumor Segmentation Using an Autoencoder-Regularized U-Net

Markus Frey$^{(\boxtimes)}$ and Matthias Nau

Kavli Institute for Systems Neuroscience, NTNU, Trondheim, Norway
markus.frey@ntnu.no

Abstract. Early diagnosis and accurate segmentation of brain tumors are imperative for successful treatment. Unfortunately, manual segmentation is time consuming, costly and despite extensive human expertise often inaccurate. Here, we present an MRI-based tumor segmentation framework using an autoencoder-regularized 3D-convolutional neural network. We trained the model on manually segmented structural T1, T1ce, T2, and Flair MRI images of 335 patients with tumors of variable severity, size and location. We then tested the model using independent data of 125 patients and successfully segmented brain tumors into three subregions: the tumor core (TC), the enhancing tumor (ET) and the whole tumor (WT). We also explored several data augmentations and preprocessing steps to improve segmentation performance. Importantly, our model was implemented on a single NVIDIA GTX1060 graphics unit and hence optimizes tumor segmentation for widely affordable hardware. In sum, we present a memory-efficient and affordable solution to tumor segmentation to support the accurate diagnostics of oncological brain pathologies.

Keywords: Brain tumor · U-Net · Autoencoder

1 Introduction

An estimated 17,760 people will die from a primary brain tumor this year in the US alone [1]. Another 23,820 will be diagnosed with having one [1]. The earlier and the more accurate this diagnosis will be, the better the patients chances are for successful treatment. In cases of doubt, patients typically undergo a brain scan either using computed tomography (CT) or magnetic resonance imaging (MRI). Both techniques acquire a 3D image of the brain, which then serves as the basis for medical examination. To understand the severity of the disease and to plan potential treatments, a critical challenge is identifying the tumor, but also to estimate its spread and growth by segmenting the affected tissue. This process still often relies on careful manual assessment by trained medical staff.

In recent years, a growing number of algorithmic solutions were proposed to aid and accelerate this process [2–4]. Most of these automatic segmentation

© Springer Nature Switzerland AG 2020
A. Crimi and S. Bakas (Eds.): BrainLes 2019, LNCS 11992, pp. 388–396, 2020.
https://doi.org/10.1007/978-3-030-46640-4_37

methods build on convolutional neural networks (CNNs) trained on manual brain segmentations of a large cohort of patients. Given enough training data, they learn to generalize across patients and allow to identify the tumor and its spread in new, previously unseen brains. However, there are at least two challenges associated with CNN's. First, they tend to overfit to the training data, making it necessary to either have large data sets to begin with, or to use a variety of data augmentations to make them generalize more robustly. Second, many current CNN implementations require powerful computational resources to be used within a reasonable time.

To solve such challenges and to promote the further development of automatic segmentation methods, the brain tumor segmentation challenge (BraTS) [2, 5–8] provides large data sets of manually segmented brains for users to test new implementations. Here, we used this data to implement a convolutional autoencoder regularized U-net for brain tumor segmentation inspired by last year's BraTS challenge winning contribution [3]. As model input, we used structural (T1) images, T1-weighted contrast-enhanced (T1ce) images, T2-weighted images and fluid-attenuated inversion recovery (Flair) MRI images of 335 patients with tumors of variable severity, size and location. As training labels, we used the corresponding manual segmentations.

The model training comprised three parts (Fig. 1). First, in an encoding stage, the model learned a low-dimensional representation of the input. Second, the variational autoencoder (VAE) stage reconstructed the input image from this low-dimensional latent space. Third, a U-Net part created the actual segmentations [9]. In this model architecture, the VAE part is supposed to act as a strong regularizer on all model weights [3] and therefore to prevent overfitting on the training data. The resulting segmentation images were compared to the manual segmentation labels. This process was repeated until the optimal model weights were found. These optimal parameters were then tested on new validation data of 125 patients, localizing and segmenting each brain tumor into three tissue categories: whole tumor, enhancing tumor and tumor core.

Importantly, all of these steps were conducted on a single NVIDIA GTX1060 graphics unit while using data exclusively from the BraTS challenge 2019. In addition, we explored various model parameters, data augmentations and preprocessing steps to improve model performance. Therefore, we address above-introduced challenges by presenting a memory-efficient and widely-affordable solution to brain tumor segmentation in line with the aims of the GreenAI initiative [10].

2   Methods

2.1   Model Architecture

As mentioned above, our model is inspired by earlier work [3], but was adapted as described in the following (see also Fig. 1). We adjusted the model architecture to incorporate a patch-wise segmentation of the input image, as the full input with a resolution of $240 \times 240 \times 155$ voxel as used in the original model is too big to fit most commercially available graphics cards (GPU). This is true even with a batch size of 1. We therefore used 3D blocks of size $80 \times 80 \times 80$ and adjusted the number of filters to make full use of the GPU memory available, leading to 32 filters in the first layer with a ratio of 2 between subsequent layers. We also replaced the rectified linear unit (ReLU) activation functions with LeakyReLU [11] as we observed an improvement in performance in a simplified version of our model.

Notably, we tested various other factors, which did not lead to an improvement in model performance, but are nevertheless included here as null-report. These included a) changing the downsampling in the convolutional layers from strides to average or max pooling, b) adjusting the ratio in the number of filters between layers (including testing non-integer steps), c) varying the number of units in the bottleneck layer, d) increasing the number of down-sampling and subsequent up-sampling steps and e) replacing the original group norm by batch norm. Due to our self-imposed computational constraints, we could not systematically test all these adjustments and possible interactions using the full model. Instead, we tested these parameters in a simplified model with only 8 filters at the input stage.

The overall model architecture follows a similar structure as a U-Net [9], with an additional variational autoencoder module [12] to regularize the segmentation of the tumor masks. As loss functions we used the mean-squared error between the reconstructed and real input image and the Kullback-Leibler loss to ensure a normal distribution in the latent space. The weights for both losses were down-weighted by a factor of 0.1. The (soft Dice) segmentation loss was averaged across all voxels belonging to the whole tumor (WT), enhancing tumor (ET) and tumor core (TC).

2.2   Optimization

For training the model we used an adjusted version of the Dice loss in [3]:

$$L_{Dice} = 1 - \left(2 * \frac{\sum (y_{true} * y_{pred}) + s}{(\sum y_{true}^2 + \sum y_{pred}^2) + s} \right) \tag{1}$$

with y_{true} being the real 3D mask and y_{pred} being the corresponding 3D prediction. This version of the Dice loss ensured that the loss estimate lies within the interval $[0,1]$. The smoothness term s ensured that the model is allowed to predict 0 tumor voxels without incurring a high loss in its overall estimate. In line with [13] we decided to use $s = 100$.

The autoencoder part of our model consisted of two loss terms. As a reconstruction loss we used the mean-squared error between the reconstructed and the real input image:

$$L_{L2} = ||y_{true} - y_{pred}||_2^2 \tag{2}$$

In addition, we used a Kullback-Leibler loss to ensure a normal distribution in our bottleneck layer, with N being the number of voxels in the input:

$$L_{KL} = \frac{1}{N} \sum \mu^2 + \sigma^2 - \log \sigma^2 - 1 \tag{3}$$

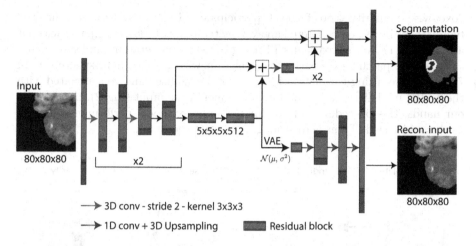

Fig. 1. Model architecture of our memory-efficient autoencoder-regularized U-Net. As input to the model we used patches of size 80 × 80 × 80 and stacked the MRI modalities in the channel dimension (n = 4). We used 3D convolutions with a kernel size of 3 × 3 × 3 throughout. We used residual blocks [14], using 3D convolutions with LeakyReLU activations, interspersed with Group Normalization [15]. For upsampling to the original image size, we used 3D bilinear upsampling and 3D convolutions with a kernel size of 1 for both the autoencoder and the segmentation part.

with μ and σ^2 the mean and variance of the estimated distribution. In line with [3] we weighted the autoencoder losses by 0.1, resulting in an overall loss according to:

$$L = 0.1 * L_{L2} + 0.1 * L_{KL} + 0.33 * L_{Dice_{wt}} + 0.33 * L_{Dice_{tc}} + 0.33 * L_{Dice_{et}} \quad (4)$$

We tested different weighting for the tumor subregions, but did not observe a clear change in model performance using the smaller test model. We therefore used the average of the three regions.

For training the model, we used the Adam optimizer [16], starting out with a learning rate of 1e-4 and decreasing it according to

$$\alpha = \alpha_0 * (1 - \frac{e}{N_e})^{0.9} \quad (5)$$

with e the epoch and N_e the number of total epochs (n = 50). We evaluated 2101 samples in each epoch, stopping early when the validation loss did not decrease further for 2 subsequent epochs.

2.3 Data Augmentation

In line with [3] we used a random scaling between 0.9 and 1.1 on each image patch, and applied random axis mirror flip for all 3 axes with a probability of 0.5. We experimented with additional augmentations. In particular, we computed a voxel-wise similarity score for each participant's T1 comparing it to a healthy template brain. We co-registered an average template of 305 participants without any tumors [17] to each patient's T1 using translations, rotations and scaling and calculated a patch-wise Pearson's correlation with a searchlight-sphere size of 7 mm. The resulting correlation images were normalized and concatenated with the 4 MRI modalities as an additional channel in the input (Fig. 2). However, in our hands, this procedure did not further improve model performance. Future work could test different across-image similarity measures.

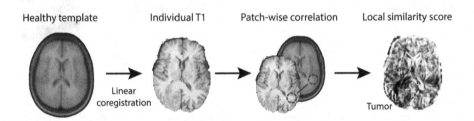

Fig. 2. Local similarity score. To aid model performance, we computed a local similarity score image, which served as additional input. We linearly co-registered a healthy template brain [17] to each participant's T1, and computed a patch-wise Pearson correlation between the two. Patch-size was 7 mm. The correlation between healthy and pathological brain drops in tumor regions.

A shortcoming of using discretized patch-input is the lack of information about the anatomical symmetry of the tested brain images. Strong asymmetry in MRI images can indicate the presence of a tumor, which is in most cases limited to one hemisphere. The other hemisphere should hence rather approximate how the healthy brain once looked like. Therefore, for each patch we also provided the mirrored patch from the opposite hemisphere as an additional input. This mirroring of image patches was only done on the sagittal plane of the MRI images. Looking forward, we believe this approach has the potential to benefit the model performance if measures other than Pearson's correlation are explored and mirror symmetry is factored in.

We used test time augmentation to make our segmentation results more robust, for this we mirrored the input on all three axes and flipped the corresponding prediction to match the original mask orientation. This gave us 16 model estimates $(2 * 2 * 2 * 2)$, which we averaged and thresholded to obtain our segmentation masks. We decided to directly optimize for tumor regions instead of the intra-tumoral regions as this resulted in better estimates during training of how well our model will perform on the BraTS 2019 competition benchmark. We optimized the values at which we thresholded our masks on the training data and used 0.55, 0.5 and 0.4 for whole tumor, tumor core and enhanced tumor respectively.

3   Results

Here, we present an autoencoder regularized U-net for brain tumor segmentation. The model was trained on the BraTS 2019 training data, which consisted of 335 patients separated into high-grade glioma and low-grade glioma cases. Initial shape of the input data was $240 \times 240 \times 155$, with multi-label segmentation masks of the same size, indicating NCR & NET (label 1), edema (label 2), and enhancing tumor (label 4). We created an average tumor template from all segmentation masks to locate the most prominent tumor regions via visual inspection. Based on that, we created our initial slice resulting in image dimensions of $160 \times 190 \times 140$. We then used a sliding window approach to create patches of size $80 \times 80 \times 80$, feeding these patches through the model while using a sampling procedure that increased the likelihood of sampling patches with a corresponding tumor (positive samples).

Table 1. Validation results. ET: enhancing tumor, WT: whole tumor, TC: tumor core.

	Dice score			Sensitivity			Specificity			Hausdorff95		
	ET	WT	TC	ET	WT	TC	ET	WT	TC	ET	WT	TC
Mean	0.787	0.896	0.800	0.782	0.907	0.787	0.998	0.994	0.997	6.005	8.171	8.241
Std	0.252	0.085	0.215	0.271	0.088	0.246	0.003	0.008	0.004	14.55	15.37	11.53
Median	0.870	0.922	0.896	0.884	0.934	0.895	0.999	0.997	0.999	2.000	3.162	3.605

We used an ensemble of two separately trained models to segment the MRI images of validation and testing set into different tumor tissue types. This allowed us to test the model on previously unseen data (Table 1, Figs. 3, 4, team-name: CYHSM). The mean Dice scores of our model on the validation dataset ($n = 125$) are 0.787 for enhanced tumor, 0.896 for whole tumor and 0.800 for tumor core.

In Fig. 3, we show the model segmentations for one exemplary patient from the validation set overlayed on the patient's T1 scan. For this patient we obtained Dice scores of 0.923 for whole tumor, 0.944 for tumor core and 0.869 for enhancing tumor from the online evaluation platform: https://ipp.cbica.upenn.edu. The distribution of Dice scores across patients can be seen in Fig. 4 [18]. The ensemble model performed well on most patients (~ 0.9 median Dice score), but failed completely in a few.

To examine why the model performed poorly in some few patients, we examined the model error pattern as a function of brain location. We calculated the average voxel-wise Dice score for the whole tumor for all 125 validation subsets and registered them to the Colin27-MNI-template (Fig. 5). We found that our model performed well in superficial gray matter (average Dice-score >0.9), but failed to segment the tumors accurately in white matter, predominantly in deeper structures in the temporal lobes. Moreover, our model segmented the whole tumor most accurately, but struggled to differentiate the enhancing tumor from the tumor core. It especially misclassified low-grade glioma cases in which no enhancing tumor was present (Dice score of 0).

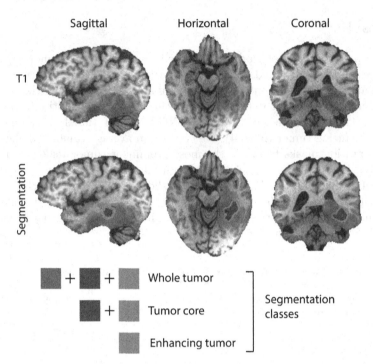

Fig. 3. Tumor segmentations. Validation data shown for one exemplary patient. We depict the T1 scan (upper panel) as well as the segmentation output of our model overlaid on the respective T1 scan (bottom panel) for sagittal, horizontal and coronal slices. Segmentations were color-coded.

4   Discussion

Tumor segmentation still often relies on manual segmentation by trained medical staff. Here, we present a fast, automated and accurate solution to this problem. Our segmentations can be used to inform physicians and aid the diagnostic process. We successfully segmented various brain tumors into three tissue types: whole tumor, enhancing tumor and tumor core in 125 patients provided by the BraTS challenge [2]. Importantly, our model was implemented and optimized on a single GTX1060 graphics unit with 6GB memory. To meet these low graphics memory demands, we split the input images into multiple 3D patches. The model iterated through these patches and converged on the most likely brain segmentation given all iterations in the end. We hence present a memory efficient and widely affordable solution to brain segmentation. Naturally, one limitation of this low-cost approach is that the model is still relatively slow. Naturally, more computational resources would alleviate this problem. In addition, more graphics memory would allow to upscale the input patch size further, in turn likely also benefiting the model performance greatly.

In addition, we implemented the model using data provided for this year's BraTS 2019 challenge alone. No other data was used. Earlier work including

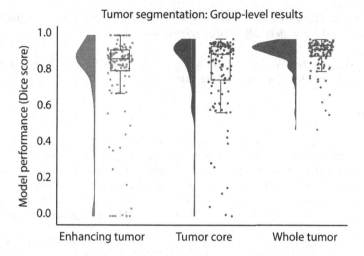

Fig. 4. Group-level segmentation performance (Dice score) for enhancing tumor (green, left), tumor core (blue, middle) and whole tumor (petrol, right) for the validation data set. We plot single-patient data overlaid on group-level whisker-boxplots (center, median; box, 25th to 75th percentiles; whiskers, 1.5 interquartile range) as well as the smoothed data distribution. (Color figure online)

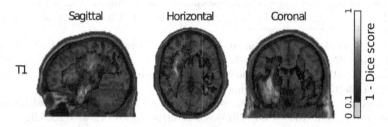

Fig. 5. Model error (1-Dice Score) overlaid on structural T1-template in MNI-space. Hot colors indicate high errors. The model performed well in superficial gray matter, but failed in deeper structures, especially in white matter tracts in the temporal lobe. (Color figure online)

previous BraTS challenges showed that incorporating additional data, hence increasing the training data set, greatly improves model performance [4]. Here, we aimed at optimizing brain tumor segmentation explicitly in the light of these common computational and data resource constraints. One interesting observation was that the model performed well on most patients (3), but failed completely in a few. The reasons for this remain unclear and need to be explored in the future.

Taken together, our results demonstrate the wide-ranging applicability of U-Nets to improve tissue segmentation and medical diagnostics. We show that dedicated memory efficient model architectures can overcome computational and data resource limitations and that fast and efficient brain tumor segmentation can be achieved on widely-affordable hardware.

Acknowledgements. We are grateful to Christian F. Doeller and the Kavli Institute for Systems Neuroscience for supporting this work.

References

1. Siegel, R.L., et al.: Cancer statistics, 2019. CA. Cancer J. Clin. **69**, 7–34 (2019)
2. Menze, B.H., et al.: The multimodal brain tumor image segmentation benchmark (BRATS). IEEE Trans. Med. Imaging **34**, 1993–2024 (2015)
3. Myronenko, A.: 3D MRI brain tumor segmentation using autoencoder regularization. In: Crimi, A., Bakas, S., Kuijf, H., Keyvan, F., Reyes, M., van Walsum, T. (eds.) BrainLes 2018. LNCS, vol. 11384, pp. 311–320. Springer, Cham (2019). https://doi.org/10.1007/978-3-030-11726-9_28
4. Isensee, F., Kickingereder, P., Wick, W., Bendszus, M., Maier-Hein, K.H.: No new-net. In: Crimi, A., Bakas, S., Kuijf, H., Keyvan, F., Reyes, M., van Walsum, T. (eds.) BrainLes 2018. LNCS, vol. 11384, pp. 234–244. Springer, Cham (2019). https://doi.org/10.1007/978-3-030-11726-9_21
5. Bakas, S., et al.: Advancing the cancer genome atlas glioma MRI collections with expert segmentation labels and radiomic features. Sci. Data **4** (2017)
6. Bakas, S., et al.: Identifying the best machine learning algorithms for brain tumor segmentation, progression assessment, and overall survival prediction in the BRATS challenge. ArXiv181102629 (2018). Cs Stat at http://arxiv.org/abs/1811.02629
7. Bakas, S., et al.: Segmentation labels for the pre-operative scans of the TCGA-LGG collection. Cancer Imaging Arch. (2017)
8. Bakas, S., et al.: Segmentation labels for the pre-operative scans of the TCGA-GBM collection. Cancer Imaging Arch. (2017)
9. Ronneberger, O., et al.: U-Net: convolutional networks for biomedical image segmentation. ArXiv150504597 (2015). Cs at http://arxiv.org/abs/1505.04597
10. Schwartz, R., et al.: Green AI. ArXiv190710597 (2019). Cs Stat at http://arxiv.org/abs/1907.10597
11. Xu, B., et al.: Empirical evaluation of rectified activations in convolutional network. ArXiv150500853 (2015). Cs Stat at http://arxiv.org/abs/1505.00853
12. Kingma, D.P., Welling, M.: Auto-encoding variational bayes. ArXiv13126114 (2013). Cs Stat at http://arxiv.org/abs/1312.6114
13. Cahall, D.E., et al.: Inception modules enhance brain tumor segmentation. Front. Comput. Neurosci. **13** (2019)
14. He, K., et al.: Deep residual learning for image recognition. ArXiv151203385 (2015). Cs at http://arxiv.org/abs/1512.03385
15. Wu, Y., Kaiming, H.: Group normalization. In: Proceedings of the European Conference on Computer Vision (ECCV), pp. 3–19 (2018)
16. Kingma, D.P., Ba, J.: Adam: a method for stochastic optimization. ArXiv14126980 (2014). Cs at http://arxiv.org/abs/1412.6980
17. Evans, A.C., et al.: 3D statistical neuroanatomical models from 305 MRI volumes. In: IEEE Conference Record Nuclear Science Symposium and Medical Imaging Conference, San Francisco, CA, USA, pp. 1813–1817 (1993)
18. Allen, M., et al.: Raincloud plots: a multi-platform tool for robust data visualization. Wellcome Open Res. **4**, 63 (2019)

Author Index

Printed in the United States
By Bookmasters